SIXTH EDITION

Student Solutions Manual for Kaufmann/Schwitters's

Elementary Algebra

Karen Schwitters
Seminole Community College

Jesse Turner
Seminole Community College

Brooks/Cole
Thomson Learning

Australia • Canada • Denmark • Japan • Mexico • New Zealand • Philippines
Puerto Rico • Singapore • Spain • United Kingdom • United States

Project Development Editor: *Michelle Paolucci*
Senior Editorial Assistant: *Erin Wickersham*
Marketing Team: *Leah Thomson, Debra Johnston*
Production Coordinator: *Dorothy Bell*
Cover Design: *Lisa Thompson*
Cover Illustration: *Laura Militzer Bryant*
Print Buyer: *Micky Lawler*
Printing and Binding: *Globus Printing*

For more information, contact:
BROOKS/COLE
511 Forest Lodge Road
Pacific Grove, CA 93950 USA
www.brookscole.com

Printed in the United States of America

10 9 8 7 6 5 4 3 2

ISBN 0-534-37343-7

Table of Contents

Chapter 1 Basic Concepts of Arithmetic and Algebra

PROBLEM SET 1.1 Numerical and Algebraic Expressions

1. True

3. True

5. 0, 12

7. $-8,\ -3,\ -1,\ 0,\ \frac{1}{5},\ 2\frac{3}{8},\ \frac{7}{2},\ 12$

9. 0, 12

11. $-8,\ -\sqrt{19},\ -3,\ -1,\ 0,\ \frac{1}{5},\ \sqrt{5},\ 2\frac{3}{8},\ \pi,\ \frac{7}{2},\ 6.23,\ 12$

13. $9+14-7$
$23-7$
16

15. $7(14-9)$
$7(5)$
35

17. $16+5\cdot 7$
$16+35$
51

19. $4(12+9)-3(8-4)$
$4(21)-3(4)$
$84-12$
72

21. $6\cdot 7+5\cdot 8-3\cdot 9$
$42+40-27$
$82-27$
55

23. $(6+9)(8-4)$
$(15)(4)$
60

25. $6+4[3(9-4)]$
$6+4[3(5)]$
$6+4(15)$
$6+60$
66

27. $16\div 8\cdot 4+36\div 4\cdot 2$
$2\cdot 4+36\div 4\cdot 2$
$8+36\div 4\cdot 2$
$8+9\cdot 2$
$8+18$
26

29. $\frac{8+12}{4}-\frac{9+15}{8}$
$\frac{20}{4}-\frac{24}{8}$
$5-3$
2

Problem Set 1.1

31. $56 - [3(9-6)]$
$56 - [3(3)]$
$56 - (9)$
$56 - 9$
47

33. $7 \cdot 4 \cdot 2 \div 8 + 14$
$28 \cdot 2 \div 8 + 14$
$56 \div 8 + 14$
$7 + 14$
21

35. $4 \cdot 9 \div 12 + 18 \div 2 + 3$
$36 \div 12 + 18 \div 2 + 3$
$3 + 18 \div 2 + 3$
$3 + 9 + 3$
$12 + 3$
15

37. $\frac{6(8-3)}{3} + \frac{12(7-4)}{9}$
$\frac{6(5)}{3} + \frac{12(3)}{9}$
$\frac{30}{3} + \frac{36}{9}$
$10 + 4$
14

39. $83 - \frac{4(12-7)}{5}$
$83 - \frac{4(5)}{5}$
$83 - \frac{20}{5}$
$83 - 4$
79

41. $\frac{4 \cdot 6 + 5 \cdot 3}{7 + 2 \cdot 3} + \frac{7 \cdot 9 + 6 \cdot 5}{3 \cdot 5 + 8 \cdot 2}$
$\frac{24+15}{7+6} + \frac{63+30}{15+16}$
$\frac{39}{13} + \frac{93}{31}$
$3 + 3$
6

43. $7x + 4y$, for $x = 6$, $y = 8$
$7(6) + 4(8)$
$42 + 32$
74

45. $16a - 9b$, for $a = 3$, $b = 4$
$16(3) - 9(4)$
$48 - 36$
12

47. $4x + 7y + 3xy$, for $x = 4$, $y = 9$
$4(4) + 7(9) + 3(4)(9)$
$16 + 63 + 108$
$79 + 108$
187

49. $14xz + 6xy - 4yz$, for $x = 8$, $y = 5$, $z = 7$
$14(8)(7) + 6(8)(5) - 4(5)(7)$
$784 + 240 - 140$
$1024 - 140$
884

51. $\frac{54}{n} + \frac{n}{3}$, for $n = 9$
$\frac{54}{9} + \frac{9}{3}$
$6 + 3$
9

53. $\frac{y+16}{6}+\frac{50-y}{3}$, for $y=8$

$\frac{8+16}{6}+\frac{50-8}{3}$

$\frac{24}{6}+\frac{42}{3}$

$4+14$

18

55. $(x+y)(x-y)$, for $x=8,\ y=3$

$(8+3)(8-3)$

$(11)(5)$

55

57. $(5x-2y)(3x+4y)$, for $x=3,\ y=6$

$[5(3)-2(6)][3(3)+4(6)]$

$[15-12][9+24]$

$(3)(33)$

99

59. $81-2[5(n+4)]$, for $n=3$

$81-2[5(3+4)]$

$81-2[5(7)]$

$81-2[35]$

$81-70$

11

61. $\frac{bh}{2}$, for $b=8,\ h=12$

$\frac{8(12)}{2}$

$\frac{96}{2}$

48

63. $\frac{bh}{2}$, for $b=7,\ h=6$

$\frac{7(6)}{2}$

$\frac{42}{2}$

21

65. $\frac{Bh}{3}$, for $B=25,\ h=12$

$\frac{25(12)}{3}$

$\frac{300}{3}$

100

67. $\frac{Bh}{3}$, for $B=36,\ h=7$

$\frac{36(7)}{3}$

$\frac{252}{3}$

84

69. $\frac{h(b_1+b_2)}{2}$, for $h=17,\ b_1=14,\ b_2=6$

$\frac{17(14+6)}{2}$

$\frac{17(20)}{2}$

$\frac{340}{2}$

170

71. $\frac{h(b_1+b_2)}{2}$, for $h=8,\ b_1=17,\ b_2=24$

$\frac{8(17+24)}{2}$

$\frac{8(41)}{2}$

$\frac{328}{2}$

164

PROBLEM SET 1.2 Prime and Composite Numbers

1. Since $8(7) = 56$, "8 divides 56" is a true statement.

3. Since $6(9) = 54$, "6 does not divide 54" is a false statement.

5. Since $8(12) = 96$, "96 is a multiple of 8" is a true statement.

7. Since there is no whole number, k, such that $4(k) = 54$, "54 is not a multiple of 4" is a true statement.

9. Since $4(36)=144$, "144 is divisible by 4" is a true statement.

11. Since there is no whole number, k, such that $3(k) = 173$, "173 is divisible by 3" is a false statement.

13. Since $11(13)=143$, "11 is a factor of 143" is a true statement.

15. Since there is no whole number, k, such that $9(k) = 119$, "9 is a factor of 119" is a false statement.

17. Since $3(19) = 57$ and 3 is a prime number, "3 is a prime factor of 57" is a true statement.

19. Since $2(2) = 4$, 4 is not a prime number, therefore "4 is a prime factor of 48" is a false statement.

21. Since 53 is only divisible by itself and 1, it is a prime number.

23. Since 59 is only divisible by itself and 1, it is a prime number.

25. Since $91 = 7(13)$, it is a composite number.

27. Since 89 is only divisible by itself and 1, it is a prime number.

29. Since $111 = 3(37)$, it is a composite number.

31. $26 = 2 \cdot 13$

33. $36 = 4 \cdot 9 = 2 \cdot 2 \cdot 3 \cdot 3$

35. $49 = 7 \cdot 7$

37. $56 = 8 \cdot 7 = 2 \cdot 4 \cdot 7 = 2 \cdot 2 \cdot 2 \cdot 7$

39. $120 = 3 \cdot 40 = 3 \cdot 4 \cdot 10 =$
$3 \cdot 2 \cdot 2 \cdot 2 \cdot 5 = 2 \cdot 2 \cdot 2 \cdot 3 \cdot 5$

41. $135 = 5 \cdot 27 = 5 \cdot 3 \cdot 9 =$
$5 \cdot 3 \cdot 3 \cdot 3 = 3 \cdot 3 \cdot 3 \cdot 5$

43. $12 = 2 \cdot 2 \cdot 3$
$16 = 2 \cdot 2 \cdot 2 \cdot 2$
The greatest common factor is $2 \cdot 2 = 4$

45. $56 = 2 \cdot 2 \cdot 2 \cdot 7$
$64 = 2 \cdot 2 \cdot 2 \cdot 2 \cdot 2 \cdot 2$
The greatest common factor is $2 \cdot 2 \cdot 2 = 8$

47. $63 = 3 \cdot 3 \cdot 7$
$81 = 3 \cdot 3 \cdot 3 \cdot 3$
The greatest common factor is $3 \cdot 3 = 9$

49. $84 = 2 \cdot 2 \cdot 3 \cdot 7$
$96 = 2 \cdot 2 \cdot 2 \cdot 2 \cdot 2 \cdot 3$
The greatest common factor is $2 \cdot 2 \cdot 3 = 12$

51. $36 = 2 \cdot 2 \cdot 3 \cdot 3$
$72 = 2 \cdot 2 \cdot 2 \cdot 3 \cdot 3$
$90 = 2 \cdot 3 \cdot 3 \cdot 5$
The greatest common factor is $2 \cdot 3 \cdot 3 = 18$

53. $48 = 2 \cdot 2 \cdot 2 \cdot 2 \cdot 3$
$60 = 2 \cdot 2 \cdot 3 \cdot 5$
$84 = 2 \cdot 2 \cdot 3 \cdot 7$
The greatest common factor is $2 \cdot 2 \cdot 3 = 12$

55. $6 = 2 \cdot 3$
$8 = 2 \cdot 2 \cdot 2$
The least common multiple is $2 \cdot 2 \cdot 2 \cdot 3 = 24$

57. $12 = 2 \cdot 2 \cdot 3$
$16 = 2 \cdot 2 \cdot 2 \cdot 2$
The least common multiple is $2 \cdot 2 \cdot 2 \cdot 2 \cdot 3 = 48$

59. $28 = 2 \cdot 2 \cdot 7$
$35 = 5 \cdot 7$
The least common multiple is $2 \cdot 2 \cdot 5 \cdot 7 = 140$

61. $49 = 7 \cdot 7$
$56 = 2 \cdot 2 \cdot 2 \cdot 7$
The least common multiple is $2 \cdot 2 \cdot 2 \cdot 7 \cdot 7 = 392$

63. $8 = 2 \cdot 2 \cdot 2$
$12 = 2 \cdot 2 \cdot 3$
$28 = 2 \cdot 2 \cdot 7$
The least common multiple is $2 \cdot 2 \cdot 2 \cdot 3 \cdot 7 = 168$

65. $9 = 3 \cdot 3$
$15 = 3 \cdot 5$
$18 = 2 \cdot 3 \cdot 3$
The least common multiple is $2 \cdot 3 \cdot 3 \cdot 5 = 90$

PROBLEM SET 1.3 Operations with Fractions

1. $\frac{8}{12} = \frac{4 \cdot 2}{4 \cdot 3} = \frac{2}{3}$

3. $\frac{16}{24} = \frac{8 \cdot 2}{8 \cdot 3} = \frac{2}{3}$

5. $\frac{15}{9} = \frac{3 \cdot 5}{3 \cdot 3} = \frac{5}{3}$

7. $\frac{24x}{44x} = \frac{6 \cdot 4x}{11 \cdot 4x} = \frac{6}{11}$

9. $\frac{9x}{21y} = \frac{3 \cdot 3x}{3 \cdot 7y} = \frac{3x}{7y}$

11. $\frac{14xy}{35y} = \frac{7y \cdot 2x}{7y \cdot 5} = \frac{2x}{5}$

13. $\frac{65abc}{91ac} = \frac{13ac \cdot 5b}{13ac \cdot 7} = \frac{5b}{7}$

15. $\frac{3}{4} \cdot \frac{5}{7} = \frac{3 \cdot 5}{4 \cdot 7} = \frac{15}{28}$

17. $\frac{2}{7} \div \frac{3}{5} = \frac{2}{7} \cdot \frac{5}{3} = \frac{10}{21}$

19. $\frac{3}{8} \cdot \frac{12}{15} = \frac{\overset{1}{\cancel{3}} \cdot \overset{3}{\cancel{12}}}{\underset{2}{\cancel{8}} \cdot \underset{5}{\cancel{15}}} = \frac{3}{10}$

21. $\frac{5x}{9y} \cdot \frac{7y}{3x} = \frac{5 \cdot 7 \cdot \cancel{x} \cdot \cancel{y}}{9 \cdot 3 \cdot \cancel{x} \cdot \cancel{y}} = \frac{35}{27}$

23. $\frac{6a}{14b} \cdot \frac{16b}{18a} = \frac{\overset{1}{\cancel{6}} \cdot \overset{8}{\cancel{16}} \cdot \cancel{a} \cdot \cancel{b}}{\underset{7}{\cancel{14}} \cdot \underset{3}{\cancel{18}} \cdot \cancel{a} \cdot \cancel{b}} = \frac{8}{21}$

25. $ab \cdot \frac{2}{b} = \frac{a \cdot \cancel{b} \cdot 2}{\cancel{b}} = 2a$

27. $\frac{3}{x} \div \frac{6}{y} = \frac{\overset{1}{\cancel{3}}}{x} \cdot \frac{y}{\underset{2}{\cancel{6}}} = \frac{y}{2x}$

29. $\frac{5x}{9y} \div \frac{13x}{36y} = \frac{5x}{9y} \cdot \frac{36y}{13x} =$
$\frac{5 \cdot \overset{4}{\cancel{36}} \cdot \cancel{x} \cdot \cancel{y}}{\underset{1}{\cancel{9}} \cdot 13 \cdot \cancel{x} \cdot \cancel{y}} = \frac{20}{13}$

Problem Set 1.3

31. $\begin{pmatrix}\text{Maria's}\\ \text{share of}\\ \text{department}\end{pmatrix} \cdot \begin{pmatrix}\text{Department's}\\ \text{share of}\\ \text{Agency}\end{pmatrix} = \begin{pmatrix}\text{Maria's}\\ \text{share of}\\ \text{Agency}\end{pmatrix}$

$\left(\frac{1}{3}\right) \cdot \left(\frac{3}{4}\right) = \frac{1}{4}$

33. $\frac{2}{7} + \frac{3}{7} = \frac{2+3}{7} = \frac{5}{7}$

35. $\frac{7}{9} - \frac{2}{9} = \frac{7-2}{9} = \frac{5}{9}$

37. $\frac{3}{4} + \frac{9}{4} = \frac{3+9}{4} = \frac{12}{4} = 3$

39. $\frac{11}{12} - \frac{3}{12} = \frac{11-3}{12} = \frac{8}{12} = \frac{2}{3}$

41. $\frac{5}{24} + \frac{11}{24} = \frac{5+11}{24} = \frac{16}{24} = \frac{2}{3}$

43. $\frac{8}{x} + \frac{7}{x} = \frac{8+7}{x} = \frac{15}{x}$

45. $\frac{5}{3y} + \frac{1}{3y} = \frac{5+1}{3y} = \frac{6}{3y} = \frac{2}{y}$

47. $\frac{1}{3} + \frac{1}{5} = \left(\frac{1 \cdot 5}{3 \cdot 5}\right) + \left(\frac{1 \cdot 3}{5 \cdot 3}\right) =$

$\frac{5}{15} + \frac{3}{15} = \frac{8}{15}$

49. $\frac{15}{16} - \frac{3}{8} = \frac{15}{16} - \left(\frac{3 \cdot 2}{8 \cdot 2}\right) =$

$\frac{15}{16} - \frac{6}{16} = \frac{9}{16}$

51. $\frac{7}{10} + \frac{8}{15} = \left(\frac{7 \cdot 3}{10 \cdot 3}\right) + \left(\frac{8 \cdot 2}{15 \cdot 2}\right) =$

$\frac{21}{30} + \frac{16}{30} = \frac{37}{30}$

53. $\frac{11}{24} + \frac{5}{32} = \left(\frac{11 \cdot 4}{24 \cdot 4}\right) + \left(\frac{5 \cdot 3}{32 \cdot 3}\right) =$

$\frac{44}{96} + \frac{15}{96} = \frac{59}{96}$

55. $\frac{3}{x} + \frac{4}{y} = \left(\frac{3 \cdot y}{x \cdot y}\right) + \left(\frac{4 \cdot x}{y \cdot x}\right) =$

$\frac{3y}{xy} + \frac{4x}{xy} = \frac{3y+4x}{xy}$

57. $\frac{7}{a} - \frac{2}{b} = \left(\frac{7 \cdot b}{a \cdot b}\right) - \left(\frac{2 \cdot a}{b \cdot a}\right) =$

$\frac{7b}{ab} - \frac{2a}{ab} = \frac{7b-2a}{ab}$

59. $\frac{2}{x} + \frac{7}{2x} = \left(\frac{2 \cdot 2}{x \cdot 2}\right) + \frac{7}{2x} =$

$\frac{4}{2x} + \frac{7}{2x} = \frac{11}{2x}$

61. $\frac{10}{3x} - \frac{2}{x} = \frac{10}{3x} - \left(\frac{2 \cdot 3}{x \cdot 3}\right) =$

$\frac{10}{3x} - \frac{6}{3x} = \frac{4}{3x}$

63. $\frac{5}{3x} + \frac{7}{3y} = \left(\frac{5 \cdot y}{3x \cdot y}\right) + \left(\frac{7 \cdot x}{3y \cdot x}\right) =$

$\frac{5y}{3xy} + \frac{7x}{3xy} = \frac{5y+7x}{3xy}$

65. $\frac{8}{5x} + \frac{3}{4y} = \left(\frac{8 \cdot 4y}{5x \cdot 4y}\right) + \left(\frac{3 \cdot 5x}{4y \cdot 5x}\right) =$

$\frac{32y}{20xy} + \frac{15x}{20xy} = \frac{32y+15x}{20xy}$

67. $\frac{7}{4x} - \frac{5}{9y} = \left(\frac{7 \cdot 9y}{4x \cdot 9y}\right) - \left(\frac{5 \cdot 4x}{9y \cdot 4x}\right) =$

$\frac{63y}{36xy} - \frac{20x}{36xy} = \frac{63y-20x}{36xy}$

69. $\frac{1}{4} - \frac{3}{8} + \frac{5}{12} - \frac{1}{24} =$

$\frac{6}{24} - \frac{9}{24} + \frac{10}{24} - \frac{1}{24} = \frac{6}{24} = \frac{1}{4}$

71. $\frac{5}{6}+\frac{2}{3}\cdot\frac{3}{4}-\frac{1}{4}\cdot\frac{2}{5}=$

$\frac{5}{6}+\frac{1}{2}-\frac{1}{10}=$

$\frac{25}{30}+\frac{15}{30}-\frac{3}{30}=\frac{37}{30}$

73. $\frac{3}{4}\cdot\frac{6}{9}-\frac{5}{6}\cdot\frac{8}{10}+\frac{2}{3}\cdot\frac{6}{8}=$

$\frac{1}{2}-\frac{2}{3}+\frac{1}{2}=$

$\frac{3}{6}-\frac{4}{6}+\frac{3}{6}=\frac{2}{6}=\frac{1}{3}$

75. $\frac{7}{13}\left(\frac{2}{3}-\frac{1}{6}\right)=\frac{7}{13}\left(\frac{2\cdot 2}{3\cdot 2}-\frac{1}{6}\right)=$

$\frac{7}{13}\left(\frac{4}{6}-\frac{1}{6}\right)=\frac{7}{13}\left(\frac{3}{6}\right)=\frac{7}{26}$

77. $11\frac{3}{4}+1\frac{1}{2}-\frac{3}{8}-\frac{1}{4}+\frac{1}{2}-\frac{5}{8}=$

$\left(11+\frac{6}{8}\right)+\left(1+\frac{4}{8}\right)-\frac{3}{8}-\frac{2}{8}+\frac{4}{8}-\frac{5}{8}=$

$12+\frac{4}{8}=12+\frac{1}{2}=12\frac{1}{2}=\$12.50$

79. $\begin{pmatrix}\text{Amount walk}\\\text{shortened}\end{pmatrix}=\begin{pmatrix}\text{Daily}\\\text{walk}\end{pmatrix}\text{less}\begin{pmatrix}\text{Amount}\\\text{walked}\end{pmatrix}$

$2\frac{1}{2}-\frac{3}{4}=\left(2+\frac{1}{2}\right)-\frac{3}{4}=$

$\frac{8}{4}+\frac{2}{4}-\frac{3}{4}=\frac{7}{4}=1\frac{3}{4}$ miles

PROBLEM SET 1.4 Addition and Subtraction of Real Numbers

1. $5+(-3)=2$

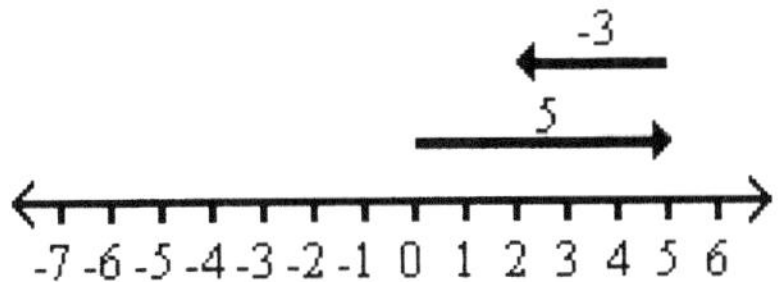

3. $-6+2=-4$

5. $-3+(-4)=-7$

7. $8+(-2)=6$

9. $5+(-11)=-6$

11. $17+(-9)=|17|-|-9|=17-9=8$

13. $8+(-19)=-(|-19|-|8|)=$
$-(19-8)=-11$

15. $-7+(-8)=-(|-7|+|-8|)=$
$-(7+8)=-15$

17. $-15+8=-(|-15|-|8|)=$
$-(15-8)=-7$

Problem Set 1.4

19. $-13+(-18)=$
$-(|-13|+|-18|)=$
$-(13+18)=-31$

21. $-27+8=-(|-27|-|8|)=$
$-(27-8)=-19$

23. $32+(-23)=+(|32|-|-23|)=$
$+(32-23)=9$

25. $\left(-\frac{3}{5}\right)+\left(-\frac{8}{5}\right)=$
$-\left(\left|-\frac{3}{5}\right|+-\left|-\frac{8}{5}\right|\right)=$
$-\left(\frac{3}{5}+\frac{8}{5}\right)=-\left(\frac{11}{5}\right)=-\frac{11}{5}$

27. $-\frac{3}{8}+\frac{1}{2}=+\left(\left|\frac{1}{2}\right|-\left|-\frac{3}{8}\right|\right)=$
$+\left(\frac{1}{2}-\frac{3}{8}\right)=+\left(\frac{4}{8}-\frac{3}{8}\right)=+\frac{1}{8}$

29. $9.38+(-16.42)=$
$-(|-16.42|-|9.38|)=$
$-(16.42-9.38)=-7.04$

31. $3-8=3+(-8)=-5$

33. $-4-9=-4+(-9)=-13$

35. $5-(-7)=5+7=12$

37. $-6-(-12)=-6+12=6$

39. $-11-(-10)=-11+10=-1$

41. $-18-27=-18+(-27)=-45$

43. $34-63=34+(-63)=-29$

45. $45-18=45+(-18)=27$

47. $-21-44=-21+(-44)=-65$

49. $-53-(-24)=-53+24=-29$

51. $6-8-9=6+(-8)+(-9)=-11$

53. $-4-(-6)+5-8=$
$-4+6+5+(-8)=-1$

55. $5+7-8-12=$
$5+7+(-8)+(-12)=-8$

57. $-\frac{1}{3}-\frac{1}{6}-\left(-\frac{2}{3}\right)+\left(-\frac{5}{6}\right)=$
$\left(-\frac{1}{3}\right)+\left(-\frac{1}{6}\right)+\frac{2}{3}+\left(-\frac{5}{6}\right)=$
$\left(-\frac{2}{6}\right)+\left(-\frac{1}{6}\right)+\frac{4}{6}+\left(-\frac{5}{6}\right)=$
$\left(-\frac{3}{6}\right)+\frac{4}{6}+\left(-\frac{5}{6}\right)=$
$\frac{1}{6}+\left(-\frac{5}{6}\right)=-\frac{4}{6}=-\frac{2}{3}$

59. $-6.4-5.32-9.17-8.6-(-1.56)=$
$-6.4-5.32-9.17-8.6+1.56=$
$-29.49+1.56=-27.93$

61. $7\frac{1}{2}-12\frac{5}{8}+2\frac{1}{8}=$
$7\frac{4}{8}-12\frac{5}{8}+2\frac{1}{8}=$
$-5\frac{1}{8}+2\frac{1}{8}=-3$

63. $-11-(-14)+(-17)-18=$
$-11+14+(-17)+(-18)=-32$

65. $16-21+(-15)-(-22)=$
$16+(-21)+(-15)+22=2$

67. $\begin{array}{r} 5 \\ -\ 9 \\ \hline -4 \end{array}$ This problem in horizontal format is $5+(-9)=-4$

69. $\begin{array}{r} -13 \\ -18 \\ \hline -31 \end{array}$ This problem in horizontal format is (-13) + (-18) = -31

71. $\begin{array}{r} -18 \\ 9 \\ \hline -9 \end{array}$ This problem in horizontal format is $(-18)+9=-9$

73. $\begin{array}{r} -21 \\ 39 \\ \hline 18 \end{array}$ This problem in horizontal format is $(-21)+39=18$

75. $\begin{array}{r} 27 \\ -19 \\ \hline 8 \end{array}$ This problem in horizontal format is $27+(-19)=8$

77. $\begin{array}{r} -53 \\ 24 \\ \hline -29 \end{array}$ This problem in horizontal format is $(-53)+24=-29$

79. $\begin{array}{r} 5 \\ 12 \\ \hline -7 \end{array}$ Change the sign of the bottom number and add $\begin{array}{r} 5 \\ -12 \\ \hline -7 \end{array}$

81. $\begin{array}{r} 6 \\ -9 \\ \hline 15 \end{array}$ Change the sign of the bottom number and add $\begin{array}{r} 6 \\ 9 \\ \hline 15 \end{array}$

83. $\begin{array}{r} -7 \\ -8 \\ \hline 1 \end{array}$ Change the sign of the bottom number and add $\begin{array}{r} -7 \\ 8 \\ \hline 1 \end{array}$

85. $\begin{array}{r} 17 \\ -19 \\ \hline 36 \end{array}$ Change the sign of the bottom number and add $\begin{array}{r} 17 \\ 19 \\ \hline 36 \end{array}$

87. $\begin{array}{r} -23 \\ 16 \\ \hline -39 \end{array}$ Change the sign of the bottom number and add $\begin{array}{r} -23 \\ -16 \\ \hline -39 \end{array}$

89. $\begin{array}{r} -12 \\ 12 \\ \hline -24 \end{array}$ Change the sign of the bottom number and add $\begin{array}{r} -12 \\ -12 \\ \hline -24 \end{array}$

91. $x=-6, y=-13: x-y$
$(-6)-(-13)$
$-6+13$
7

93. $x=3,\ y=-4,\ z=-6:\ -x+y-z$
$-(3)+(-4)-(-6)$
$-3+(-4)+6$
-1

95. $-x-y-z$ for $x=-\frac{1}{5},\ y=\frac{1}{2},\ z=-\frac{3}{10}$
$-\left(-\frac{1}{5}\right)-\left(\frac{1}{2}\right)-\left(-\frac{3}{10}\right)=$
$\frac{1}{5}+\left(-\frac{1}{2}\right)+\frac{3}{10}=$
$\frac{2}{10}+\left(-\frac{5}{10}\right)+\frac{3}{10}=$
$-\frac{3}{10}+\frac{3}{10}=0$

97. $-x+y+z$ for $x=-1.4,\ y=2.7,\ z=-3.6$
$-(-1.4)+(2.7)+(-3.6)=$
$1.4+2.7+(-3.6)=$
$4.1+(-3.6)=0.5$

99. $x=-15,\ y=12,\ z=-10;\ x-y-z$
$(-15)-(12)-(-10)=$
$(-15)+(-12)+10=-17$

101. $4+(-7)=-3$ (3 yards behind original scrimmage line)

103. $-4+(-6)=-10$ (10 yards behind original scrimmage line)

105. $-5+2=-3$ (3 yards behind original scrimmage line)

107. $-4+15=11$ (11 yards ahead of original scrimmage line)

109. $-12+17=5$ (5 yards ahead of original scrimmage line)

111. $60+(-125)=-65$
(overall expense of $65)

113. $-55+(-45)=-100$
(overall expense of $100)

Problem Set 1.4

115. $-70+45=-25$
(overall expense of \$25)

117. $-120+250=130$
(overall income of \$130)

119. $145+(-65)=80$
(overall income of \$80)

121. $-17+14=-3$ (temperature at noon is $-3°$F.)

123. $(+3)+(-2)+(-3)+(-5)=-7$
(7 under par for tournament)

PROBLEM SET 1.5 Multiplication and Division of Real Numbers

1. $5(-6)=-(|5|\cdot|-6|)=$
$-(5\cdot 6)=-30$

3. $\dfrac{-27}{3}=-\left(\dfrac{|-27|}{|3|}\right)=$
$-\left(\dfrac{27}{3}\right)=-9$

5. $\dfrac{-42}{-6}=\dfrac{|-42|}{|-6|}=\dfrac{42}{6}=7$

7. $\left(-\dfrac{2}{3}\right)\left(\dfrac{6}{5}\right)=-\left(\left|-\dfrac{2}{3}\right|\cdot\left|\dfrac{6}{5}\right|\right)=$
$-\left(\dfrac{2}{3}\cdot\dfrac{6}{5}\right)=-\left(\dfrac{12}{15}\right)=$
$-\left(\dfrac{4}{5}\right)=-\dfrac{4}{5}$

9. $\left(-\dfrac{1}{2}\right)(-12)=\left|-\dfrac{1}{2}\right|\cdot|-12|$
$\dfrac{1}{2}\cdot 12=6$

11. $\dfrac{7.2}{-8}=-\left(\dfrac{|7.2|}{|-8|}\right)=$
$-\left(\dfrac{7.2}{8}\right)=-0.9$

13. $14(-9)=-(|14|\cdot|-9|)=$
$-(14\cdot 9)=-126$

15. $(-11)(-14)=|-11|\cdot|-14|=$
$11\cdot 14=154$

17. $\dfrac{135}{-15}=-\left(\dfrac{|135|}{|-15|}\right)=$
$-\left(\dfrac{135}{15}\right)=-9$

19. $\dfrac{-121}{-11}=\dfrac{|-121|}{|-11|}=\dfrac{121}{11}=11$

21. $(-15)(-15)=|-15|\cdot|-15|=$
$15\cdot 15=225$

23. $\dfrac{112}{-8}=-\left(\dfrac{|112|}{|-8|}\right)=$
$-\left(\dfrac{112}{8}\right)=-14$

25. $\dfrac{0}{-8}=0$ because $(-8)(0)=0$

27. $\dfrac{-138}{-6}=\dfrac{|-138|}{|-6|}=\dfrac{138}{6}=23$

29. $\dfrac{2.48}{-0.4}=-\left(\dfrac{|2.48|}{|-0.4|}\right)=$
$-\left(\dfrac{2.48}{0.4}\right)=-(6.2)=-6.2$

31. $\left(-6\frac{1}{2}\right)\left(-1\frac{3}{5}\right)=\left|-6\frac{1}{2}\right|\cdot\left|-1\frac{3}{5}\right|=$
$\left(6\frac{1}{2}\right)\left(1\frac{3}{5}\right)=\left(\frac{13}{2}\right)\left(\frac{8}{5}\right)=\frac{52}{5}$

33. $\left(-\frac{3}{8}\right)\div\left(-\frac{1}{4}\right)=\left|-\frac{3}{8}\right|\div\left|-\frac{1}{4}\right|=$
$\frac{3}{8}\div\frac{1}{4}=\frac{3}{8}\cdot\frac{4}{1}=\frac{3}{2}$

35. $\left(-\frac{1}{3}\right)\div 0$ is undefined because no number times zero produces $-\frac{1}{3}$.

37. $(-72)\div 18=-(|-72|\div|18|)=$
$-(72\div 18)=-4$

39. $(-36)(27)=-(|-36|\cdot|27|)=$
$-(36\cdot 27)=-972$

41. $3(-4)+5(-7)=$
$-12+(-35)=-47$

43. $7(-2)-4(-8)=-14-(-32)=$
$-14+32=18$

45. $\left(-\frac{1}{2}\right)\left(-\frac{3}{2}\right)+\left(-\frac{1}{6}\right)\left(-\frac{5}{2}\right)=$
$\frac{3}{4}+\frac{5}{12}=\frac{9}{12}+\frac{5}{12}=\frac{14}{12}=\frac{7}{6}$

47. $10\left(-\frac{1}{5}\right)-4\left(-\frac{1}{2}\right)+3\left(-\frac{2}{3}\right)=$
$-2+2-2=-2$

49. $\frac{1.3+(-2.5)}{-0.3}=\frac{-1.2}{-0.3}=4$

51. $\frac{12-48}{0.6}=\frac{-36}{0.6}=-60$

53. $\frac{-7(10)+6(-9)}{-4}=\frac{-70+(-54)}{-4}=$
$\frac{-124}{-4}=31$

55. $\frac{4(-7)-8(-9)}{11}=\frac{-28-(-72)}{11}=$
$\frac{-28+72}{11}=\frac{44}{11}=4$

57. $-2(3)-3(-4)+4(-5)-6(-7)=$
$-6-(-12)+(-20)-(-42)=$
$-6+12+(-20)+42=28$

59. $-1(-6)-4+6(-2)-7(-3)-18=$
$6-4+(-12)-(-21)-18=$
$6-4+(-12)+21-18=-7$

61. $x=-5,\ y=9:\ 7x+5y=$
$7(-5)+5(9)=$
$-35+45=10$

63. $a=-5,\ b=7:\ 9a-2b=$
$9(-5)-2(7)=-45-14=$
$-45+(-14)=-59$

65. $-6x-7y,$ for $x=-\frac{1}{8},\ y=-\frac{1}{2}$
$-6\left(-\frac{1}{8}\right)-7\left(-\frac{1}{2}\right)=$
$\frac{6}{8}+\frac{7}{2}=\frac{3}{4}+\frac{7}{2}=\frac{3}{4}+\frac{14}{4}=\frac{17}{4}$

67. $x=-6,\ y=4:\ \frac{5x-3y}{-6}=$
$\frac{5(-6)-3(4)}{-6}=\frac{-30-12}{-6}=$
$\frac{-30+(-12)}{-6}=\frac{-42}{-6}=7$

69. $a=-1,\ b=-5:\ 3(2a-5b)=$
$3[2(-1)-5(-5)]=$
$3[-2-(-25)]=$
$3[-2+25]=3(23)=69$

71. $-2x+6y-xy;\ x=1.6,\ y=-1.2$
$-2(1.6)+6(-1.2)-(1.6)(-1.2)=$

$-3.2 - 7.2 - (-1.92) =$
$-3.2 - 7.2 + 1.92 = -8.48$

73. $a = 2,\ b = -14:\ -4ab - b =$
$-4(2)(-14) - (-14) =$
$-8(-14) + 14 =$
$112 + 14 = 126$

75. $a = -2,\ b = -3,\ c = 4:$
$(ab + c)(b - c) =$
$[(-2)(-3) + (4)][(-3) - (4)] =$
$(6 + 4)(-7) = 10(-7) = -70$

77. $\text{F} = 59:\ \dfrac{5(\text{F} - 32)}{9} = \dfrac{5[(59) - 32]}{9} =$

$\dfrac{5(27)}{9} = \dfrac{135}{9} = 15$

79. $\text{F} = 14:\ \dfrac{5(\text{F} - 32)}{9} = \dfrac{5[(14) - 32]}{9} =$

$\dfrac{5(-18)}{9} = \dfrac{-90}{9} = -10$

81. $\text{F} = -13:\ \dfrac{5(\text{F} - 32)}{9} =$

$\dfrac{5[(-13) - 32]}{9} =$

$\dfrac{5(-45)}{9} = \dfrac{-225}{9} = -25$

83. $\text{C} = 25:\ \dfrac{9\text{C}}{5} + 32 =$

$\dfrac{9(25)}{5} + 32 = \dfrac{225}{5} + 32 = 45 + 32 = 77$

85. $\text{C} = 40:\ \dfrac{9\text{C}}{5} + 32 =$

$\dfrac{9(40)}{5} + 32 = \dfrac{360}{5} + 32 = 72 + 32 = 104$

87. $\text{C} = -10:\ \dfrac{9\text{C}}{5} + 32 =$

$\dfrac{9(-10)}{5} + 32 = \dfrac{-90}{5} + 32 =$

$-18 + 32 = 14$

89. $\left(\begin{array}{c}\text{Value of}\\ \text{800 shares}\end{array}\right) = \left(\begin{array}{c}\text{800 times the price}\\ \text{of shares at closing}\end{array}\right):$

$$\begin{aligned}\text{Value} &= 800\left[\text{price} + \left(\begin{array}{c}\text{1 day's}\\ \text{increase}\end{array}\right) + \left(\begin{array}{c}\text{4 day's}\\ \text{decrease}\end{array}\right)\right]\\ &= 800[19 + 2 + 4(-1)]\\ &= 800[19 + 2 + (-4)]\\ &= 800(17)\\ &= \$13,600\end{aligned}$$

PROBLEM SET 1.6 Use of Properties

1. $3(7 + 8) = 3(7) + 3(8):$
Distributive property

3. $-2 + (5 + 7) = (-2 + 5) + 7:$
Associative property of addition

5. $143(-7) = -7(143):$
Commutative property of multiplication

7. $-119 + 119 = 0:$
Additive inverse property

9. $-56 + 0 = -56:$
Identity property of addition

11. $[5(-8)]4 = 5[-8(4)]:$
Associative property of multiplication

13. $(-18 + 56) + 18 =$
$[56 + (-18)] + 18 =$
$56 + (-18 + 18) =$
$56 + 0 = 56$

15. $36 - 48 - 22 + 41 =$
$-12 - 22 + 41 =$
$-34 + 41 = 7$

17. $(25)(-18)(-4) =$
$(25)(-4)(-18) =$
$(-100)(-18) = 1,800$

19. $(4)(-16)(-9)(-25) =$
$(4)(-25)(-16)(-9) =$
$(-100)(144) = -14{,}400$

21. $37(-42-58) = 37(-100) = -3{,}700$

23. $59(36) + 59(64) = 59(36+64) =$
$59(100) = 5{,}900$

25. $15(-14) + 16(-8) =$
$-210 - 128 = -338$

27. $17 + (-18) - 19 - 14 + 13 - 17 =$
$[17 + (-17)] + (-18) - 19 - 14 + 13 =$
$0 + (-37) + (-14 + 13) =$
$-37 - 1 = -38$

29. $-21 + 22 - 23 + 27 + 21 - 19 =$
$(-21 + 22) + (-23 + 27) + (21 - 19) =$
$1 + 4 + 2 = 7$

31. $9x - 14x = (9 - 14)x - 5x$

33. $4m + m - 8m = (4 + 1 - 8)m = -3m$

35. $-9y + 5y - 7y =$
$(-9 + 5 - 7)y = -11y$

37. $4x - 3y - 7x + y =$
$4x - 7x - 3y + y =$
$(4 - 7)x + (-3 + 1)y =$
$-3x - 2y$

39. $-7a - 7b - 9a + 3b =$
$-7a - 9a - 7b + 3b =$
$(-7 - 9)a + (-7 + 3)b =$
$-16a - 4b$

41. $6xy - x - 13xy + 4x =$
$6xy - 13xy - x + 4x =$
$-7xy + 3x$

43. $5x - 4 + 7x - 2x + 9 =$
$5x + 7x - 2x - 4 + 9 =$
$10x + 5$

45. $-2xy + 12 + 8xy - 16 =$
$-2xy + 8xy + 12 - 16 =$
$6xy - 4$

47. $-2a + 3b - 7b - b + 5a - 9a =$
$-2a + 5a - 9a + 3b - 7b - b =$
$-6a - 5b$

49. $13ab + 2a - 7a - 9ab + ab - 6a =$
$13ab - 9ab + ab + 2a - 7a - 6a =$
$5ab - 11a$

51. $3(x + 2) + 5(x + 6) =$
$3(x) + 3(2) + 5(x) + 5(6) =$
$3x + 6 + 5x + 30 =$
$3x + 5x + 6 + 30$
$8x + 36$

53. $5(x - 4) + 6(x + 8) =$
$5x - 20 + 6x + 48 =$
$11x + 28$

55. $9(x + 4) - (x - 8) =$
$9(x + 4) - 1(x - 8) =$
$9x + 36 - x + 8 =$
$8x + 44$

57. $3(a - 1) - 2(a - 6) + 4(a + 5) =$
$3a - 3 - 2a + 12 + 4a + 20 =$
$5a + 29$

59. $-2(m + 3) - 3(m - 1) + 8(m + 4) =$
$-2m - 6 - 3m + 3 + 8m + 32 =$
$3m + 29$

61. $(y + 3) - 1(y - 2) - 1(y + 6) - 7(y - 1) =$
$y + 3 - y + 2 - y - 6 - 7y + 7 =$
$-8y + 6$

63. $3x + 5y + 4x - 2y = 7x + 3y =$
$7(-2) + 3(3)$ for $x = -2$ and $y = 3$
$-14 + 9 = -5$

65. $5(x - 2) + 8(x + 6) =$
$5x - 10 + 8x + 48 = 13x + 38 =$

$13(-6)+38$ for $x=-6$,
$-78+38=-40$

67. $8(x+4)-10(x-3)=$
$8x+32-10x+30=-2x+62=$
$-2(-5)+62$ for $x=-5$
$10+62=72$

69. $(x-6)-(x+12)=$
$x-6-x-12=-18$
-18 for $x=-3$

71. $2(x+y)-3(x-y)=$
$2x+2y-3x+3y=-x+5y=$
$-(-2)+5(7)$ for $x=-2$ and $y=7$
$2+35=37$

73. $2xy+6+7xy-8=9xy-2=$

$9(2)(-4)-2$ for $x=2$ and $y=-4$
$-72-2=-74$

75. $5x-9xy+3x+2xy=8x-7xy=$
$8(12)-7(12)(-1)$,
for $x=12$ and $y=-1$
$96+84=180$

77. $(a-b)-(a+b)=$
$a-b-a-b=-2b=$
$-2(-17)$ for $a=19$ and $b=-17$
34

79. $-3x+7x+4x-2x-x=5x=$
$5(-13)=-65$ for $x=-13$

PROBLEM SET 1.7 Exponents

1. $2^6=2\cdot2\cdot2\cdot2\cdot2\cdot2=64$

3. $3^4=3\cdot3\cdot3\cdot3=81$

5. $(-2)^3=(-2)(-2)(-2)=-8$

7. $-3^2=-(3\cdot3)=-9$

9. $(-4)^2=(-4)(-4)=16$

11. $\left(\frac{2}{3}\right)^4=\left(\frac{2}{3}\right)\left(\frac{2}{3}\right)\left(\frac{2}{3}\right)\left(\frac{2}{3}\right)=\frac{16}{81}$

13. $-\left(\frac{1}{2}\right)^3=-\left(\frac{1}{2}\right)\left(\frac{1}{2}\right)\left(\frac{1}{2}\right)=-\frac{1}{8}$

15. $\left(-\frac{3}{2}\right)^2=\left(-\frac{3}{2}\right)\left(-\frac{3}{2}\right)=\frac{9}{4}$

17. $(0.3)^3=(0.3)(0.3)(0.3)=0.027$

19. $-(1.2)^2=-(1.2)(1.2)=-1.44$

21. $3^2+2^3-4^3=9+8-64=-47$

23. $(-2)^3-2^4-3^2=$
$-8-16-9=-33$

25. $5(2)^2-4(2)-1=5(4)-8-1=$
$20-8-1=11$

27. $-2(3)^3-3(3)^2+4(3)-6=$
$-2(27)-3(9)+12-6=$
$-54-27+12-6=-75$

29. $-7^2-6^2+5^2=$
$-49-36+25=-60$

31. $-3(-4)^2-2(-3)^3+(-5)^2=$
$-3(16)-2(-27)+25=$
$-48+54+25=31$

33. $\dfrac{-3(2)^4}{12}+\dfrac{5(-3)^3}{15}=$
$\dfrac{-3(16)}{12}+\dfrac{5(-27)}{15}=$

$$\frac{-48}{12} + \frac{-135}{15} =$$
$$-4 - 9 = -13$$

35. $9 \cdot x \cdot x = 9x^2$

37. $3 \cdot 4 \cdot x \cdot y \cdot y = 12xy^2$

39. $-2 \cdot 9 \cdot x \cdot x \cdot x \cdot x \cdot y = -18x^4y$

41. $(5x)(3y) = 5 \cdot 3 \cdot x \cdot y = 15xy$

43. $(6x^2)(2x^2) = 6 \cdot 2 \cdot x \cdot x \cdot x \cdot x = 12x^4$

45. $(-4a^2)(-2a^3) =$
$(-4)(-2) \cdot a \cdot a \cdot a \cdot a \cdot a =$
$8a^5$

47. $3x^2 - 7x^2 - 4x^2 =$
$(3 - 7 - 4)x^2 = -8x^2$

49. $-12y^3 + 17y^3 - y^3 =$
$(-12 + 17 - 1)y^3 = 4y^3$

51. $7x^2 - 2y^2 - 9x^2 + 8y^2 =$
$(7 - 9)x^2 + (-2 + 8)y^2 =$
$-2x^2 + 6y^2$

53. $\frac{2}{3}n^2 - \frac{1}{4}n^2 - \frac{3}{5}n^2 =$
$\left(\frac{2}{3} - \frac{1}{4} - \frac{3}{5}\right)n^2 =$
$\left(\frac{40}{60} - \frac{15}{60} - \frac{36}{60}\right)n^2 =$
$-\frac{11}{60}n^2$

55. $5x^2 - 8x - 7x^2 + 2x =$
$(5 - 7)x^2 + (-8 + 2)x =$
$-2x^2 - 6x$

57. $x^2 - 2x - 4 + 6x^2 - x + 12 =$
$x^2 + 6x^2 - 2x - x - 4 + 12 =$
$7x^2 - 3x + 8$

59. $\frac{9xy}{15x} = \frac{\overset{3}{\cancel{9}} \cdot \cancel{x} \cdot y}{\underset{5}{\cancel{15}} \cdot \cancel{x}} = \frac{3y}{5}$

61. $\frac{22xy^2}{6xy^3} = \frac{\overset{11}{\cancel{22}} \cdot \cancel{x} \cdot \cancel{y} \cdot \cancel{y}}{\underset{3}{\cancel{6}} \cdot \cancel{x} \cdot \cancel{y} \cdot \cancel{y} \cdot y} = \frac{11}{3y}$

63. $\frac{7a^2b^3}{17a^3b} = \frac{7 \cdot \cancel{a} \cdot \cancel{a} \cdot \cancel{b} \cdot b \cdot b}{17 \cdot \cancel{a} \cdot \cancel{a} \cdot a \cdot \cancel{b}} = \frac{7b^2}{17a}$

65. $\frac{-24abc^2}{32bc} = \frac{\overset{3}{\cancel{24}} \cdot a \cdot \cancel{b} \cdot \cancel{c} \cdot c}{\underset{4}{\cancel{32}} \cdot \cancel{b} \cdot \cancel{c}} = -\frac{3ac}{4}$

67. $\frac{-5x^4y^3}{-20x^2y} = \frac{\cancel{5} \cdot \cancel{x} \cdot \cancel{x} \cdot x \cdot x \cdot \cancel{y} \cdot y \cdot y}{4 \cdot \cancel{5} \cdot \cancel{x} \cdot \cancel{x} \cdot \cancel{y}} =$
$\frac{x^2y^2}{4}$

69. $\left(\frac{7x^2}{9y}\right)\left(\frac{12y}{21x}\right) = \frac{\cancel{7} \cdot \overset{4}{\cancel{12}} \cdot \cancel{x} \cdot x \cdot \cancel{y}}{\underset{3}{\cancel{9}} \cdot \underset{3}{\cancel{21}} \cdot \cancel{x} \cdot \cancel{y}} =$
$\frac{4x}{9}$

71. $\left(\frac{5c}{a^2b^2}\right) \div \left(\frac{12c}{ab}\right) = \frac{5c}{a^2b^2} \cdot \frac{ab}{12c} =$
$\frac{5 \cdot \cancel{a} \cdot \cancel{b} \cdot \cancel{c}}{12 \cdot \cancel{a} \cdot a \cdot \cancel{b} \cdot b \cdot \cancel{c}} = \frac{5}{12ab}$

73. $\frac{6}{x} + \frac{5}{y^2} = \frac{6 \cdot y^2}{x \cdot y^2} + \frac{5 \cdot x}{y^2 \cdot x} =$
$\frac{6y^2}{xy^2} + \frac{5x}{xy^2} = \frac{6y^2 + 5x}{xy^2}$

75. $\frac{5}{x^4} - \frac{7}{x^2} = \frac{5}{x^4} - \frac{7 \cdot x \cdot x}{x \cdot x \cdot x \cdot x} =$
$\frac{5}{x^4} - \frac{7x^2}{x^4} = \frac{5 - 7x^2}{x^4}$

Problem Set 1.7

77. $\frac{3}{2x^3}+\frac{6}{x}=\frac{3}{2x^3}+\frac{6\cdot 2\cdot x\cdot x}{x\cdot 2\cdot x\cdot x}=$
$\frac{3}{2x^3}+\frac{12x^2}{2x^3}=\frac{3+12x^2}{2x^3}$

79. $\frac{-5}{4x^2}+\frac{7}{3x^2}=\frac{-5\cdot 3}{4x^2\cdot 3}+\frac{7\cdot 4}{3x^2\cdot 4}=$
$\frac{-15}{12x^2}+\frac{28}{12x^2}=\frac{13}{12x^2}$

81. $\frac{11}{a^2}-\frac{14}{b^2}=\frac{11\cdot b^2}{a^2\cdot b^2}-\frac{14\cdot a^2}{b^2\cdot a^2}=$
$\frac{11b^2}{a^2b^2}-\frac{14a^2}{a^2b^2}=\frac{11b^2-14a^2}{a^2b^2}$

83. $\frac{1}{2x^3}-\frac{4}{3x^2}=\frac{1\cdot 3}{2x^3\cdot 3}-\frac{4\cdot 2x}{3x^2\cdot 2x}=$
$\frac{3}{6x^3}-\frac{8x}{6x^3}=\frac{3-8x}{6x^3}$

85. $\frac{3}{x}-\frac{4}{y}-\frac{5}{xy}=\frac{3\cdot y}{x\cdot y}-\frac{4\cdot x}{y\cdot x}-\frac{5}{xy}=$
$\frac{3y}{xy}-\frac{4x}{xy}-\frac{5}{xy}=\frac{3y-4x-5}{xy}$

87. $x=-2,\ y=-3:\ 4x^2+7y^2=$
$4(-2)^2+7(-3)^2=$
$4(4)+7(9)=$
$16+63=79$

89. $x=\frac{1}{2},\ y=-\frac{1}{3}:\ 3x^2-y^2=$
$3\left(\frac{1}{2}\right)^2-\left(-\frac{1}{3}\right)^2=$
$3\left(\frac{1}{4}\right)-\left(\frac{1}{9}\right)=$
$\frac{3}{4}-\frac{1}{9}=\frac{3\cdot 9}{4\cdot 9}-\frac{1\cdot 4}{9\cdot 4}=$
$\frac{27}{36}-\frac{4}{36}=\frac{23}{36}$

91. $x=-\frac{1}{2},\ y=2:\ x^2-2xy+y^2=$
$\left(-\frac{1}{2}\right)^2-2\left(-\frac{1}{2}\right)(2)+(2)^2=$
$\left(\frac{1}{4}\right)+2+4=\frac{1}{4}+\frac{2\cdot 4}{1\cdot 4}+\frac{4\cdot 4}{1\cdot 4}=$
$\frac{1}{4}+\frac{8}{4}+\frac{16}{4}=\frac{25}{4}$

93. $x=-8:\ -x^2=-(-8)^2=$
$-(-8)(-8)=-(64)=-64$

95. $x=-3,\ y=-4:\ -x^2-y^2=$
$-(-3)^2-(-4)^2=$
$-9-16=-25$

97. $a=-6,\ b=-1:\ -a^2-3b^3=$
$-(-6)^2-3(-1)^3=$
$-36-3(-1)=$
$-36+3=-33$

99. $x=0.4,\ y=-0.3:\ y^2-3xy=$
$(-0.3)^2-3(0.4)(-0.3)=$
$0.09-3(-0.12)=$
$0.09+0.36=0.45$

PROBLEM SET 1.8 Translating from English to Algebra

For problems, 1-12, the answers may vary.

1. The difference of a and b
3. One-third of the product of B and h
5. Two times the quantity, l plus w
7. The quotient of A divided by w
9. The quantity, a plus b, divided by 2
11. Two more than three times y

13. $l+w$

15. ab

17. $\frac{\text{d}}{\text{t}}$

19. lwh

21. $y - x$

23. $xy + 2$

25. $7 - y^2$

27. $\frac{x-y}{4}$

29. $10 - x$

31. $10(n + 2)$

33. $xy - 7$

35. $xy - 12$

For problems 37-68, it may help to do a specific example before trying to formulate the general expression.

37. Suppose that the sum of two numbers is 35 and one of the numbers is 14. Then we subtract to find the other number, $35 - 14$. Thus, if one of the numbers is n, then the other number is $35 - n$.

39. Since the smaller number plus the difference must equal the larger number, the other number must be $n + 45$.

41. In ten years, Janet's age will be increased by 10, so it is represented as $y + 10$.

43. Twice Debra's age is $2x$, so Debra's mother is 3 years less than $2x$ or $2x - 3$.

45. Three dimes and five quarters is $10(3) + 25(2) = 80$ cents. Thus d dimes and q quarters is $10d + 25q$ cents.

47. If a car travels 200 miles in 5 hours, then it would be traveling at a rate of $\frac{\text{200 miles}}{\text{5 hours}} = 40$ miles per hour.
Therefore, the rate of the car is $\frac{d}{t}$.

49. If 5 pounds of candy cost \$15, then to find the price per pound we divide 15 by 5.
Thus, if p pounds cost d dollars, $\frac{d}{f}$ represents the price per pound.

51. To find a monthly salary, the annual salary is divided by 12, so Larry's monthly salary is $\frac{d}{12}$.

53. If 6 is the whole number, then the next larger whole number would be $6 + 1 = 7$. Thus, if n is the whole number, the next larger whole number is $n + 1$.

55. Suppose 7 is the odd number, then the next larger odd number would be $7 + 2 = 9$. Thus, if n is the odd number, the next larger odd number would be $n + 2$.

57. If Willie is y years old, then twiceWillie's age is $2y$. Since his father is 2 years less than twice Willie's age, the father's age is $2y - 2$. The sum of their ages is $y + (2y - 2) = 3y - 2$.

59. To convert yards to inches, the yards must be multiplied by 36. To convert feet to inches, the number of feet must be multiplied by 12. If the perimeter of a rectangle is 5 yards and 2 feet, then the perimeter in inches is $36(5) + 12(2) = 204$ inches. Thus, the perimeter of a rectangle that is y yards and f feet is $36y + 12f$ inches.

Problem Set 1.8

61. To change feet to yards, we divide the number of feet by 3. Therefore, f feet equals $\frac{f}{3}$ yards.

63. The width of a rectangle is w feet and the length is three times the width, or $3w$ feet. The perimeter of the rectangle is the sum of the lengths of the four sides. There are 2 widths and 2 lengths in a rectangle so the perimeter is 2 times the width plus 2 times the length or $2(w) + 2(3w) = 8w$ feet.

3w = length

w

w = width

3w

65. See problem 63. The length is l inches and the width is $\left(\frac{l}{2} - 2\right)$ inches. The perimeter is

$2l + 2\left(\frac{l}{2} - 2\right) = 3l - 4$ inches.

l = length

$\frac{l}{2}$ - 2 = width

67. The first side of a triangle is f feet long. The second side is 2 feet longer, or $f + 2$ feet long. The third side is twice as long as the second side, or $2(f + 2)$. The perimeter is the sum of the lengths of the sides, or $f + (f + 2) + 2(f + 2) = 4f + 6$ feet. To convert to inches, multiply the number of feet by 12. The perimeter of the triangle in inches is $12(4f + 6) = 48f + 72$ inches.

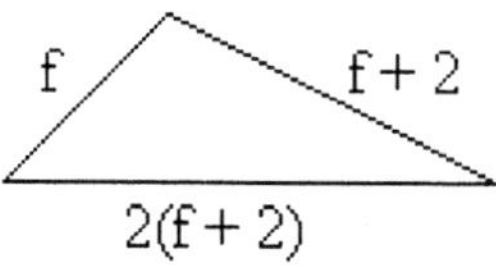

CHAPTER 1 | Review Problem Set

1. 73, prime number

2. Because $87 = 3(29)$, it is a composite number

3. Because $63 = 9(7)$, it is a composite number

4. Because $81 = 9(9)$, it is a composite number

5. Because $91 = 7(13)$, it is a composite number

6. $24 = 4 \cdot 6 = 2 \cdot 2 \cdot 2 \cdot 3$

7. $63 = 9 \cdot 7 = 3 \cdot 3 \cdot 7$

8. $57 = 3 \cdot 19$

9. $64 = 16 \cdot 4 = 4 \cdot 4 \cdot 4 = 2 \cdot 2 \cdot 2 \cdot 2 \cdot 2 \cdot 2$

10. $84 = 4 \cdot 21 = 2 \cdot 2 \cdot 3 \cdot 7$

11. Associative Property of Addition

12. Commutative Property of Multiplication

13. Distributive Property

14. Additive Identity

15. Commutative Property of Addition

16. Associative Property of Multiplication

17. Additive Inverse

18. Distributive Property

19. $8+(-9)+(-16)+(-14)+17+12=$
$37+(-39)=-2$

20. $19-23-14+21+14-13=4$

21. $3(-4)-6=-12-6=-18$

22. $(-5)(-4)-8=20-8=12$

23. $5(-2)+6(-4)=-10-24=-34$

24. $(-6)(8)+(-7)(-3)=$
$-48+21=-27$

25. $\frac{4}{5}\div\frac{1}{5}\cdot\frac{2}{3}-\frac{1}{4}=\left(\frac{4}{5}\cdot\frac{5}{1}\right)\cdot\frac{2}{3}-\frac{1}{4}=$
$\frac{4}{1}\cdot\frac{2}{3}-\frac{1}{4}=\frac{8}{3}-\frac{1}{4}=\frac{8\cdot 4}{3\cdot 4}-\frac{1\cdot 3}{4\cdot 3}=$
$\frac{32}{12}-\frac{3}{12}=\frac{29}{12}$

26. $\frac{2}{3}\cdot\frac{1}{4}\div\frac{1}{2}+\frac{2}{3}\cdot\frac{1}{4}=$
$\frac{2}{12}\div\frac{1}{2}+\frac{2}{12}=$
$\left(\frac{2}{12}\cdot\frac{2}{1}\right)+\frac{2}{12}=$
$\frac{4}{12}+\frac{2}{12}=\frac{6}{12}=\frac{1}{2}$

27. $0.48+0.72-0.35-0.18=$
$1.20-0.53=0.67$

28. $0.81+(0.6)(0.4)-(0.7)(0.8)=$
$0.81+0.24-0.56=$
$1.05-0.56=0.49$

29. $2^6=2\cdot2\cdot2\cdot2\cdot2\cdot2=64$

30. $(-3)^3=(-3)(-3)(-3)=-27$

31. $-4^2=-(4\cdot4)=-(16)=-16$

32. $\left(\frac{3}{4}\right)^2=\left(\frac{3}{4}\right)\left(\frac{3}{4}\right)=\frac{9}{16}$

33. $\left(\frac{1}{2}+\frac{2}{3}\right)^2=\left(\frac{3}{6}+\frac{4}{6}\right)^2=$
$\left(\frac{7}{6}\right)^2=\left(\frac{7}{6}\right)\left(\frac{7}{6}\right)=\frac{49}{36}$

34. $(0.6)^3=(0.6)(0.6)(0.6)=0.216$

35. $(0.12)^2=(0.12)(0.12)=0.0144$

36. $(0.06)^2=(0.06)(0.06)=0.0036$

37. $\left(-\frac{2}{3}\right)^3=$
$\left(-\frac{2}{3}\right)\left(-\frac{2}{3}\right)\left(-\frac{2}{3}\right)=-\frac{8}{27}$

38. $\left(-\frac{1}{2}\right)^4=$
$\left(-\frac{1}{2}\right)\left(-\frac{1}{2}\right)\left(-\frac{1}{2}\right)\left(-\frac{1}{2}\right)=\frac{1}{16}$

39. $3ab-4ab-2a=-ab-2a$

40. $5xy-9xy+xy-y=-3xy-y$

41. $3(x+6)+7(x+8)=$
$3x+18+7x+56=10x+74$

42. $5(x-4)-3(x-9)=$
$5x-20-3x+27=2x+7$

43. $-3(x-2)-4(x+6)=$
$-3x+6-4x-24=-7x-18$

44. $-2x-3(x-4)+2x=$
$-2x-3x+12+2x=-3x+12$

45. $2(a-1)-a-3(a-2)=$
$2a-2-a-3a+6=-2a+4$

46. $-(a-1)+3(a-2)-4a+1=$
$-a+1+3a-6-4a+1=-2a-4$

47. $\frac{1}{2}x+\frac{3}{4}x-\frac{5}{6}x+\frac{1}{24}x=$

$\frac{12}{24}x+\frac{18}{24}x-\frac{20}{24}x+\frac{1}{24}x=$

$\frac{31}{24}x-\frac{20}{24}x=\frac{11}{24}x$

48. $1.4a-1.9b+0.8a+3.6b=2.2a+1.7b$

49. $\frac{2}{5}n+\frac{1}{3}n-\frac{5}{6}n=$

$\frac{12}{30}n+\frac{10}{30}n-\frac{25}{30}n=$

$\frac{22}{30}n-\frac{25}{30}n=-\frac{3}{30}n=-\frac{1}{10}n$

50. $n-\frac{3}{4}n+2n-\frac{1}{5}n=$

$\left(1-\frac{3}{4}+2-\frac{1}{5}\right)n=$

$\left(\frac{20}{20}-\frac{15}{20}+\frac{40}{20}-\frac{4}{20}\right)n=\frac{41}{20}n$

51. $\frac{7}{x}+\frac{9}{2y}=\frac{7\cdot 2y}{x\cdot 2y}+\frac{9\cdot x}{2y\cdot x}=$

$\frac{14y}{2xy}+\frac{9x}{2xy}=\frac{14y+9x}{2xy}$

52. $\frac{5}{xy}-\frac{8}{x^2}=\frac{5\cdot x}{xy\cdot x}-\frac{8\cdot y}{x^2\cdot y}=$

$\frac{5x}{x^2y}-\frac{8y}{x^2y}=\frac{5x-8y}{x^2y}$

53. $\left(\frac{7y}{8x}\right)\left(\frac{14x}{35}\right)=\frac{\overset{1}{7y}}{\underset{4}{\cancel{8}\cancel{x}}}\cdot\frac{\overset{7}{\cancel{14}\cancel{x}}}{\underset{5}{\cancel{35}}}=\frac{7y}{20}$

54. $\left(\frac{6xy}{9y^2}\right)\div\left(\frac{15y}{18x^2}\right)=\frac{6xy}{9y^2}\cdot\frac{18x^2}{15y}=$

$\frac{108x^3y}{135y^3}=\frac{4x^3}{5y^2}$

55. $5x+8y$ for $x=-7$ and $y=-3$
$5(-7)+8(-3)=-35-24=-59$

56. $7x-9y$ for $x=-3$ and $y=4$
$7(-3)-9(4)=-21-36=-57$

57. $\frac{-5x-2y}{-2x-7y}$ for $x=6$ and $y=4$

$\frac{-5(6)-2(4)}{-2(6)-7(4)}=\frac{-30-8}{-12-28}=\frac{-38}{-40}=\frac{19}{20}$

58. $x=-4,\ y=-6:\ \frac{-3x+4y}{3x}=$

$\frac{-3(-4)+4(-6)}{3(-4)}=$

$\frac{12+(-24)}{-12}=\frac{-12}{-12}=1$

59. $5a+6b-7a-2b$ for $a=-1$ and $b=5$
$5(-1)+6(5)-7(-1)-2(5)=$
$-5+30+7-10=$
$-15+37=22$

60. $3x+7y-5x+y=$
$-2x+8y$ for $x=-4$ and $y=3$
$-2(-4)+8(3)=8+24=32$

61. $\frac{1}{4}x-\frac{2}{5}y$ for $x=\frac{2}{3}$ and $y=-\frac{5}{7}$

$\frac{1}{4}\left(\frac{2}{3}\right)-\frac{2}{5}\left(-\frac{5}{7}\right)=$

$\frac{1}{6}+\frac{2}{7}=\frac{7}{42}+\frac{12}{42}=\frac{19}{42}$

62. $a=-\frac{1}{2},\ b=\frac{1}{3}:\ a^3+b^2=$

$\left(-\frac{1}{2}\right)\left(-\frac{1}{2}\right)\left(-\frac{1}{2}\right)+\left(\frac{1}{3}\right)\left(\frac{1}{3}\right)=$

$-\frac{1}{8}+\frac{1}{9}=-\frac{9}{72}+\frac{8}{72}=-\frac{1}{72}$

63. $2x^2-3y^2$ for $x=0.6$ and $y=0.7$
$2(0.6)^2-3(0.7)^2=$
$2(0.36)-3(0.49)=$
$0.72-1.47=-0.75$

64. $w = 0.4,\ z = -0.7 : 0.7w + 0.9z =$
$0.7(0.4) + 0.9(-0.7) =$
$0.28 - 0.63 = -0.35$

65. If one number is n,
the other number is $72 - n$

66. If Joan has 3 pennies and 4 dimes,
she has $3 + 10(4) = 43$ cents.
Thus, if she has p pennies and d dimes,
she has $p + 10d$ cents.

67. 1 hour = 60 minutes
$$\frac{x \text{ words}}{1 \text{ hour}} \cdot \frac{1 \text{ hour}}{60 \text{ minutes}} = \frac{x}{60} \text{ words/minute}$$

68. Twice Harry's age is 2y. If his brother is
3 years less than twice Harry's age,
his brother is $2y - 3$ years old.

69. Larry chose n. Cindy chose 3 more than
5 times Larry's number or $5n + 3$.

70. To convert yards to inches, the
yards must be multiplied by 36.
To convert feet to inches, the
feet must be multiplied by 12. If
the file cabinet is y yards and f
feet tall, then it would be
$36y + 12f$ inches tall.

71. $100\,\text{cm} = 1\text{m}$
m meters $\cdot\ 100\text{cm/m} = 100\text{m}$

72. If Corinne has 3 nickels, 4 dimes,
and 2 quarters, she has
$5(3) + 10(4) + 25(2) = 105$ cents.
Thus, if she has n nickels, d dimes, and
q quarters she has $5n + 10d + 25q$ cents.

73. $n - 5$

74. $5 - n$

75. $10(x - 2)$

76. $10x - 2$

77. $x - 3$

78. $\frac{d}{r}$

79. $x^2 + 9$

80. $(x + 9)^2$

81. $x^3 + y^3$

82. $xy - 4$

CHAPTER 1 Test

1. $16 - 18 - 14 + 21 - 14 + 19 =$
$56 - 46 = 10$

2. $6 - [3 - (10 - 12)] =$
$6 - [3 - (-2)] =$
$6 - [3 + 2] =$
$6 - (5) =$
$6 - 5 = 1$

3. $\frac{-7(-4) - 5(-6)}{-2} =$
$\frac{28 + 30}{-2} = \frac{58}{-2} = -29$

4. $\frac{3}{4} + \frac{1}{3} \div \frac{4}{3} - \frac{1}{2} =$
$\frac{3}{4} + \frac{1}{3} \cdot \frac{3}{4} - \frac{1}{2} =$
$\frac{3}{4} + \frac{1}{4} - \frac{1}{2} =$
$\frac{4}{4} - \frac{1}{2} = 1 - \frac{1}{2} = \frac{1}{2}$

Chapter 1 Test

5. $-4^3 = -(4^3) = -(4 \cdot 4 \cdot 4) =$
$-(64) = -64$

6. $(0.2)^2 - (0.3)^3 + (0.4)^2 =$
$0.04 - 0.27 + 0.16 =$
0.173

7. $3xy - 2x - 4y$ for $x = -6$ and $y = 7$
$3(-6)(7) - 2(-6) - 4(7) =$
$-126 + 12 - 28 =$
$-154 + 12 = -142$

8. $-4x^2y - 2xy^2 + xy$ for $x = -2$ and $y = -4$
$-4(-2)^2(-4) - 2(-2)(-4)^2 + (-2)(-4) =$
$-4(4)(-4) - 2(-2)(16) + 8 =$
$64 + 64 + 8 = 136$

9. $\frac{1}{5}n - \frac{1}{3}n + n - \frac{1}{6}n =$
$\frac{6}{30}n - \frac{10}{30}n + \frac{30}{30}n - \frac{5}{30}n =$
$\frac{36}{30}n - \frac{15}{30}n = \frac{21}{30}n = \frac{7}{50}$

10. $42 = 2 \cdot 3 \cdot 7$, $70 = 2 \cdot 2 \cdot 3 \cdot 5$ } The greatest common factor is $2 \cdot 3 = 6$

11.
$12 = 2 \cdot 2 \cdot 3 = 2^2 \cdot 3$
$18 = 2 \cdot 3 \cdot 3 = 2 \cdot 3^2$
$27 = 3 \cdot 3 \cdot 3 = 3^3$
$\text{LCM} = 2^2 \cdot 3^3 = 4 \cdot 27 = 108$

12. $-x + 4(x-1) - 3(x+2) - (x+5) =$
$-x + 4x - 4 - 3x - 6 - x - 5 =$
$-x + 4x - 3x - x - 4 - 6 - 5 =$
$-x - 15$

13. $-(n-1) + 2(n-2) - 3(n-3) =$
$-n + 1 + 2n - 4 - 3n + 9 =$
$-2n + 6$

14. $\frac{2}{3}x - \frac{1}{4}y - \frac{3}{4}x - \frac{2}{3}y =$
$\frac{2}{3}x - \frac{3}{4}x - \frac{1}{4}y - \frac{2}{3}y =$
$\frac{8}{12}x - \frac{9}{12}x - \frac{3}{12}y - \frac{8}{12}y =$
$-\frac{1}{12}x - \frac{11}{12}y$

15. $\frac{5}{xy} - \frac{2}{x} + \frac{3}{y} =$
$\frac{5}{xy} - \frac{2(y)}{x(y)} + \frac{3(x)}{y(x)} =$
$\frac{5}{xy} - \frac{2y}{xy} + \frac{3x}{xy} =$
$\frac{5 - 2y + 3x}{xy}$

16. $\left(\frac{6x^2y}{11}\right) \div \left(\frac{9y^2}{22}\right) =$
$\frac{\overset{2}{\cancel{6}}x^2y}{\underset{1}{\cancel{11}}} \cdot \frac{\overset{2}{\cancel{22}}}{\underset{3}{\cancel{9}}y^2} =$
$\frac{4x^2y}{3y^2} = \frac{4x^2}{3y}$

17. Associative Property of Addition

18. Distributive Property

19. πd

20. $y - 3$

21. $2(m + 5)$

22. $\frac{15}{x}$

23. p pennies and n nickels and d dimes is $p + 5n + 10d$ cents.

24. $4n - 5$

25. y yards and f feet and i inches is $36y + 12f + i$ inches.

Chapter 2 First-Degree Equations and Inequalities of One Variable

PROBLEM SET 2.1 Solving First-Degree Equations

1. $x+9=17$
$x+9-9=17-9$
Subtract 9 from both sides.
$x=8$
The solution set is $\{8\}$.

3. $x+11=5$
$x+11-11=5-11$
Subtract 11 from both sides.
$x=-6$
The solution set is $\{-6\}$.

5. $-7=x+2$
$-7-2=x+2-2$
Subtract 2 from both sides.
$-9=x$
The solution set is $\{-9\}$.

7. $8=n+14$
$8-14=n+14-14$
Subtract 14 from both sides.
$-6=n$
The solution set is $\{-6\}$.

9. $21+y=34$
$21+y-21=34-21$
Subtract 21 from both sides.
$y=13$
The solution set is $\{13\}$.

11. $x-17=31$
$x-17+17=31+17$
Add 17 to both sides.
$x=48$
The solution set is $\{48\}$.

13. $14=x-9$
$14+9=x-9+9$
Add 9 to both sides.
$23=x$
The solution set is $\{23\}$.

15. $-26=n-19$
$-26+19=n-19+19$
Add 19 to both sides.
$-7=n$
The solution set is $\{-7\}$.

17. $y-\frac{2}{3}=\frac{3}{4}$
$y-\frac{2}{3}+\frac{2}{3}=\frac{3}{4}+\frac{2}{3}$
Add $\frac{2}{3}$ to both sides.
$y=\frac{9}{12}+\frac{8}{12}=\frac{17}{12}$
The solution set is $\left\{\frac{17}{12}\right\}$.

19. $x+\frac{3}{5}=\frac{1}{3}$
$x+\frac{3}{5}-\frac{3}{5}=\frac{1}{3}-\frac{3}{5}$
Subtract $\frac{3}{5}$ from both sides.
$x=\frac{5}{15}-\frac{9}{15}=-\frac{4}{15}$
The solution set is $\left\{-\frac{4}{15}\right\}$.

21. $b+0.19=0.46$
$b+0.19-0.19=0.46-0.19$
Subtract 0.19 from both sides.
$b=0.27$
The solution set is $\{0.27\}$.

23. $n - 1.7 = -5.2$
$n - 1.7 + 1.7 = -5.2 + 1.7$
Add 1.7 to both sides.
$n = -3.5$
The solution set is $\{-3.5\}$.

25. $15 - x = 32$
$15 - x - 15 = 32 - 15$
Subtract 15 from both sides.
$-x = 17$
$-1(-x) = -1(17)$
Multiply both sides by -1.
$x = -17$
The solution set is $\{-17\}$.

27. $-14 - n = 21$
$-14 - n + 14 = 21 + 14$
Add 14 to both sides.
$-n = 35$
$-1(-n) = -1(35)$
Multiply both sides by -1.
$n = -35$
The solution set is $\{-35\}$.

31. $7x = -56$
$\frac{7x}{7} = \frac{-56}{7}$
Divide both sides by 7.
$x = -8$
The solution set is $\{-8\}$.

31. $-6x = 102$
$\frac{-6x}{-6} = \frac{102}{-6}$
Divide both sides by -6.
$x = -17$
The solution set is $\{-17\}$.

33. $5x = 37$
$\frac{5x}{5} = \frac{37}{5}$
Divide both sides by 5.
$x = \frac{37}{5}$
The solution set is $\left\{\frac{37}{5}\right\}$.

35. $-18 = 6n$
$\frac{-18}{6} = \frac{6n}{6}$
Divide both sides by 6.
$-3 = n$
The solution set is $\{-3\}$.

37. $-26 = -4n$
$\frac{-26}{-4} = \frac{-4n}{-4}$
Divide both sides by -4.
$\frac{13}{2} = n$
The solution set is $\left\{\frac{13}{2}\right\}$.

39. $\frac{t}{9} = 16$
$9\left(\frac{t}{9}\right) = 9(16)$
Multiply both sides by 9.
$t = 144$
The solution set is $\{144\}$.

41. $\frac{n}{-8} = -3$
$-8\left(\frac{n}{-8}\right) = -8(-3)$
Multiply both sides by -8.
$n = 24$
The solution set is $\{24\}$.

43. $-x = 15$
$-1(-x) = -1(15)$
Multiply both sides by -1.
$x = -15$
The solution set is $\{-15\}$.

45. $\frac{3}{4}x = 18$

$\frac{4}{3}\left(\frac{3}{4}x\right) = \frac{4}{3}(18)$

Multiply both sides by $\frac{4}{3}$.

$x = 24$

The solution set is $\{24\}$.

47. $-\frac{2}{5}n = 14$

$-\frac{5}{2}\left(-\frac{2}{5}n\right) = -\frac{5}{2}(14)$

Multiply both sides by $-\frac{5}{2}$.

$n = -35$

The solution set is $\{-35\}$.

49. $\frac{2}{3}n = \frac{1}{5}$

$\frac{3}{2}\left(\frac{2}{3}n\right) = \frac{3}{2}\left(\frac{1}{5}\right)$

Multiply both sides by $\frac{3}{2}$.

$n = \frac{3}{10}$

The solution set is $\left\{\frac{3}{10}\right\}$.

51. $\frac{5}{6}n = -\frac{3}{4}$

$\frac{6}{5}\left(\frac{5}{6}n\right) = \frac{6}{5}\left(-\frac{3}{4}\right)$

Multiply both sides by $\frac{6}{5}$.

$n = -\frac{9}{10}$

The solution set is $\left\{-\frac{9}{10}\right\}$.

53. $\frac{3x}{10} = \frac{3}{20}$

$\frac{10}{3}\left(\frac{3x}{10}\right) = \frac{10}{3}\left(\frac{3}{20}\right)$

Multiply both sides by $\frac{10}{3}$.

$x = \frac{1}{2}$

The solution set is $\left\{\frac{1}{2}\right\}$.

55. $\frac{-y}{2} = \frac{1}{6}$

$-2\left(\frac{-y}{2}\right) = -2\left(\frac{1}{6}\right)$

Multiply both sides by -2.

$y = -\frac{1}{3}$

The solution set is $\left\{-\frac{1}{3}\right\}$.

57. $-\frac{4}{3}x = -\frac{9}{8}$

$-\frac{3}{4}\left(-\frac{4}{3}x\right) = -\frac{3}{4}\left(-\frac{9}{8}\right)$

Multiply both sides by $-\frac{3}{4}$.

$x = \frac{27}{32}$

The solution set is $\left\{\frac{27}{32}\right\}$.

59. $-\frac{5}{12} = \frac{7}{6}x$

$\frac{6}{7}\left(-\frac{5}{12}\right) = \frac{6}{7}\left(\frac{7}{6}x\right)$

Multiply both sides by $\frac{6}{7}$.

$-\frac{5}{14} = x$

The solution set is $\left\{-\frac{5}{14}\right\}$.

61. $-\frac{5}{7}x = 1$

$-\frac{7}{5}\left(-\frac{5}{7}x\right) = -\frac{7}{5}(1)$

Multiply both sides by $-\frac{7}{5}$.

$x = -\frac{7}{5}$

The solution set is $\left\{-\frac{7}{5}\right\}$.

63. $-4n = \frac{1}{3}$

$-\frac{1}{4}(-4n) = -\frac{1}{4}\left(\frac{1}{3}\right)$

Multiply both sides by $-\frac{1}{4}$.

$n = -\frac{1}{12}$

The solution set is $\left\{-\frac{1}{12}\right\}$.

65. $-8n = \frac{6}{5}$

$-\frac{1}{8}(-8n) = -\frac{1}{8}\left(\frac{6}{5}\right)$

Multiply both sides by $-\frac{1}{8}$.

$n = -\frac{3}{20}$

The solution set is $\left\{-\frac{3}{20}\right\}$.

67. $1.2x = 0.36$

$\frac{1.2x}{1.2} = \frac{0.36}{1.2}$

Divide both sides by 1.2.

$x = 0.3$

The solution set is $\{0.3\}$.

69. $30.6 = 3.4n$

$\frac{30.6}{3.4} = \frac{3.4n}{3.4}$

Divide both sides by 3.4.

$9 = n$

The solution set is $\{9\}$.

71. $-3.4x = 17$

$\frac{-3.4x}{-3.4} = \frac{17}{-3.4}$

Divide both sides by -3.4.

$x = -5$

The solution set is $\{-5\}$.

PROBLEM SET 2.2 Equations and Problem Solving

1. $2x + 5 = 13$

$2x + 5 - 5 = 13 - 5$

Subtract 5 from both sides.

$2x = 8$

$\frac{2x}{2} = \frac{8}{2}$

Divide both sides by 2.

$x = 4$

The solution set is $\{4\}$.

3. $5x + 2 = 32$

$5x + 2 - 2 = 32 - 2$

Subtract 2 from both sides.

$5x = 30$

$\frac{5x}{5} = \frac{30}{5}$

Divide both sides by 5.

$x = 6$

The solution set is $\{6\}$.

5. $3x - 1 = 23$

$3x - 1 + 1 = 23 + 1$

Add 1 to both sides.

$3x = 24$

$\frac{3x}{3} = \frac{24}{3}$

Divide both sides by 3.

$x = 8$

The solution set is $\{8\}$.

7. $4n - 3 = 41$

$4n - 3 + 3 = 41 + 3$

Add 3 to both sides.

$4n = 44$

$\frac{4n}{4} = \frac{44}{4}$

Divide both sides by 4.

$x = 11$

The solution set is $\{11\}$.

9. $6y - 1 = 16$

$6y - 1 + 1 = 16 + 1$

Add 1 to both sides.

$6y = 17$

$\frac{6y}{6} = \frac{17}{6}$

Divide both sides by 6.

$x = \frac{17}{6}$

The solution set is $\left\{\frac{17}{6}\right\}$.

11. $2x + 3 = 22$

$2x + 3 - 3 = 22 - 3$

Subtract 3 from both sides.

$2x = 19$

$\frac{2x}{2} = \frac{19}{2}$

Divide both sides by 2.

$x = \frac{19}{2}$

The solution set is $\left\{\frac{19}{2}\right\}$.

13. $10 = 3t - 8$

$10 + 8 = 3t - 8 + 8$

Add 8 to both sides.

$18 = 3t$

$\frac{18}{3} = \frac{3t}{3}$

Divide both sides by 3.

$6 = t$

The solution set is $\{6\}$.

15. $5x + 14 = 9$

$5x + 14 - 14 = 9 - 14$

Subtract 14 from both sides.

$5x = -5$

$\frac{5x}{5} = \frac{-5}{5}$

Divide both sides by 5.

$x = -1$

The solution set is $\{-1\}$.

17. $18 - n = 23$

$18 - n - 18 = 23 - 18$

Subtract 18 from both sides.

$-n = 5$

$-1(-n) = -1(5)$

Multiply both sides by -1.

$n = -5$

The solution set is $\{-5\}$.

19. $-3x + 2 = 20$

$-3x + 2 - 2 = 20 - 2$

Subtract 2 from both sides.

$-3x = 18$

$\frac{-3x}{-3} = \frac{18}{-3}$

Divide both sides by -3.

$x = -6$

The solution set is $\{-6\}$.

Problem Set 2.2

21. $7+4x=29$

$7+4x-7=29-7$

Subtract 7 from both sides.

$4x=22$

$\frac{4x}{4}=\frac{22}{4}$

Divide both sides by 4.

$x=\frac{11}{2}$

The solution set is $\left\{\frac{11}{2}\right\}$.

23. $16=-2-9a$

$16+2=-2-9a+2$

Add 2 to both sides.

$18=-9a$

$\frac{18}{-9}=\frac{-9a}{-9}$

Divide both sides by -9.

$-2=a$

The solution set is $\{-2\}$.

25. $-7x+3=-7$

$-7x+3-3=-7-3$

Subtract 3 from both sides.

$-7x=-10$

$\frac{-7x}{-7}=\frac{-10}{-7}$

Divide both sides by -7.

$x=\frac{10}{7}$

The solution set is $\left\{\frac{10}{7}\right\}$.

27. $17-2x=-19$

$17-2x-17=-19-17$

Subtract 17 from both sides.

$-2x=-36$

$\frac{-2x}{-2}=\frac{-36}{-2}$

Divide both sides by -2.

$x=18$

The solution set is $\{18\}$.

29. $-16-4x=9$

$-16-4x+16=9+16$

Add 16 to both sides.

$-4x=25$

$\frac{-4x}{-4}=\frac{25}{-4}$

Divide both sides by -4.

$x=-\frac{25}{4}$

The solution set is $\left\{-\frac{25}{4}\right\}$.

31. $-12t+4=88$

$-12t+4-4=88-4$

Subtract 4 from both sides.

$-12t=84$

$\frac{-12t}{-12}=\frac{84}{-12}$

Divide both sides by -12.

$t=-7$

The solution set is $\{-7\}$.

33. $14y+15=-33$

$14y+15-15=-33-15$

Subtract 15 from both sides.

$14y=-48$

$\frac{14y}{14}=\frac{-48}{14}$

Divide both sides by 14.

$y=-\frac{24}{7}$

The solution set is $\left\{-\frac{24}{7}\right\}$.

35. $32-16n=-8$

$32-16n-32=-8-32$

Subtract 32 from both sides.

$-16n=-40$

$\frac{-16n}{-16}=\frac{-40}{-16}$

Divide both sides by -16.

$n=\frac{5}{2}$

The solution set is $\left\{\frac{5}{2}\right\}$.

37. $17x - 41 = -37$
$17x - 41 + 41 = -37 + 41$
Add 41 to both sides.
$17x = 4$
$\frac{17x}{17} = \frac{4}{17}$
Divide both sides by 17.
$x = \frac{4}{17}$
The solution set is $\left\{\frac{4}{17}\right\}$.

39. $29 = -7 - 15x$
$29 + 7 = -7 - 15x + 7$
Add 7 to both sides.
$36 = -15x$
$\frac{36}{-15} = \frac{-15x}{-15}$
Divide both sides by -15.
$-\frac{12}{5} = x$
The solution set is $\left\{-\frac{12}{5}\right\}$.

41. Let n represent the number.
$n + 12 = 21$
$n = 9$
The number is 9.

43. Let n represent the number.
$n - 9 = 13$
$n = 22$
The number is 22.

45. Let c represent the cost of the other item.
$c + 25 = 43$
$c = 18$
The cost of the other item is $18.

47. Let x represent Nora's age now; therefore, $x + 6$ represents her age 6 years from now.
$x + 6 = 41$
$x = 35$
Nora is presently 35 years old.

49. Let h represent his hourly rate. Then the product of the number of hours times the hourly rate equals total amount earned.
$6h = 39$
$h = \frac{39}{6} = 6.5$
His hourly rate was $6.50.

51. Let x represent the number.
Then 3 times the number is $3x$.
$3x + 6 = 24$
$3x = 18$
$x = 6$
The number is 6.

53. Let n represent the number.
Then 3 times the number is $3n$.
$19 = 3n + 4$
$15 = 3n$
$5 = n$
The number is 5.

55. Let x represent the number.
Then 5 times the number is $5x$.
$49 = 5x - 6$
$55 = 5x$
$11 = x$
The number is 11.

57. Let n represent the number.
$6n - 1 = 47$
$6n = 48$
$n = 8$
The number is 8.

59. Let x represent the number.
$27 - 8x = 3$
$-8x = -24$
$x = 3$
The number is 3.

61. Let c represent the cost of the ring.
$550 = 2c - 50$
$600 = 2c$
$300 = c$
The cost of the ring was $300.

63. Let w represent the width of the floor.
$18 = 5w - 2$
$20 = 5w$
$4 = w$
The width of the floor is 4 meters.

65. Let n represent the number of cars sold during December of 1983.
$32 = 2n + 4$
$28 = 2n$
$14 = n$
They sold 14 cars during December of 1983.

67. Let h represent the hours of labor.
$156 = 36 + 24h$
$120 = 24h$
$5 = h$
There were 5 hours of labor in the repair bill.

PROBLEM SET 2.3 More on Solving Equations and Problem Solving

1.
$$2x + 7 + 3x = 32$$
$$5x + 7 = 32$$
$$5x + 7 - 7 = 32 - 7$$
$$5x = 25$$
$$\frac{5x}{5} = \frac{25}{5}$$
$$x = 5$$
The solution set is $\{5\}$.

3.
$$7x - 4 - 3x = -36$$
$$4x - 4 = -36$$
$$4x - 4 + 4 = -36 + 4$$
$$4x = -32$$
$$\frac{4x}{4} = \frac{-32}{4}$$
$$x = -8$$
The solution set is $\{-8\}$.

5.
$$3y - 1 + 2y - 3 = 4$$
$$5y - 4 = 4$$
$$5y - 4 + 4 = 4 + 4$$
$$5y = 8$$
$$\frac{5y}{5} = \frac{8}{5}$$
$$y = \frac{8}{5}$$
The solution set is $\left\{\frac{8}{5}\right\}$.

7.
$$5n - 2 - 8n = 31$$
$$-3n - 2 = 31$$
$$-3n - 2 + 2 = 31 + 2$$
$$-3n = 33$$
$$\frac{-3n}{-3} = \frac{33}{-3}$$
$$n = -11$$
The solution set is $\{-11\}$.

9.
$$\begin{aligned} -2n+1-3n+n-4&=7\\ -4n-3&=7\\ -4n-3+3&=7+3\\ -4n&=10\\ \frac{-4n}{-4}&=\frac{10}{-4}\\ n&=-\frac{5}{2}\end{aligned}$$
The solution set is $\left\{-\frac{5}{2}\right\}$.

11.
$$\begin{aligned} 3x+4&=2x-5\\ 3x+4-2x&=2x-5-2x\\ x+4-4&=-5-4\\ x&=-9\end{aligned}$$
The solution set is $\{-9\}$.

13.
$$\begin{aligned} 5x-7&=6x-9\\ 5x-7-6x&=6x-9-6x\\ -x-7+7&=-9+7\\ -x&=-2\\ x&=2\end{aligned}$$
The solution set is $\{2\}$.

15.
$$\begin{aligned} 6x+1&=3x-8\\ 6x+1-3x&=3x-8-3x\\ 3x+1-1&=-8-1\\ 3x&=-9\\ \frac{3x}{3}&=\frac{-9}{3}\\ x&=-3\end{aligned}$$
The solution set is $\{-3\}$.

17.
$$\begin{aligned} 7y-3&=5y+10\\ 7y-3-5y&=5y+10-5y\\ 2y-3+3&=10+3\\ 2y&=13\\ \frac{2y}{2}&=\frac{13}{2}\\ y&=\frac{13}{2}\end{aligned}$$
The solution set is $\left\{\frac{13}{2}\right\}$.

19.
$$\begin{aligned} 8n-2&=8n-7\\ 8n-2-8n&=8n-7-8n\\ -2&=-7\end{aligned}$$
Since $-2\neq -7$, this is a contradiction.
The solution set is $\emptyset$.

21.
$$\begin{aligned} -2x-7&=-3x+10\\ -2x-7+3x&=-3x+10+3x\\ x-7+7&=10+7\\ x&=17\end{aligned}$$
The solution set is $\{17\}$.

23.
$$\begin{aligned} -3x+5&=-5x-8\\ -3x+5-5&=-5x-8-5\\ -3x+5x&=-5x-13+5x\\ 2x&=-13\\ \frac{2x}{2}&=\frac{-13}{2}\\ x&=-\frac{13}{2}\end{aligned}$$
The solution set is $\left\{-\frac{13}{2}\right\}$.

25.
$$\begin{aligned} -7-6x&=9-9x\\ -7-6x+9x&=9-9x+9x\\ -7+3x&=9\\ -7+3x+7&=9+7\\ 3x&=16\\ \frac{3x}{3}&=\frac{16}{3}\\ x&=\frac{16}{3}\end{aligned}$$
The solution set is $\left\{\frac{16}{3}\right\}$.

27.
$$\begin{aligned} 2x-1-x&=x-1\\ x-1&=x-1\\ x-1-x&=x-1-x\\ -1&=-1\end{aligned}$$
Since $-1=-1$, this is an identity.
The solution set is {all real numbers}.

Problem Set 2.3

29.
$$\begin{aligned} 5n-4-n &= -3n-6+n \\ 4n-4 &= -2n-6 \\ 4n-4+2n &= -2n-6+2n \\ 6n-4+4 &= -6+4 \\ 6n &= -2 \\ \frac{6n}{6} &= \frac{-2}{6} = -\frac{1}{3} \end{aligned}$$
The solution set is $\left\{-\frac{1}{3}\right\}$.

31.
$$\begin{aligned} -7-2n-6n &= 7n-5n+12 \\ -7-8n &= 2n+12 \\ -7-8n-2n &= 2n+12-2n \\ -7-10n+7 &= 12+7 \\ -10n &= 19 \\ \frac{-10n}{-10} &= \frac{19}{-10} = -\frac{19}{10} \end{aligned}$$
The solution set is $\left\{-\frac{19}{10}\right\}$.

33. Let n represent the number. Then $4n$ represents four times the number.
$$\begin{aligned} n+4n &= 85 \\ 5n &= 85 \\ n &= 17 \end{aligned}$$
The number is 17.

35. Let n represent the first odd number, then $n+2$ represents the next odd number..
$$\begin{aligned} n+(n+2) &= 72 \\ 2n+2 &= 72 \\ 2n &= 70 \\ n &= 35 \end{aligned}$$
The two consecutive odd numbers are 35 and 37.

37. Let n, $n+2$, and $n+4$ represent the three consecutive even numbers.
$$\begin{aligned} n+(n+2)+(n+4) &= 114 \\ 3n+6 &= 114 \\ 3n &= 108 \\ n &= 36 \end{aligned}$$
The numbers are 36, 38, and 40.

39. Let n represent the number.
$$\begin{aligned} 3n+2 &= 7n-4 \\ -4n+2 &= -4 \\ -4n &= -6 \\ n &= \frac{-6}{-4} = \frac{3}{2} \end{aligned}$$
The number is $\frac{3}{2}$.

41. Let n represent the number.
$$\begin{aligned} n+5n &= 3n-18 \\ 6n &= 3n-18 \\ 3n &= -18 \\ n &= -6 \end{aligned}$$
The number is -6.

43. Let a represent the first angle. Then the other angle is represented by $2a-6$. Since the angles are complementary, the sum of their measures is $90°$.
$$\begin{aligned} a+(2a-6) &= 90 \\ 3a-6 &= 90 \\ 3a &= 96 \\ a &= 32 \end{aligned}$$
The measures of the angles are $32°$ and $2(32)-6=58°$.

45. Let a represent the smaller angle. Then $3a-20$ represents the larger angle. Since they are supplementary angles, the sum of their measures is $180°$.
$$\begin{aligned} a+(3a-20) &= 180 \\ 4a-20 &= 180 \\ 4a &= 200 \\ a &= 50 \end{aligned}$$
The measures of the angles are $50°$, and $3(50)-20=130°$.

47. Let a represent the smaller of the other two angles; then $a+10$ represents the larger angle. The sum of the measures of the three angles of a triangle is $180°$.

$40+a+a+10=180$
$50+2a=180$
$2a=130$
$a=65$

The other two angles of the triangle would be $65°$ and $75°$.

49. Let x represent the price per share of the stock.

$2x-17=35$
$2x=52$
$x=26$

He paid \$26 per share for the stock.

51. Let h represent Bob's normal hourly rate. Bob worked 8 hours at the higher rate.

$40h+8(2h)=504$
$40h+16h=504$
$56h=504$
$h=9$

Bob's normal hourly rate is \$9.00 per hour.

53. Let m represent the number of males; then $3m$ represents the number of females.

$m+3m=600$
$4m=600$
$m=150$

Therefore, 150 males and $3(150)=450$ females attended the concert.

55. Let x represent the number of votes that Melton received; then Sanchez received $2x+10$.

$x+(2x+10)=1030$
$3x+10=1030$
$3x=1020$
$x=340$

Sanchez received $2(340)+10=690$ votes.

57. Let x represent the length of the shorter piece; then $x+8$ represents the length of the other piece.

$x+(x+8)=20$
$2x+8=20$
$2x=12$
$x=6$

The shorter piece would be 6 feet long.

PROBLEM SET 2.4 Equations Involving Parentheses and Fractional Forms

1. $7(x+2)=21$
$7x+14=21$ Apply distributive property.
$7x=7$ Subtract 14 from both sides.
$x=1$ Divide both sides by 7.
The solution set is $\{1\}$.

3. $5(x-3)=35$
$5x-15=35$ Apply distributive property.
$5x=50$ Add 15 to both sides.
$x=10$ Divide both sides by 5.
The solution set is $\{10\}$.

5. $-3(x+5)=12$
$-3x-15=12$ Apply distributive property.
$-3x=27$ Add 15 to both sides.
$x=-9$ Divide both sides by -3.
The solution set is $\{-9\}$.

7. $4(n-6)=5$
$4n-24=5$ Apply distributive property.
$4n=29$ Add 24 to both sides.
$n=\frac{29}{4}$ Divide both sides by 4.
The solution set is $\left\{\frac{29}{4}\right\}$.

Problem Set 2.4

9. $6(n+7)=8$

$6n+42=8$ Apply distributive property.

$6n=-34$ Subtract 42 from both sides.

$n=\frac{-34}{6}$ Divide both sides by 6.

$n=-\frac{17}{3}$ Reduce

The solution set is $\left\{-\frac{17}{3}\right\}$.

11. $-10=-5(t-8)$

$-10=-5t+40$ Apply distributive property.

$-50=-5t$ Subtract 40 from both sides.

$10=t$ Divide both sides by -5.

The solution set is $\{10\}$.

13. $5(x-4)=4(x+6)$

$5x-20=4x+24$ Apply distributive property.

$x-20=24$ Subtract $4x$ from both sides.

$x=44$ Add 20 to both sides.

The solution set is $\{44\}$.

We will dicontinue giving reasons for each step but will continue to show enough of the work so that you can follow the steps. If a new technique is introduced, then we will indicate some of the reasons again.

15. $8(x+1)=9(x-2)$

$8x+8=9x-18$

$-x+8=-18$

$-x=-26$

$x=26$

The solution set is $\{26\}$.

17. $8(t+5)=6(t-6)$

$8t+40=6t-36$

$2t+40=-36$

$2t=-76$

$t=-38$

The solution set is $\{-38\}$.

19. $3(2t+1)=4(3t-2)$

$6t+3=12t-8$

$-6t+3=-8$

$-6t=-11$

$t=\frac{-11}{-6}=\frac{11}{6}$

The solution set is $\left\{\frac{11}{6}\right\}$.

21. $-2(x-6)=-(x-9)$

$-2x+12=-x+9$

$-x+12=9$

$-x=-3$

$x=3$

The solution set is $\{3\}$.

23. $-3(t-4)-2(t+4)=9$

$-3t+12-2t-8=9$

$-5t+4=9$

$-5t=5$

$t=-1$

The solution set is $\{-1\}$.

25. $3(n-10)-5(n+12)=-86$

$3n-30-5n-60=-86$

$-2n-90=-86$

$-2n=4$

$n=-2$

The solution set is $\{-2\}$.

27. $3(x+1)+4(2x-1)=5(2x+3)$

$3x+3+8x-4=10x+15$

$11x-1=10x+15$

$x-1=15$

$x=16$

The solution set is $\{16\}$.

29. $$\begin{aligned} -(x+2)+2(x-3) &= -2(x-7) \\ -x-2+2x-6 &= -2x+14 \\ x-8 &= -2x+14 \\ 3x-8 &= 14 \\ 3x &= 22 \\ x &= \frac{22}{3} \end{aligned}$$
The solution set is $\left\{\frac{22}{3}\right\}$.

31. $$\begin{aligned} 5(2x-1)-(3x+4) &= 4(x+3)-27 \\ 10x-5-3x-4 &= 4x+12-27 \\ 7x-9 &= 4x-15 \\ 3x-9 &= -15 \\ 3x &= -6 \\ x &= -2 \end{aligned}$$
The solution set is $\{-2\}$.

33. $$\begin{aligned} -(a-1)-(3a-2) &= 6+2(a-1) \\ -a+1-3a+2 &= 6+2a-2 \\ -4a+3 &= 2a+4 \\ -6a+3 &= 4 \\ -6a &= 1 \\ a &= -\frac{1}{6} \end{aligned}$$
The solution set is $\left\{-\frac{1}{6}\right\}$.

35. $$\begin{aligned} 3(x-1)+2(x-3) &= -4(x-2)+10(x+4) \\ 3x-3+2x-6 &= -4x+8+10x+40 \\ 5x-9 &= 6x+48 \\ -x-9 &= 48 \\ -x &= 57 \\ x &= -57 \end{aligned}$$
The solution set is $\{-57\}$.

37. $$\begin{aligned} 3-7(x-1) &= 9-6(2x+1) \\ 3-7x+7 &= 9-12x-6 \\ -7x+10 &= -12x+3 \\ 5x+10 &= 3 \\ 5x &= -7 \\ x &= -\frac{7}{5} \end{aligned}$$
The solution set is $\left\{-\frac{7}{5}\right\}$.

For Problems 39 – 60, we begin each solution by multiplying both sides of the given equation by the least common denominator of all of the denominators in the equation. This has the effect of "clearing the equation of all fractions."

39. $$\begin{aligned} \frac{3}{4}x-\frac{2}{3} &= \frac{5}{6} \\ 12\left(\frac{3}{4}x-\frac{2}{3}\right) &= 12\left(\frac{5}{6}\right) \\ 9x-8 &= 10 \\ 9x &= 18 \\ x &= 2 \end{aligned}$$
The solution set is $\{2\}$.

41. $$\begin{aligned} \frac{5}{6}x+\frac{1}{4} &= -\frac{9}{4} \\ 12\left(\frac{5}{6}x+\frac{1}{4}\right) &= 12\left(-\frac{9}{4}\right) \\ 10x+3 &= -27 \\ 10x &= -30 \\ x &= -3 \end{aligned}$$
The solution set is $\{-3\}$.

43. $$\begin{aligned} \frac{1}{2}x-\frac{3}{5} &= \frac{3}{4} \\ 20\left(\frac{1}{2}x-\frac{3}{5}\right) &= 20\left(\frac{3}{4}\right) \\ 10x-12 &= 15 \\ 10x &= 27 \\ x &= \frac{27}{10} \end{aligned}$$
The solution set is $\left\{\frac{27}{10}\right\}$.

45. $$\begin{aligned} \frac{n}{3}+\frac{5n}{6} &= \frac{1}{8} \\ 24\left(\frac{n}{3}+\frac{5n}{6}\right) &= 24\left(\frac{1}{8}\right) \\ 8n+20n &= 3 \\ 28n &= 3 \\ n &= \frac{3}{28} \end{aligned}$$
The solution set is $\left\{\frac{3}{28}\right\}$.

47.
$$\frac{5y}{6}-\frac{3}{5}=\frac{2y}{3}$$
$$30\left(\frac{5y}{6}-\frac{3}{5}\right)=30\left(\frac{2y}{3}\right)$$
$$25y-18=20y$$
$$5y-18=0$$
$$5y=18$$
$$y=\frac{18}{5}$$
The solution set is $\left\{\frac{18}{5}\right\}$.

49.
$$\frac{h}{6}+\frac{h}{8}=1$$
$$24\left(\frac{h}{6}+\frac{h}{8}\right)=24(1)$$
$$4h+3h=24$$
$$7h=24$$
$$h=\frac{24}{7}$$
The solution set is $\left\{\frac{24}{7}\right\}$.

51.
$$\frac{x+2}{3}+\frac{x+3}{4}=\frac{13}{3}$$
$$12\left(\frac{x+2}{3}+\frac{x+3}{4}\right)=12\left(\frac{13}{3}\right)$$
$$4(x+2)+3(x+3)=52$$
$$4x+8+3x+9=52$$
$$7x+17=52$$
$$7x=35$$
$$x=5$$
The solution set is $\{5\}$.

53.
$$\frac{x-1}{5}-\frac{x+4}{6}=-\frac{13}{15}$$
$$30\left(\frac{x-1}{5}-\frac{x+4}{6}\right)=30\left(-\frac{13}{15}\right)$$
$$6(x-1)-5(x+4)=-26$$
$$6x-6-5x-20=-26$$
$$x-26=-26$$
$$x=0$$
The solution set is $\{0\}$.

55.
$$\frac{x+8}{2}-\frac{x+10}{7}=\frac{3}{4}$$
$$28\left(\frac{x+8}{2}-\frac{x+10}{7}\right)=28\left(\frac{3}{4}\right)$$
$$14(x+8)-4(x+10)=21$$
$$14x+112-4x-40=21$$
$$10x+72=21$$
$$10x=-51$$
$$x=-\frac{51}{10}$$
The solution set is $\left\{-\frac{51}{10}\right\}$.

57.
$$\frac{x-2}{8}-1=\frac{x+1}{4}$$
$$8\left(\frac{x-2}{8}-1\right)=8\left(\frac{x+1}{4}\right)$$
$$1(x-2)-8=2(x+1)$$
$$x-2-8=2x+2$$
$$x-10=2x+2$$
$$-10=x+2$$
$$-12=x$$
The solution set is $\{-12\}$.

59.
$$\frac{x+1}{4}=\frac{x-3}{6}+2$$
$$12\left(\frac{x+1}{4}\right)=12\left(\frac{x-3}{6}+2\right)$$
$$3(x+1)=2(x-3)+24$$
$$3x+3=2x-6+24$$
$$3x+3=2x+18$$
$$x+3=18$$
$$x=15$$
The solution set is $\{15\}$.

61. Let n and $n+1$ represent the consecutive whole numbers.
$$n+4(n+1)=39$$
$$n+4n+4=39$$
$$5n=35$$
$$n=7$$
The numbers are 7 and 8.

63. Let n, $n+1$, and $n+2$ represent the consecutive whole numbers.

$$2(n+n+1) = 3(n+2)+10$$
$$2n+2n+2 = 3n+6+10$$
$$4n+2 = 3n+16$$
$$n = 14$$

The numbers are 14, 15, and 16.

65. Let n represent the smaller number; then, $17-n$ represents the larger number.

$$2n = (17-n)+1$$
$$2n = 17-n+1$$
$$2n = 18-n$$
$$3n = 18$$
$$n = 6$$

The numbers are 6 and 11.

67. Let n represent the number.

$$\frac{1}{3}n+20 = \frac{3}{4}n$$
$$12\left(\frac{1}{3}n+20\right) = 12\left(\frac{3}{4}n\right)$$
$$4n+240 = 9n$$
$$240 = 5n$$
$$48 = n$$

The number is 48.

69. Let x represent the time waiting in line at the grocery store.

$$4 = \frac{1}{2}x-3$$
$$2(4) = 2\left(\frac{1}{2}x-3\right)$$
$$8 = 2\left(\frac{1}{2}x\right)-2(3)$$
$$8 = x-6$$
$$14 = x$$

Mrs. Nelson waited 14 minutes in line at the grocery store.

71. Let x represent the shorter piece of board; then $20-x$ represents the longer piece.

$$4x = 3(20-x)-4$$
$$4x = 60-3x-4$$
$$4x = 56-3x$$
$$7x = 56$$
$$x = 8$$

The length of the pieces are 8 feet and 12 feet.

73. Let x represent the number of nickels; then $35-x$ is the number of quarters.

$$.05x+.25(35-x) = 5.75$$
$$.05x+8.75-.25x = 5.75$$
$$-.20x+8.75 = 5.75$$
$$-.20x = -3$$
$$x = \frac{-3}{-.20}$$
$$x = 15$$

There are 15 nickels and $35-15 = 20$ quarters.

75. Let n represent the number of nickels, $2n$ the number of dimes, and $2n+10$ the number of quarters.

$$n+2n+2n+10 = 210$$
$$5n+10 = 210$$
$$5n = 200$$
$$n = 40$$

Max has 40 nickels, $2(40) = 80$ dimes, and $2(40)+10 = 90$ quarters.

77. Let d represent the number of dimes and $18-d$ the number of quarters. The value in cents of the dimes is $10d$ and $25(18-d)$ is the value in cents of the quarters.

$$10d+25(18-d) = 330$$
$$10d+450-25d = 330$$
$$-15d+450 = 330$$
$$-15d = -120$$
$$d = 8$$

Maida has 8 dimes and $18-8 = 10$ quarters.

79. Let x represent the number of crabs; then $3x$ is the number of fish; then $x+2$ is the number of plants.

$$x + 3x + x + 2 = 22$$
$$5x + 2 = 22$$
$$5x = 20$$
$$x = 4$$

There are 4 crabs, $3(4) = 12$ fish, and $4 + 2 = 6$ plants.

81. Let a represent the measure of the angle; then $180 - a$ represents its supplement and $90 - a$ represents its complement.

$$180 - a = 2(90 - a) + 30$$
$$180 - a = 180 - 2a + 30$$
$$180 - a = -2a + 210$$
$$a = 30$$

The angle has a measure of $30°$.

83. Let c represent the measure of angle C. Then $\frac{1}{5}c - 2$ represents angle A and $\frac{1}{2}c - 5$ represents angle B. The sum of the measures of all angles in a triangle is $180°$.

$$c + \frac{1}{5}c - 2 + \frac{1}{2}c - 5 = 180$$
$$10\left[c + \frac{1}{5}c - 2 + \frac{1}{2}c - 5\right] = 10(180)$$
$$10c + 2c - 20 + 5c - 50 = 1800$$
$$17c - 70 = 1800$$
$$17c = 1870$$
$$c = 110$$

Angle A measures $\frac{1}{5}(110) - 2 = 20°$, angle B measures $\frac{1}{2}(110) - 5 = 50°$, and angle C measures $110°$.

85. Let a represent the measure of the angle; then $180 - a$ represents its supplement and $90 - a$ represents its complement.

$$180 - a = 3(90 - a) - 10$$
$$180 - a = 270 - 3a - 10$$
$$180 - a = -3a + 260$$
$$2a = 80$$
$$a = 40$$

The angle has a measure of $40°$.

PROBLEM SET 2.5 Inequalities

1. The left side simplifies to $2(3) - 4(5) = 6 - 20 = -14$ and the right side to $5(3) - 2(-1) + 4 = 15 + 2 + 4 = 21$. Since $-14 < 21$, the given inequality is true.

3. The left side simplifies to $\frac{2}{3} - \frac{3}{4} + \frac{1}{6} = \frac{8}{12} - \frac{9}{12} + \frac{2}{12} = \frac{1}{12}$ and the right side to $\frac{1}{5} + \frac{3}{4} - \frac{7}{10} = \frac{4}{20} + \frac{15}{20} - \frac{14}{20} = \frac{5}{20} = \frac{1}{4}$. Since $\frac{1}{12} < \frac{1}{4}$, the given inequality is false.

5. The left side simplifies to $\left(-\frac{1}{2}\right)\left(\frac{4}{9}\right) = -\frac{2}{9}$ and the right side to $\left(\frac{3}{5}\right)\left(-\frac{1}{3}\right) = -\frac{1}{5}$. Since $-\frac{2}{9} < -\frac{1}{5}$, the given inequality is false.

7. The left side simplifies to
$\frac{3}{4}+\frac{2}{3}\div\frac{1}{5}=\frac{3}{4}+\frac{2}{3}\cdot\frac{5}{1}=\frac{3}{4}+\frac{10}{3}$
$=\frac{9}{12}+\frac{40}{12}=\frac{49}{12}$ and the right side to
$\frac{2}{3}+\frac{1}{2}\div\frac{3}{4}=\frac{2}{3}+\frac{1}{2}\cdot\frac{4}{3}=\frac{2}{3}+\frac{2}{3}=\frac{4}{3}.$
Since $\frac{49}{12}>\frac{4}{3}$, the given inequality is true.

9. The left side simplifies to
$0.16+0.34=0.50$
and the right side to
$0.23+0.17=0.40.$
Since $0.50>0.40$, the given inequality is true.

11. $\{x|x>-2\}$ or $(-2,\infty)$

13. $\{x|x\leq 3\}$ or $(-\infty, 3]$

15. $\{x|x>2\}$ or $(2,\infty)$

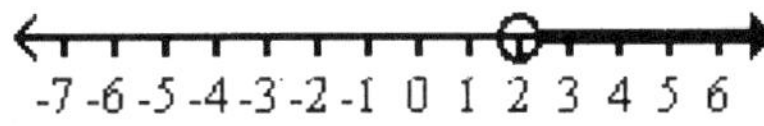

17. $\{x|x\leq -2\}$ or $(-\infty, -2]$

19. $\{x|x<-1\}$ or $(-\infty, -1)$

21. $\{x|x<2\}$ or $(-\infty, 2)$

23. $x+6<-14$
Subtract 6 from both sides.
$x+6-6<-14-6$
$x<-20$
The solution set is
$\{x|x<-20\}$ or $(-\infty, -20)$.

25. $x-4\geq -13$
Add 4 to both sides.
$x-4+4\geq -13+4$
$x\geq -9$
The solution set is
$\{x|x\geq -9\}$ or $[-9,\infty)$.

27. $4x>36$
Divide both sides by 4.
$\frac{4x}{4}>\frac{36}{4}$
$x>9$
The solution set is
$\{x|x>9\}$ or $(9,\infty)$.

29. $6x<20$
Divide both sides by 6.
$\frac{6x}{6}<\frac{20}{6}$
$x<\frac{10}{3}$
The solution set is
$\left\{x|x<\frac{10}{3}\right\}$ or $\left(-\infty, \frac{10}{3}\right)$.

31. $-5x>40$
Divide both sides by -5, which reverses the inequality.
$\frac{-5x}{-5}<\frac{40}{-5}$
$x<-8$
The solution set is
$\{x|x<-8\}$ or $(-\infty, -8)$.

Problem Set 2.5

33. $-7n \leq -56$

Divide both sides by -7, which reverses the inequality.

$\frac{-7n}{-7} \geq \frac{-56}{-7}$

$n \geq 8$

The solution set is $\{n|n \geq 8\}$ or $[8, \infty)$.

35. $48 > -14n$

Divide both sides by -14, which reverses the inequality.

$\frac{48}{-14} < \frac{-14n}{-14}$

$-\frac{24}{7} < n$

$-\frac{24}{7} < n$ means $n > -\frac{24}{7}$.

$n > -\frac{24}{7}$

The solution set is $\left\{n|n > -\frac{24}{7}\right\}$ or $\left(-\frac{24}{7}, \infty\right)$.

37. $16 < 9 + n$

Subtract 9 from both sides.

$16 - 9 < 9 + n - 9$

$7 < n$

$7 < n$ means $n > 7$.

$n > 7$

The solution set is $\{n|n > 7\}$ or $(7, \infty)$.

39. $3x + 2 > 17$

Subtract 2 from both sides.

$3x + 2 - 2 > 17 - 2$

$3x > 15$

Divide both sides by 3.

$\frac{3x}{3} > \frac{15}{3}$

$x > 5$

The solution set is $\{x|x > 5\}$ or $(5, \infty)$.

41. $4x - 3 \leq 21$

Add 3 to both sides.

$4x - 3 + 3 \leq 21 + 3$

$4x \leq 24$

Divide both sides by 4.

$\frac{4x}{4} \leq \frac{24}{4}$

$x \leq 6$

The solution set is $\{x|x \leq 6\}$ or $(-\infty, 6]$.

43. $-2x - 1 \geq 41$

Add 1 to both sides.

$-2x - 1 + 1 \geq 41 + 1$

$-2x \geq 42$

Divide both sides by -2, which reverses the inequality.

$\frac{-2x}{-2} \leq \frac{42}{-2}$

$x \leq -21$

The solution set is $\{x|x \leq -21\}$ or $(-\infty, -21]$.

45. $6x + 2 < 18$

Subtract 2 from both sides.

$6x + 2 - 2 < 18 - 2$

$6x < 16$

Divide both sides by 6.

$\frac{6x}{6} < \frac{16}{6}$

$x < \frac{8}{3}$

The solution set is $\left\{x|x < \frac{8}{3}\right\}$ or $\left(-\infty, \frac{8}{3}\right)$.

47. $3 > 4x - 2$

Add 2 to both sides.

$3 + 2 > 4x - 2 + 2$

$5 > 4x$

Divide both sides by 4.

$\frac{5}{4} > \frac{4x}{4}$

$\frac{5}{4} > x$

$\frac{5}{4} > x$ means $x < \frac{5}{4}$.

$x < \frac{5}{4}$

The solution set is

$\left\{x|x < \frac{5}{4}\right\}$ or $\left(-\infty, \frac{5}{4}\right)$.

49. $-2 < -3x + 1$

Subtract 1 from both sides.

$-2 - 1 < -3x + 1 - 1$

$-3 < -3x$

Divide both sides by -3,

which reverses the inequality.

$\frac{-3}{-3} > \frac{-3x}{-3}$

$1 > x$

$1 > x$ means $x < 1$.

$x < 1$

The solution set is

$\{x|x < 1\}$ or $(-\infty, 1)$.

51. $-38 \geq -9t - 2$

Add 2 to both sides.

$-38 + 2 \geq -9t - 2 + 2$

$-36 \geq -9t$

Divide both sides by -9,

which reverses the inequality.

$\frac{-36}{-9} \leq \frac{-9t}{-9}$

$4 \leq t$

$4 \leq t$ means $t \geq 4$.

$t \geq 4$

The solution set is

$\{t|t \geq 4\}$ or $[4, \infty)$.

53. $5x - 4 - 3x > 24$

Combine similar terms on the left side.

$2x - 4 > 24$

Add 4 to both sides.

$2x - 4 + 4 > 24 + 4$

$2x > 28$

Divide both sides by 2.

$\frac{2x}{2} > \frac{28}{2}$

$x > 14$

The solution set is

$\{x|x > 14\}$ or $(14, \infty)$.

55. $4x + 2 - 6x < -1$

Combine similar terms.

$-2x + 2 < -1$

Subtract 2 from both sides.

$-2x + 2 - 2 < -1 - 2$

$-2x < -3$

Divide both sides by -2,

which reverses the inequality.

$\frac{-2x}{-2} > \frac{-3}{-2}$

$x > \frac{3}{2}$

The solution set is

$\left\{x|x > \frac{3}{2}\right\}$ or $\left(\frac{3}{2}, \infty\right)$.

57. $-5 \geq 3t - 4 - 7t$

Combine similar terms.

$-5 \geq -4t - 4$

Add 4 to both sides.

$-5 + 4 \geq -4t - 4 + 4$

$-1 \geq -4t$

Divide both sides by -4, which reverses the inequality.

$\frac{-1}{-4} \leq \frac{-4t}{-4}$

$\frac{1}{4} \leq t$

$\frac{1}{4} \leq t$ means $t \geq \frac{1}{4}$

$t \geq \frac{1}{4}$

The solution set is

$\left\{t | t \geq \frac{1}{4}\right\}$ or $\left[\frac{1}{4}, \infty\right)$.

59. $-x - 4 - 3x > 5$

Combine similar terms.

$-4x - 4 > 5$

Add 4 to both sides.

$-4x - 4 + 4 > 5 + 4$

$-4x > 9$

Divide both sides by -4, which reverses the inequality.

$\frac{-4x}{-4} < \frac{9}{-4}$

$x < -\frac{9}{4}$

The solution set is

$\left\{x | x < -\frac{9}{4}\right\}$ or $\left(-\infty, -\frac{9}{4}\right)$.

PROBLEM SET 2.6 Inequalities, Compound Inequalities, and Problem Solving

1. $3x + 4 > x + 8$

Subtract x from both sides.

$3x + 4 - x > x + 8 - x$

$2x + 4 > 8$

Subtract 4 from both sides.

$2x + 4 - 4 > 8 - 4$

$2x > 4$

Divide both sides by 2.

$\frac{2x}{2} > \frac{4}{2}$

$x > 2$

The solution set is

$\{x | x > 2\}$ or $(2, \infty)$.

3. $7x - 2 < 3x - 6$

Subtract $3x$ from both sides.

$7x - 2 - 3x < 3x - 6 - 3x$

$4x - 2 < -6$

Add 2 to both sides.

$4x - 2 + 2 < -6 + 2$

$4x < -4$

Divide both sides by 4.

$\frac{4x}{4} < \frac{-4}{4}$

$x < -1$

The solution set is

$\{x | x < -1\}$ or $(-\infty, -1)$.

5. $6x + 7 > 3x - 3$

Subtract $3x$ from both sides.

$6x + 7 - 3x > 3x - 3 - 3x$

$3x + 7 > -3$

Subtract 7 from both sides.

$3x + 7 - 7 > -3 - 7$

$3x > -10$

Divide both sides by 3.

$\frac{3x}{3} > \frac{-10}{3}$

$x > -\frac{10}{3}$

The solution set is

$\left\{x|x > -\frac{10}{3}\right\}$ or $\left(-\frac{10}{3}, \infty\right)$.

7. $5n - 2 \leq 6n + 9$

Subtract $6n$ from both sides.

$5n - 2 - 6n \leq 6n + 9 - 6n$

$-n - 2 \leq 9$

Add 2 to both sides.

$-n - 2 + 2 \leq 9 + 2$

$-n \leq 11$

Multiply by -1, which reverses the inequality.

$n \geq -11$

The solution set is

$\{n|n \geq -11\}$ or $[-11, \infty)$.

9. $2t + 9 \geq 4t - 13$

Subtract $4t$ from both sides.

$2t + 9 - 4t \geq 4t - 13 - 4t$

$-2t + 9 \geq -13$

Subtract 9 from both sides.

$-2t + 9 - 9 \geq -13 - 9$

$-2t \geq -22$

Divide both sides by -2, which reverses the inequality.

$\frac{-2t}{-2} \leq \frac{-22}{-2}$

$t \leq 11$

The solution set is

$\{t|t \leq 11\}$ or $(-\infty, 11]$.

11. $-3x - 4 < 2x + 7$

Subtract $2x$ from both sides.

$-3x - 4 - 2x < 2x + 7 - 2x$

$-5x - 4 < 7$

Add 4 to both sides.

$-5x - 4 + 4 < 7 + 4$

$-5x < 11$

Divide both sides by -5, which reverses the inequality.

$\frac{-5x}{-5} > \frac{11}{-5}$

$x > -\frac{11}{5}$

The solution set is

$\left\{x|x > -\frac{11}{5}\right\}$ or $\left(-\frac{11}{5}, \infty\right)$.

13. $-4x + 6 > -2x + 1$

Add $2x$ to both sides.

$-4x + 6 + 2x > -2x + 1 + 2x$

$-2x + 6 > 1$

Subtract 6 from both sides.

$-2x + 6 - 6 > 1 - 6$

$-2x > -5$

Divide both sides by -2, which reverses the inequality.

$\frac{-2x}{-2} < \frac{-5}{-2}$

$x < \frac{5}{2}$

The solution set is

$\left\{x|x < \frac{5}{2}\right\}$ or $\left(-\infty, \frac{5}{2}\right)$.

Problem Set 2.6

15. $5(x-2) \le 30$

Apply distributive property.

$5x - 10 \le 30$

Add 10 to both sides.

$5x - 10 + 10 \le 30 + 10$

$5x \le 40$

Divide both sides by 5.

$\frac{5x}{5} \le \frac{40}{5}$

$x \le 8$

The solution set is

$\{x|x \le 8\}$ or $(-\infty, 8]$.

17. $2(n+3) > 9$

Apply distributive property.

$2n + 6 > 9$

Subtract 6 from both sides.

$2n + 6 - 6 > 9 - 6$

$2n > 3$

Divide both sides by 2.

$\frac{2n}{2} > \frac{3}{2}$

$n > \frac{3}{2}$

The solution set is

$\left\{n|n > \frac{3}{2}\right\}$ or $\left(\frac{3}{2}, \infty\right)$.

19. $-3(y-1) < 12$

Apply distributive property.

$-3y + 3 < 12$

Subtract 3 from to both sides.

$-3y + 3 - 3 < 12 - 3$

$-3y < 9$

Divide both sides by -3,

which reverses the inequality.

$\frac{-3y}{-3} > \frac{9}{-3}$

$y > -3$

The solution set is

$\{y|y > -3\}$ or $(-3, \infty)$.

21. $-2(x+6) > -17$

Apply distributive property.

$-2x - 12 > -17$

Add 12 to both sides.

$-2x - 12 + 12 > -17 + 12$

$-2x > -5$

Divide both sides by -2,

which reverses the inequality.

$\frac{-2x}{-2} < \frac{-5}{-2}$

$x < \frac{5}{2}$

The solution set is

$\left\{x|x < \frac{5}{2}\right\}$ or $\left(-\infty, \frac{5}{2}\right)$.

23. $3(x-2) < 2(x+1)$

Apply distributive property.

$3x - 6 < 2x + 2$

Subtract $2x$ from both sides.

$3x - 6 - 2x < 2x + 2 - 2x$

$x - 6 < 2$

Add 6 to both sides.

$x - 6 + 6 < 2 + 6$

$x < 8$

The solution set is

$\{x|x < 8\}$ or $(-\infty, 8)$.

25. $4(x+3) > 6(x-5)$

Apply distributive property.

$4x + 12 > 6x - 30$

Subtract $6x$ from both sides.

$4x + 12 - 6x > 6x - 30 - 6x$

$-2x + 12 > -30$

Subtract 12 from both sides.

$-2x + 12 - 12 > -30 - 12$

$-2x > -42$

Divide both sides by -2,

which reverses the inequality.

$\frac{-2x}{-2} < \frac{-42}{-2}$

$x < 21$

The solution set is

$\{x|x < 21\}$ or $(-\infty, 21)$.

27. $3(x-4)+2(x+3)<24$
Apply distributive property.
$3x-12+2x+6<24$
Combine similar terms.
$5x-6<24$
Add 6 to both sides.
$5x-6+6<24+6$
$5x<30$
Divide both sides by 5.
$\frac{5x}{5}<\frac{30}{5}$
$x<6$
The solution set is
$\{x|x<6\}$ or $(-\infty, 6)$.

29. $5(n+1)-3(n-1)>-9$
Apply distributive property.
$5n+5-3n+3>-9$
Combine similar terms.
$2n+8>-9$
Subtract 8 from both sides.
$2n+8-8>-9-8$
$2n>-17$
Divide both sides by 2.
$\frac{2n}{2}>\frac{-17}{2}$
$n>-\frac{17}{2}$
The solution set is
$\left\{n|n>-\frac{17}{2}\right\}$ or $\left(-\frac{17}{2}, \infty\right)$.

31. $\frac{1}{2}n-\frac{2}{3}n\geq -7$
Multiply both sides by 6.
$6\left(\frac{1}{2}n-\frac{2}{3}n\right)\geq 6(-7)$
$3n-4n\geq -42$
$-n\geq -42$
Multiply both sides by -1,
which reverses the inequality.
$n\leq 42$
The solution set is
$\{n|n\leq 42\}$ or $(-\infty, 42]$.

33. $\frac{3}{4}n-\frac{5}{6}n<\frac{3}{8}$
Multiply both sides by 24.
$24\left(\frac{3}{4}n-\frac{5}{6}n\right)<24\left(\frac{3}{8}\right)$
$18n-20n<9$
Combine similar terms.
$-2n<9$
Divide both sides by -2,
which reverses the inequality.
$\frac{-2n}{-2}>\frac{9}{-2}$
$n>-\frac{9}{2}$
The solution set is
$\left\{n|n>-\frac{9}{2}\right\}$ or $\left(-\frac{9}{2}, \infty\right)$.

Problem Set 2.6

35. $\frac{3x}{5}-\frac{2}{3}>\frac{x}{10}$

Multiply both sides by 30.

$30\left(\frac{3x}{5}-\frac{2}{3}\right)>30\left(\frac{x}{10}\right)$

$18x-20>3x$

$18x-20+20>3x+20$

$18x>3x+20$

$15x>20$

Divide both sides by 15.

$\frac{15x}{15}>\frac{20}{15}$

$x>\frac{4}{3}$

The solution set is

$\left\{x|x>\frac{4}{3}\right\}$ or $\left(\frac{4}{3},\ \infty\right)$.

37. $n\geq 3.4+0.15n$

Subtract $0.15n$ from both sides.

$n-0.15n\geq 3.4$

$0.85n\geq 3.4$

$n\geq 4$

The solution set is

$\{n|n\geq 4\}$ or $[4,\ \infty)$.

39. $0.09t+0.1(t+200)>77$

$0.09t+0.1t+20>77$

$0.19t+20>77$

$0.19t>57$

$t>300$

The solution set is

$\{t|t>300\}$ or $(300,\ \infty)$.

41. $0.06x+0.08(250-x)\geq 19$

$0.06x+20-0.08x\geq 19$

$-0.02x+20\geq 19$

$-0.02x\geq -1$

$x\leq 50$

The solution set is

$\{x|x\leq 50\}$ or $(-\infty,\ 50]$.

43. $\frac{x-1}{2}+\frac{x+3}{5}>\frac{1}{10}$

Multiply both sides by 10.

$10\left(\frac{x-1}{2}+\frac{x+3}{5}\right)>10\left(\frac{1}{10}\right)$

$5(x-1)+2(x+3)>1(1)$

$5x-5+2x+6>1$

$7x+1>1$

$7x>0$

$x>0$

The solution set is

$\{x|x>0\}$ or $(0,\ \infty)$.

45. $\frac{x+2}{6}-\frac{x+1}{5}<-2$

Multiply both sides by 30.

$30\left(\frac{x+2}{6}-\frac{x+1}{5}\right)<30(-2)$

$5(x+2)-6(x+1)<-60$

$5x+10-6x-6<-60$

$-x+4<-60$

$-x<-64$

Multiply both sides by -1, which reverses the inequality.

$x>64$

The solution set is

$\{x|x>64\}$ or $(64,\ \infty)$.

47. $\frac{n+3}{3}+\frac{n-7}{2}>3$

Multiply both sides by 6.

$6\left(\frac{n+3}{3}+\frac{n-7}{2}\right)>6(3)$

$2(n+3)+3(n-7)>18$

$2n+6+3n-21>18$

$5n-15>18$

$5n>33$

$n>\frac{33}{5}$

The solution set is

$\left\{n|n>\frac{33}{5}\right\}$ or $\left(\frac{33}{5},\ \infty\right)$.

49. $$\frac{x-3}{7}-\frac{x-2}{4}\le\frac{9}{14}$$
Multiply both sides by 28.
$$28\left(\frac{x-3}{7}-\frac{x-2}{4}\right)\le 28\left(\frac{9}{14}\right)$$
$$4(x-3)-7(x-2)\le 2(9)$$
$$4x-12-7x+14\le 18$$
$$-3x+2\le 18$$
$$-3x\le 16$$
$$x\ge -\frac{16}{3}$$
The solution set is
$\left\{x|x\ge -\frac{16}{3}\right\}$ or $\left[-\frac{16}{3},\infty\right)$.

51. Since it is an "and" statement, we need to satisfy both inequalities at the same time. Thus, all numbers between -1 and 2, but not including -1 and 2, are solutions.

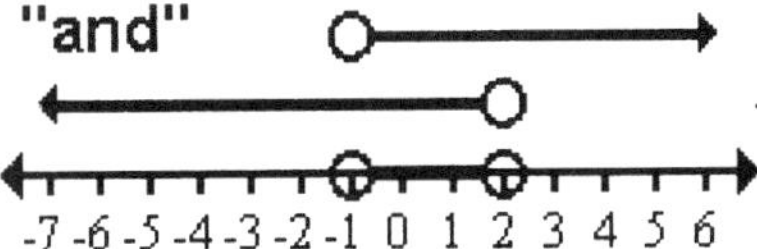

53. Since it is an "or" statement, the solution set consists of all numbers less than (but not including) -2 along with all numbers greater than (but not including) 1.

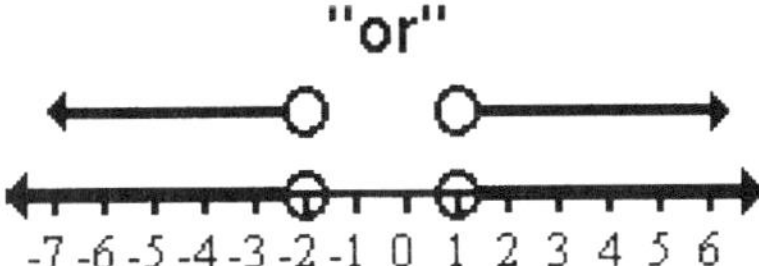

55. Since it is an "and" statement, we need to satisfy both inequalities at the same time. Thus, all numbers between -2 and 2, including 2 (but not including -2), are solutions.

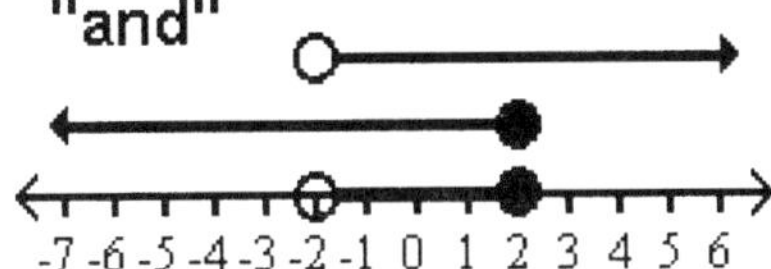

57. Since it is an "and" statement we are looking for all numbers that satisfy both inequalities at the same time. Thus, any number greater than 2 will work.

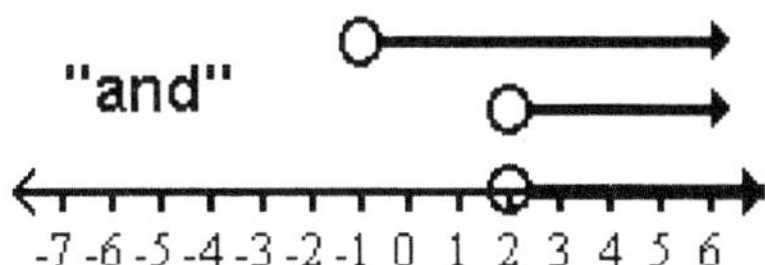

59. Since it is an "or" statement we are looking for all numbers that satisfy either inequality. Thus, any number greater than -4 will work.

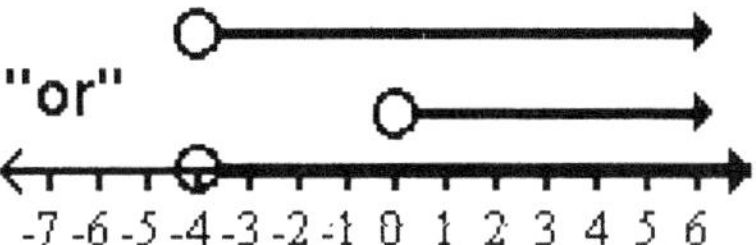

61. Since it is an "and" statement we are looking for all numbers that satisfy both inequalities at the same time. There are no numbers that are both greater than 3 and less than -1. So the solution set is $\emptyset$.

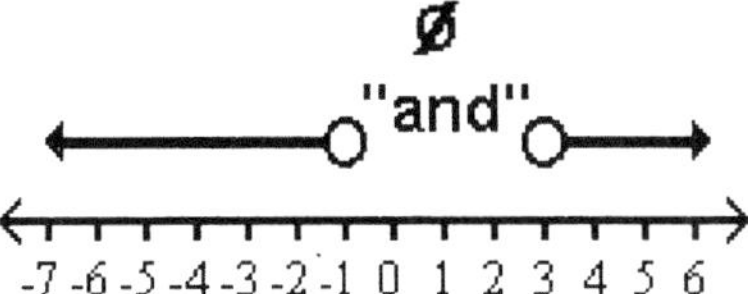

63. Since it is an "or" statement, the solution set consists of all numbers less than or equal to 0 along with all numbers greater than or equal to 2.

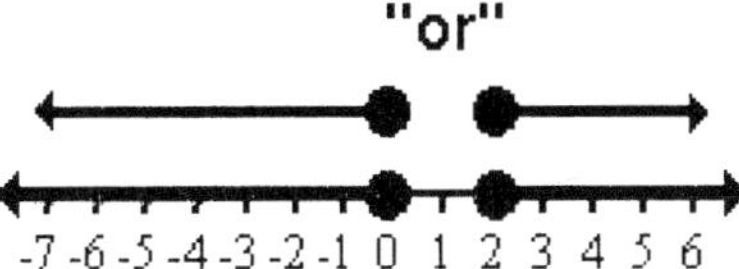

65. Since it is an "or" statement, we want all numbers greater than -4 along with all numbers less than 3. Thus, the solution set is the entire set of real numbers.

Problem Set 2.6

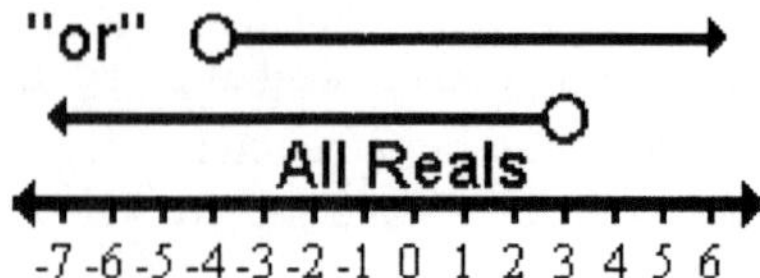

67. Let n represent the number.
$3n + 5 > 26$
$3n > 21$
$n > 7$
The numbers must be greater than 7.

69. Let w represent the width of the rectangle. Also remember that "length plus width equals one-half of the perimeter" of a rectangle.
$w + 20 \leq 35$
$w \leq 15$
Thus, 15 inches is the largest possible value for the width.

71. Let x be her score in the last game.
$\frac{132 + 160 + x}{3} \geq 150$
$292 + x \geq 450$
$x \geq 158$
She must bowl 158 or better in her last game.

73. Let x be his average on the last two exams.
$\frac{96 + 90 + 94 + 2x}{5} > 92$
$280 + 2x > 460$
$2x > 180$
$x > 90$
He must have an average better than 90 on the last two exams.

75. Let r represent the other rate.
$500(0.08) + 500r > 100$
$40 + 500r > 100$
$500r > 60$
$r > 0.12$
The other rate must be more than 12%.

77. Let x be his score on the final round.
$\frac{82 + 84 + 78 + 79 + x}{5} \leq 80$
$323 + x \leq 400$
$x \leq 77$
He must shoot 77 or less.

CHAPTER 2 Review Problem Set

1. $9x - 2 = -29$
$9x - 2 + 2 = -29 + 2$
$9x = -27$
$x = -3$
The solution set is $\{-3\}$.

2. $-3 = -4y + 1$
$-4 = -4y$
$1 = y$
The solution set is $\{1\}$.

3. $7 - 4x = 10$
$7 - 4x - 7 = 10 - 7$
$-4x = 3$
$x = -\frac{3}{4}$
The solution set is $\left\{-\frac{3}{4}\right\}$.

4. $6y - 5 = 4y + 13$
$6y - 5 - 4y = 4y + 13 - 4y$
$2y - 5 = 13$
$2y = 18$
$y = 9$
The solution set is $\{9\}$.

5.
$$4n - 3 = 7n + 9$$
$$4n - 3 - 7n = 7n + 9 - 7n$$
$$-3n - 3 = 9$$
$$-3n - 3 + 3 = 9 + 3$$
$$-3n = 12$$
$$n = -4$$
The solution set is $\{-4\}$.

6.
$$7(y - 4) = 4(y + 3)$$
$$7y - 28 = 4y + 12$$
$$3y - 28 = 12$$
$$3y = 40$$
$$y = \frac{40}{3}$$
The solution set is $\left\{\frac{40}{3}\right\}$.

7.
$$2(x + 1) + 5(x - 3) = 11(x - 2)$$
$$2x + 2 + 5x - 15 = 11x - 22$$
$$7x - 13 = 11x - 22$$
$$7x - 13 - 11x = 11x - 22 - 11x$$
$$-4x - 13 = -22$$
$$-4x - 13 + 13 = -22 + 13$$
$$-4x = -9$$
$$x = \frac{9}{4}$$
The solution set is $\left\{\frac{9}{4}\right\}$.

8.
$$-3(x + 6) = 5x - 3$$
$$-3x - 18 = 5x - 3$$
$$-8x - 18 = -3$$
$$-8x = 15$$
$$x = -\frac{15}{8}$$
The solution set is $\left\{-\frac{15}{8}\right\}$.

9.
$$\frac{2}{5}n - \frac{1}{2}n = \frac{7}{10}$$
$$10\left(\frac{2}{5}n - \frac{1}{2}n\right) = 10\left(\frac{7}{10}\right)$$
$$4n - 5n = 7$$
$$-n = 7$$
$$n = -7$$
The solution set is $\{-7\}$.

10.
$$\frac{3n}{4} + \frac{5n}{7} = \frac{1}{14}$$
$$28\left(\frac{3n}{4} + \frac{5n}{7}\right) = 28\left(\frac{1}{14}\right)$$
$$21n + 20n = 2$$
$$41n = 2$$
$$n = \frac{2}{41}$$
The solution set is $\left\{\frac{2}{41}\right\}$.

11.
$$\frac{x - 3}{6} + \frac{x + 5}{8} = \frac{11}{12}$$
$$24\left(\frac{x - 3}{6} + \frac{x + 5}{8}\right) = 24\left(\frac{11}{12}\right)$$
$$4(x - 3) + 3(x + 5) = 2(11)$$
$$4x - 12 + 3x + 15 = 22$$
$$7x + 3 = 22$$
$$7x = 19$$
$$x = \frac{19}{7}$$
The solution set is $\left\{\frac{19}{7}\right\}$.

12.
$$\frac{n}{2} - \frac{n - 1}{4} = \frac{3}{8}$$
$$8\left(\frac{n}{2} - \frac{n - 1}{4}\right) = 8\left(\frac{3}{8}\right)$$
$$4n - 2(n - 1) = 3$$
$$4n - 2n + 2 = 3$$
$$2n + 2 = 3$$
$$2n = 1$$
$$n = \frac{1}{2}$$
The solution set is $\left\{\frac{1}{2}\right\}$.

13. $$\begin{aligned} -2(x-4) &= -3(x+8) \\ -2x+8 &= -3x-24 \\ -2x+8+3x &= -3x-24+3x \\ x+8 &= -24 \\ x &= -32 \end{aligned}$$
The solution set is $\{-32\}$.

14. $$\begin{aligned} 3x-4x-2 &= 7x-14-9x \\ -x-2 &= -2x-14 \\ -x-2+2x &= -2x-14+2x \\ x-2 &= -14 \\ x &= -12 \end{aligned}$$
The solution set is $\{-12\}$.

15. $$\begin{aligned} 5(n-1)-4(n+2) &= -3(n-1)+3n+5 \\ 5n-5-4n-8 &= -3n+3+3n+5 \\ n-13 &= 8 \\ n &= 21 \end{aligned}$$
The solution set is $\{21\}$.

16. $$\begin{aligned} \frac{9}{x-3} &= \frac{8}{x+4} \\ 9(x+4) &= 8(x-3) \\ 9x+36 &= 8x-24 \\ x &= -60 \end{aligned}$$
The solution set is $\{-60\}$.

17. $$\begin{aligned} \frac{-3}{x-1} &= \frac{-4}{x+2} \\ -3(x+2) &= -4(x-1) \\ -3x-6 &= -4x+4 \\ x-6 &= 4 \\ x &= 10 \end{aligned}$$
The solution set is $\{10\}$.

18. $$\begin{aligned} -(t-3)-(2t+1) &= 3(t+5)-2(t+1) \\ -t+3-2t-1 &= 3t+15-2t-2 \\ -3t+2 &= t+13 \\ -4t &= 11 \\ t &= -\frac{11}{4} \end{aligned}$$
The solution set is $\left\{-\frac{11}{4}\right\}$.

19. $$\begin{aligned} \frac{2x-1}{3} &= \frac{3x+2}{2} \\ 2(2x-1) &= 3(3x+2) \\ 4x-2 &= 9x+6 \\ -5x-2 &= 6 \\ -5x &= 8 \\ x &= -\frac{8}{5} \end{aligned}$$
The solution set is $\left\{-\frac{8}{5}\right\}$.

20. $$\begin{aligned} 3(2t-4)+2(3t+1) &= -2(4t+3)-(t-1) \\ 6t-12+6t+2 &= -8t-6-t+1 \\ 12t-10 &= -9t-5 \\ 21t-10 &= -5 \\ 21t &= 5 \\ t &= \frac{5}{21} \end{aligned}$$
The solution set is $\left\{\frac{5}{21}\right\}$.

21. $$\begin{aligned} 3x-2 &> 10 \\ 3x &> 12 \\ x &> 4 \end{aligned}$$
The solution set is
$\{x|x>4\}$ or $(4, \infty)$.

22. $$\begin{aligned} -2x-5 &< 3 \\ -2x &< 8 \\ x &> -4 \end{aligned}$$
The solution set is
$\{x|x>-4\}$ or $(-4, \infty)$.

23. $$\begin{aligned} 2x-9 &\geq x+4 \\ x-9 &\geq 4 \\ x &\geq 13 \end{aligned}$$
The solution set is
$\{x|x\geq 13\}$ or $[13, \infty)$.

24.
$$\begin{aligned} 3x+1 &\leq 5x-10 \\ -2x+1 &\leq -10 \\ -2x &\leq -11 \\ x &\geq \frac{11}{2} \end{aligned}$$
The solution set is
$\left\{x|x \geq \frac{11}{2}\right\}$ or $\left[\frac{11}{2}, \infty\right)$.

25.
$$\begin{aligned} 6(x-3) &> 4(x+13) \\ 6x-18 &> 4x+52 \\ 2x-18 &> 52 \\ 2x &> 70 \\ x &> 35 \end{aligned}$$
The solution set is
$\{x|x > 35\}$ or $(35, \infty)$.

26.
$$\begin{aligned} 2(x+3)+3(x-6) &< 14 \\ 2x+6+3x-18 &< 14 \\ 5x-12 &< 14 \\ 5x &< 26 \\ x &< \frac{26}{5} \end{aligned}$$
The solution set is
$\left\{x|x < \frac{26}{5}\right\}$ or $\left(-\infty, \frac{26}{5}\right)$.

27.
$$\begin{aligned} \frac{2n}{5}-\frac{n}{4} &< \frac{3}{10} \\ 20\left(\frac{2n}{5}-\frac{n}{4}\right) &< 20\left(\frac{3}{10}\right) \\ 8n-5n &< 6 \\ 3n &< 6 \\ n &< 2 \end{aligned}$$
The solution set is
$\{n|n < 2\}$ or $(-\infty, 2)$.

28.
$$\begin{aligned} \frac{n+4}{5}+\frac{n-3}{6} &> \frac{7}{15} \\ 30\left(\frac{n+4}{5}+\frac{n-3}{6}\right) &> 30\left(\frac{7}{15}\right) \\ 6(n+4)+5(n-3) &> 14 \\ 6n+24+5n-15 &> 14 \\ 11n+9 &> 14 \\ 11n &> 5 \\ n &> \frac{5}{11} \end{aligned}$$
The solution set is
$\left\{n|n > \frac{5}{11}\right\}$ or $\left(\frac{5}{11}, \infty\right)$.

29.
$$\begin{aligned} -16 &< 8+2y-3y \\ -16 &< 8-y \\ -24 &< -y \\ 24 &> y \\ y &< 24 \end{aligned}$$
The solution set is
$\{y|y < 24\}$ or $(-\infty, 24)$.

30.
$$\begin{aligned} -24 &> 5x-4-7x \\ -24 &> -2x-4 \\ -20 &> -2x \\ 10 &< x \\ x &> 10 \end{aligned}$$
The solution set is
$\{x|x > 10\}$ or $(10, \infty)$.

31.
$$\begin{aligned} -3(n-4) &> 5(n+2)+3n \\ -3n+12 &> 5n+10+3n \\ -3n+12 &> 8n+10 \\ -11n+12 &> 10 \\ -11n &> -2 \\ n &< \frac{2}{11} \end{aligned}$$
The solution set is
$\left\{n|n < \frac{2}{11}\right\}$ or $\left(-\infty, \frac{2}{11}\right)$.

32.
$$\begin{aligned} -4(n-2)-(n-1) &< -4(n+6) \\ -4n+8-n+1 &< -4n-24 \\ -5n+9 &< -4n-24 \\ -n+9 &< -24 \\ -n &< -33 \\ n &> 33 \end{aligned}$$

The solution set is $\{n|n > 33\}$ or $(33, \infty)$.

33.
$$\begin{aligned} \frac{3}{4}n-6 &\le \frac{2}{3}n+4 \\ 12\left(\frac{3}{4}n-6\right) &\le 12\left(\frac{2}{3}n+4\right) \\ 9n-72 &\le 8n+48 \\ n-72 &\le 48 \\ n &\le 120 \end{aligned}$$

The solution set is $\{n|n \le 120\}$ or $(-\infty, 120]$.

34.
$$\begin{aligned} \frac{1}{2}n-\frac{1}{3}n-4 &\ge \frac{3}{5}n+2 \\ 30\left(\frac{1}{2}n-\frac{1}{3}n-4\right) &\ge 30\left(\frac{3}{5}n+2\right) \\ 15n-10n-120 &\ge 18n+60 \\ 5n-120 &\ge 18n+60 \\ -13n &\ge 180 \\ n &\le -\frac{180}{13} \end{aligned}$$

The solution set is $\left\{n|n \le -\frac{180}{13}\right\}$ or $\left(-\infty, -\frac{180}{13}\right]$.

35.
$$\begin{aligned} -12 &> -4(x-1)+2 \\ -12 &> -4x+4+2 \\ -12 &> -4x+6 \\ -18 &> -4x \\ \frac{18}{4} &< x \\ x &> \frac{9}{2} \end{aligned}$$

The solution set is $\left\{x|x > \frac{9}{2}\right\}$ or $\left(\frac{9}{2}, \infty\right)$.

36.
$$\begin{aligned} 36 &< -3(x+2)-1 \\ 36 &< -3x-6-1 \\ 36 &< -3x-7 \\ 43 &< -3x \\ -\frac{43}{3} &> x \\ x &< -\frac{43}{3} \end{aligned}$$

The solution set is $\left\{x|x < -\frac{43}{3}\right\}$ or $\left(-\infty, -\frac{43}{3}\right)$.

37. Since it is an "and" statement, we need to satisfy both inequalities at the same time. Thus, all numbers between -3 and 2, but not including -3 and 2, are solutions.

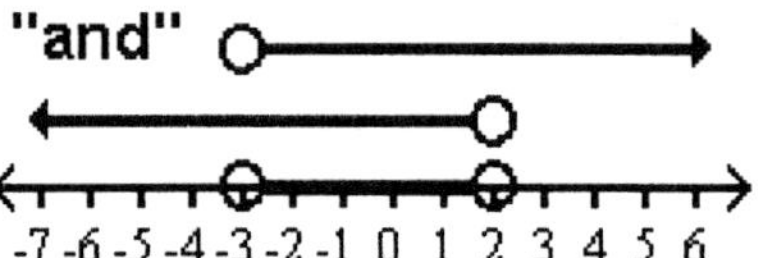

38. Since it is an "or" statement, the solution set consists of all numbers less than (but not including) -1 along with all numbers greater than (but not including) 4.

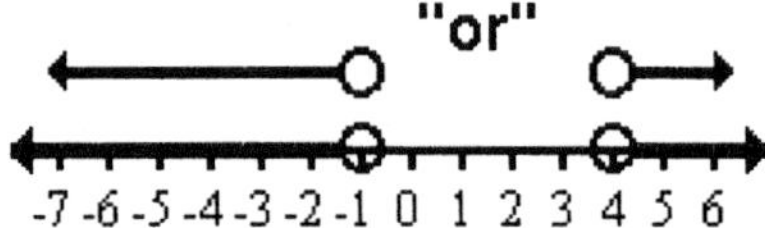

39. Since it is an "or" statement, we want all numbers less than 2 along with all numbers greater than 0. Thus, the solution set is the entire set of real numbers.

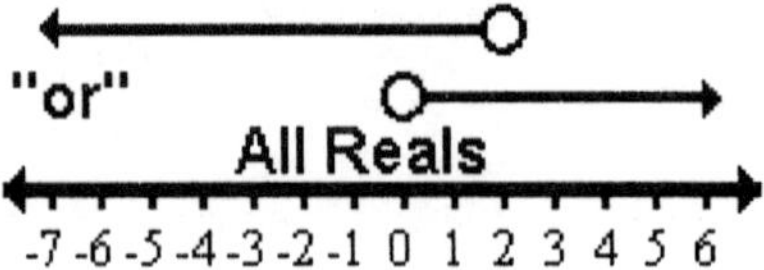

40. Since it is an "and" statement, we are looking for all numbers that satisfy both inequalities at the same time. Thus, any number greater than 1 will work.

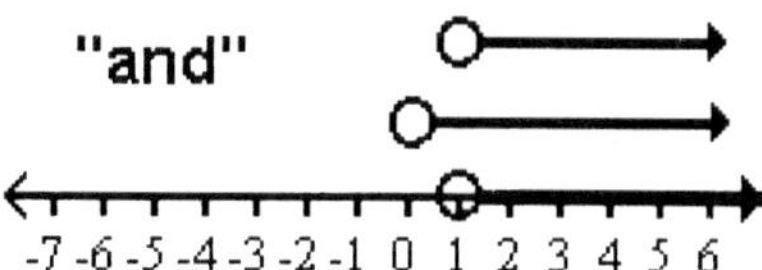

41. Let n represent the number.

$$\frac{3}{4}n = 18$$
$$4\left(\frac{3}{4}n\right) = 4(18)$$
$$3n = 72$$
$$n = 24$$

The number is 24.

42. Let n represent the number.

$$19 = 3n - 2$$
$$21 = 3n$$
$$7 = n$$

The number is 7.

43. Let n represent the larger number.

$$n - 21 = 12$$
$$n = 33$$

The larger number is 33.

44. Let n represent the number.

$$9n - 1 = 7n + 15$$
$$2n - 1 = 15$$
$$2n = 16$$
$$n = 8$$

The number is 8.

45. Let x represent the score on the fifth exam.

$$\frac{83 + 89 + 78 + 86 + x}{5} \geq 85$$
$$336 + x \geq 425$$
$$x \geq 89$$

She must have a score of 89 or better.

46. Let n represent the smaller number; then $40 - n$ represents the larger number.

$$6n = 4(40 - n)$$
$$6n = 160 - 4n$$
$$10n = 160$$
$$n = 16$$

The smaller number is 16 and the larger number is 24.

47. Let n represent the number.

$$\frac{2}{3}n - 2 = \frac{1}{2}n + 1$$
$$6\left(\frac{2}{3}n - 2\right) = 6\left(\frac{1}{2}n + 1\right)$$
$$4n - 12 = 3n + 6$$
$$n - 12 = 6$$
$$n = 18$$

The number is 18.

48. Let x represent the score on the fourth exam. The total score from the first three exams was $3(84) = 252$.

$$\frac{252 + x}{4} \geq 85$$
$$252 + x \geq 340$$
$$x \geq 88$$

She must have a score of 88 or better.

49. Let x represent the number of nickels, then $30 - x$ represents the number of dimes. The value in cents of the nickels is $5x$ and $10(30 - x)$ is the value in cents of the dimes.

$$5x + 10(30 - x) = 260$$
$$5x + 300 - 10x = 260$$
$$-5x + 300 = 260$$
$$-5x = -40$$
$$x = 8$$

There would be 8 nickels and $30 - (8) = 22$ dimes.

50. Let n represent the number of nickels, $3n+1$ the number of dimes, and $2(3n+1)$ the number of quarters. The value in cents of the nickels is $5n$, $10(3n+1)$ is the value in cents of the dimes, and $25(6n+2)$ is the value in cents of the quarters.

$$5n+10(3n+1)+25(6n+2)=1540$$
$$5n+30n+10+150n+50=1540$$
$$185n+60=1540$$
$$185n=1480$$
$$n=8$$

There would be 8 nickels, $3(8)+1=25$ dimes, and $2(25)=50$ quarters.

51. Let a represent the measure of the angle; then $180-a$ represents its supplement and $90-a$ represents its complement.

$$180-a=3(90-a)+14$$
$$180-a=270-3a+14$$
$$180-a=284-3a$$
$$180+2a=284$$
$$2a=104$$
$$a=52$$

The measure of the angle is 52°.

52. Let m represent the number of miles. that she drove.

$$3(25)+0.20m=215$$
$$75+0.20m=215$$
$$0.20m=140$$
$$20m=14,000$$
$$m=700$$

She drove 700 miles.

CHAPTER 2 Test

1. $$7x-3=11$$
$$7x=14$$
$$x=2$$

The solution set is $\{2\}$.

2. $$-7=-3x+2$$
$$-9=-3x$$
$$3=x$$

The solution set is $\{3\}$.

3. $$4n+3=2n-15$$
$$2n+3=-15$$
$$2n=-18$$
$$n=-9$$

The solution set is $\{-9\}$.

4. $$3n-5=8n+20$$
$$-5n-5=20$$
$$-5n=25$$
$$n=-5$$

The solution set is $\{-5\}$.

5. $$4(x-2)=5(x+9)$$
$$4x-8=5x+45$$
$$-x-8=45$$
$$-x=53$$
$$x=-53$$

The solution set is $\{-53\}$.

6. $$9(x+4)=6(x-3)$$
$$9x+36=6x-18$$
$$3x+36=-18$$
$$3x=-54$$
$$x=-18$$

The solution set is $\{-18\}$.

7.
$$\begin{aligned} 5(y-2)+2(y+1) &= 3(y-6) \\ 5y-10+2y+2 &= 3y-18 \\ 7y-8 &= 3y-18 \\ 4y-8 &= -18 \\ 4y &= -10 \\ y &= -\frac{10}{4} = -\frac{5}{2} \end{aligned}$$
The solution set is $\left\{-\frac{5}{2}\right\}$.

8.
$$\begin{aligned} \frac{3}{5}x-\frac{2}{3} &= \frac{1}{2} \\ 30\left(\frac{3}{5}x-\frac{2}{3}\right) &= 30\left(\frac{1}{2}\right) \\ 18x-20 &= 15 \\ 18x &= 35 \\ x &= \frac{35}{18} \end{aligned}$$
The solution set is $\left\{\frac{35}{18}\right\}$.

9.
$$\begin{aligned} \frac{x-2}{4} &= \frac{x+3}{6} \\ 12\left(\frac{x-2}{4}\right) &= 12\left(\frac{x+3}{6}\right) \\ 3(x-2) &= 2(x+3) \\ 3x-6 &= 2x+6 \\ x-6 &= 6 \\ x &= 12 \end{aligned}$$
The solution set is $\{12\}$.

10.
$$\begin{aligned} \frac{x+2}{3}+\frac{x-1}{2} &= 2 \\ 6\left(\frac{x+2}{3}+\frac{x-1}{2}\right) &= 6(2) \\ 2(x+2)+3(x-1) &= 12 \\ 2x+4+3x-3 &= 12 \\ 5x+1 &= 12 \\ 5x &= 11 \\ x &= \frac{11}{5} \end{aligned}$$
The solution set is $\left\{\frac{11}{5}\right\}$.

11.
$$\begin{aligned} \frac{x-3}{6}-\frac{x-1}{8} &= \frac{13}{24} \\ 24\left(\frac{x-3}{6}-\frac{x-1}{8}\right) &= 24\left(\frac{13}{24}\right) \\ 4(x-3)-3(x-1) &= 13 \\ 4x-12-3x+3 &= 13 \\ x-9 &= 13 \\ x &= 22 \end{aligned}$$
The solution set is $\{22\}$.

12.
$$\begin{aligned} -5(n-2) &= -3(n+7) \\ -5n+10 &= -3n-21 \\ -2n+10 &= -21 \\ -2n &= -31 \\ n &= \frac{31}{2} \end{aligned}$$
The solution set is $\left\{\frac{31}{2}\right\}$.

13.
$$\begin{aligned} 3x-2 &< 13 \\ 3x &< 15 \\ x &< 5 \end{aligned}$$
The solution set is
$\{x|x<5\}$ or $(-\infty, 5)$.

14.
$$\begin{aligned} -2x+5 &\geq 3 \\ -2x &\geq -2 \\ x &\leq 1 \end{aligned}$$
The solution set is
$\{x|x\leq 1\}$ or $(-\infty, 1]$.

15.
$$\begin{aligned} 3(x-1) &\leq 5(x+3) \\ 3x-3 &\leq 5x+15 \\ -2x-3 &\leq 15 \\ -2x &\leq 18 \\ x &\geq -9 \end{aligned}$$
The solution set is
$\{x|x\geq -9\}$ or $[-9, \infty)$.

16. $-4 > 7(x-1)+3$
$-4 > 7x - 7 + 3$
$-4 > 7x - 4$
$0 > 7x$
$0 > x$
$x < 0$
The solution set is
$\{x|x < 0\}$ or $(-\infty, 0)$.

17. $-2(x-1)+5(x-2) < 5(x+3)$
$-2x + 2 + 5x - 10 < 5x + 15$
$3x - 8 < 5x + 15$
$-2x - 8 < 15$
$-2x < 23$
$x > -\frac{23}{2}$
The solution set is
$\left\{x|x > -\frac{23}{2}\right\}$ or $\left(-\frac{23}{2}, \infty\right)$.

18. $\frac{1}{2}n + 2 \leq \frac{3}{4}n - 1$
$8\left(\frac{1}{2}n + 2\right) \leq 8\left(\frac{3}{4}n - 1\right)$
$4n + 16 \leq 6n - 8$
$-2n + 16 \leq -8$
$-2n \leq -24$
$n \geq 12$
The solution set is
$\{n|n \geq 12\}$ or $[12, \infty)$.

19. Since it is an "and" statement, we need to satisfy both inequalities at the same time. Thus, all numbers between and including -2 and 4 are solutions.

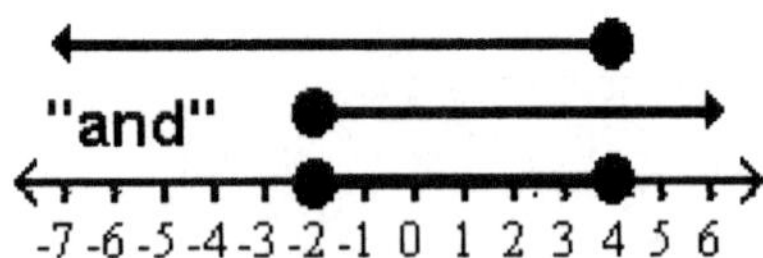

20. Since it is an "or" statement, the solution set consists of all numbers less than (but not including) 1 along with all numbers greater than (but not including) 3.

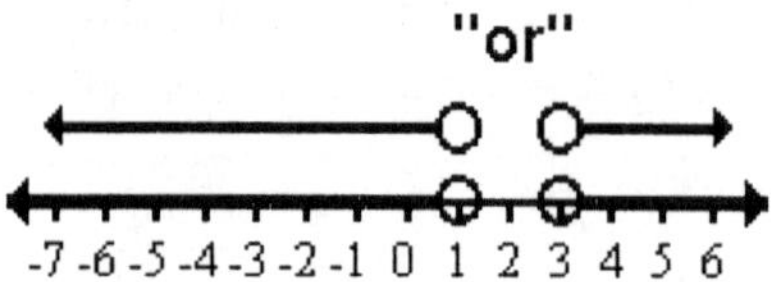

21. Let h represent the hourly rate for the labor.
$53 + 4h = 127$
$4h = 74$
$h = 18.50$
The hourly rate was \$18.50 per hour.

22. Let s represent the length of the shortest side, $3s$ the longest side, and $s + 10$ the third side.
$s + 3s + (s + 10) = 85$
$5s + 10 = 85$
$5s = 75$
$s = 15$
The shortest side is 15 meters, the longest side is 45 meters and the third side is 25 meters.

23. Let x represent the score on the fifth exam.
$\frac{86 + 88 + 89 + 91 + x}{5} \geq 90$
$354 + x \geq 450$
$x \geq 96$
She would need a score of 96 or better.

24. Let n represent the number of nickels, $2n - 1$ the number of dimes, and $3n + 2$ the number of quarters.
$n + (2n - 1) + (3n + 2) = 103$
$6n + 1 = 103$
$6n = 102$
$n = 17$
There are 17 nickels, $2(17) - 1 = 33$ dimes, and $3(17) + 2 = 53$ quarters.

25. Let a represent the measure of the angle; then $180 - a$ represents its supplement and $90 - a$ represents its complement.

$$90 - a = \frac{1}{2}(180 - a) - 10$$
$$2(90 - a) = 2\left[\frac{1}{2}(180 - a) - 10\right]$$
$$180 - 2a = (180 - a) - 20$$
$$180 - 2a = 160 - a$$
$$180 - a = 160$$
$$-a = -20$$
$$a = 20$$

The measure of the angle is 20°.

Chapter 3 Formulas and Problem Solving

PROBLEM SET 3.1 Ratio, Proportion, and Percent

1. $\frac{x}{6} = \frac{3}{2}$

$2x = 18$ Cross products are equal.

$x = 9$

The solution set is $\{9\}$.

3. $\frac{5}{12} = \frac{n}{24}$

$120 = 12n$ Cross products are equal.

$10 = n$

The solution set is $\{10\}$.

5. $\frac{x}{3} = \frac{5}{2}$

$2x = 15$ Cross products are equal.

$x = \frac{15}{2}$

The solution set is $\left\{\frac{15}{2}\right\}$.

7. $\frac{x-2}{4} = \frac{x+4}{3}$

Cross products are equal.

$3(x-2) = 4(x+4)$

$3x - 6 = 4x + 16$

$-x = 22$

$x = -22$

The solution set is $\{-22\}$.

9. $\frac{x+1}{6} = \frac{x+2}{4}$

Cross products are equal.

$4(x+1) = 6(x+2)$

$4x + 4 = 6x + 12$

$-2x = 8$

$x = -4$

The solution set is $\{-4\}$.

11. $\frac{h}{2} - \frac{h}{3} = 1$

Be careful, this is not a proportion.

Multiply both sides by 6.

$6\left(\frac{h}{2} - \frac{h}{3}\right) = 6(1)$

$3h - 2h = 6$

$h = 6$

The solution set is $\{6\}$.

13. $\frac{x+1}{3} - \frac{x+2}{2} = 4$

Be careful, this is not a proportion.

Multiply both sides by 6.

$6\left(\frac{x+1}{3} - \frac{x+2}{2}\right) = 6(4)$

$2(x+1) - 3(x+2) = 24$

$2x + 2 - 3x - 6 = 24$

$-x - 4 = 24$

$-x = 28$

$x = -28$

The solution set is $\{-28\}$.

15. $\frac{-4}{x+2} = \frac{-3}{x-7}$

Cross products are equal.

$-4(x-7) = -3(x+2)$

$-4x + 28 = -3x - 6$

$-x = -34$

$x = 34$

The solution set is $\{34\}$.

17. $\dfrac{-1}{x-7}=\dfrac{5}{x-1}$

Cross products are equal.

$-1(x-1)=5(x-7)$

$-x+1=5x-35$

$-6x=-36$

$x=6$

The solution set is $\{6\}$.

19. $\dfrac{3}{2x-1}=\dfrac{2}{3x+2}$

Cross products are equal.

$3(3x+2)=2(2x-1)$

$9x+6=4x-2$

$5x=-8$

$x=-\dfrac{8}{5}$

The solution set is $\left\{-\dfrac{8}{5}\right\}$.

21. $\dfrac{n+1}{n}=\dfrac{8}{7}$

Cross products are equal.

$7(n+1)=8n$

$7n+7=8n$

$7=n$

The solution set is $\{7\}$.

23. $\dfrac{x-1}{2}-1=\dfrac{3}{4}$

Be careful, this is not a proportion.

Multiply both sides by 8.

$8\left(\dfrac{x-1}{2}-1\right)=8\left(\dfrac{3}{4}\right)$

$4(x-1)-8=2(3)$

$4x-4-8=6$

$4x-12=6$

$4x=18$

$x=\dfrac{18}{4}=\dfrac{9}{2}$

The solution set is $\left\{\dfrac{9}{2}\right\}$.

25. $-3-\dfrac{x+4}{5}=\dfrac{3}{2}$

Be careful, this is not a proportion.

Multiply both sides by 10.

$10\left(-3-\dfrac{x+4}{5}\right)=10\left(\dfrac{3}{2}\right)$

$-30-2(x+4)=5(3)$

$-30-2x-8=15$

$-2x-38=15$

$-2x=53$

$x=-\dfrac{53}{2}$

The solution set is $\left\{\dfrac{-53}{2}\right\}$.

27. $\dfrac{n}{150-n}=\dfrac{1}{2}$

Cross products are equal.

$2n=150-n$

$3n=150$

$n=50$

The solution set is $\{50\}$.

29. $\dfrac{300-n}{n}=\dfrac{3}{2}$

Cross products are equal.

$2(300-n)=3n$

$600-2n=3n$

$600=5n$

$120=n$

The solution set is $\{120\}$.

31. $\dfrac{-1}{5x-1}=\dfrac{-2}{3x+7}$

Cross products are equal.

$-1(3x+7)=-2(5x-1)$

$-3x-7=-10x+2$

$7x-7=2$

$7x=9$

$x=\dfrac{9}{7}$

The solution set is $\left\{\dfrac{9}{7}\right\}$.

Problem Set 3.1

33. $\frac{11}{20} = \frac{n}{100}$

$20n = 1100$

$n = 55$

Therefore, $\frac{11}{20} = \frac{55}{100} = 55\%$

35. $\frac{3}{5} = \frac{n}{100}$

$5n = 300$

$n = 60$

Therefore, $\frac{3}{5} = \frac{60}{100} = 60\%$

37. $\frac{1}{6} = \frac{n}{100}$

$6n = 100$

$n = \frac{100}{6} = 16\frac{2}{3}$

Therefore, $\frac{1}{6} = \frac{16\frac{2}{3}}{100} = 16\frac{2}{3}\%$

39. $\frac{3}{8} = \frac{n}{100}$

$8n = 300$

$n = \frac{300}{8} = 37\frac{1}{2}$

Therefore, $\frac{3}{8} = \frac{37\frac{1}{2}}{100} = 37\frac{1}{2}\%$

41. $\frac{3}{2} = \frac{n}{100}$

$2n = 300$

$n = 150$

Therefore, $\frac{3}{2} = \frac{150}{100} = 150\%$

43. $\frac{12}{5} = \frac{n}{100}$

$5n = 1200$

$n = 240$

Therefore, $\frac{12}{5} = \frac{240}{100} = 240\%$

45. Let n represent the number.

$n = (7\%)(38)$

$n = 0.07(38)$

$n = 2.66$

Therefore, 2.66 is 7% of 38.

47. Let n represent the number.

$(15\%)(n) = 6.3$

$0.15n = 6.3$

Multiply both sides by 100.

$15n = 630$

$n = 42$

Therefore, 15% of 42 is 6.3.

49. Let r represent the percent to be found.

$76 = r(95)$

$\frac{76}{95} = r$

$0.80 = r$

Therefore, 76 is 80% of 95.

51. Let n represent the number.

$n = (120\%)(50)$

$n = 1.2(50)$

$n = 60$

Therefore, 60 is 120% of 50.

53. Let r represent the percent to be found.

$46 = r(40)$

$\frac{46}{40} = r$

$1.15 = r$

Therefore, 46 is 115% of 40.

55. Let n represent the number.

$(160\%)(n) = 144$

$1.6n = 144$

Multiply both sides by 10.

$16n = 1440$

$n = \frac{1440}{16} = 90$

Therefore, 160% of 90 is 144.

57. Let l and w represent the length and width of the room measured in feet.

$$\frac{\text{map}}{\text{room}} \quad \frac{1}{2\frac{1}{2}} = \frac{6}{w} \quad (\text{inches} \;\; \text{feet})$$

$$w = 6\left(2\frac{1}{2}\right)$$

$$w = 6\left(\frac{5}{2}\right)$$

$$w = 15$$

$$\frac{\text{map}}{\text{room}} \quad \frac{1}{3\frac{1}{4}} = \frac{6}{l} \quad (\text{inches} \;\; \text{feet})$$

$$l = 6\left(3\frac{1}{4}\right)$$

$$l = 6\left(\frac{13}{4}\right)$$

$$l = \frac{39}{2} = 19\frac{1}{2}$$

The room measures 15 feet by $19\frac{1}{2}$ feet.

59. Let m represent the number of miles traveled.

$$\frac{\text{miles}}{\text{gallons}} \quad \frac{264}{12} = \frac{m}{15}$$

$$12m = 3960$$

$$m = 330$$

The car will travel 330 miles.

61. Let l represent the length of the rectangle.

$$\frac{\text{length}}{\text{width}} \quad \frac{5}{2} = \frac{l}{24}$$

$$2l = 120$$

$$l = 60$$

The length of the rectangle is 60 centimeters.

63. Let s represent the number of pounds of salt needed.

$$\frac{\text{salt}}{\text{water}} \quad \frac{3}{10} = \frac{s}{25}$$

$$10s = 75$$

$$s = 7.5$$

It will take 7.5 pounds of salt.

65. Let p represent the number of pounds of fertilizer need.

$$\frac{\text{fertilizer}}{\text{lawn}} \quad \frac{20}{1500} = \frac{p}{2500}$$

$$1500p = 50,000$$

$$p = 33\frac{1}{3}$$

It will take $33\frac{1}{3}$ pounds of fertilizer to cover the lawn.

67. Let n represent the number of people who are expected to vote.

$$\frac{3}{7} = \frac{n}{210,000}$$

$$7n = 630,000$$

$$n = 90,000$$

It is expected that $90,000$ people will vote in the election.

69. Let x represent the length of the rectangle. The sum of the length and the width is equal to one-half of the perimeter. Therefore, $25 - x$ represents the width.

$$\frac{\text{length}}{\text{width}} \quad \frac{3}{2} = \frac{x}{25 - x}$$

$$2x = 3(25 - x)$$

$$2x = 75 - 3x$$

$$5x = 75$$

$$x = 15$$

The dimensions of the rectangle are 15 inches by 10 inches.

71. Let x represent the additional money to be invested.

$$\frac{\text{investment}}{\text{earnings}} \quad \frac{500}{45} = \frac{500 + x}{72}$$

$$45(500 + x) = 500(72)$$

$$22,500 + 45x = 36,000$$

$$45x = 13,500$$

$$x = 300$$

The additional investment would be \$300.

73. Let x represent the money the child will receive; then, the cancer fund will receive $180,000 - x$.

$$\frac{\text{child}}{\text{fund}} \quad \frac{5}{1} = \frac{x}{180,000 - x}$$
$$x = 5(180,000 - x)$$
$$x = 900,000 - 5x$$
$$6x = 900,000$$
$$x = 150,000$$
The child would receive $\$150,000$.

PROBLEM SET 3.2 More on Percents and Problem Solving

1. $x - 0.36 = 0.75$
Add 0.36 to both sides.
$x - 0.36 + 0.36 = 0.75 + 0.36$
$x = 1.11$
The solution set is $\{1.11\}$.

3. $x + 7.6 = 14.2$
Subtract 7.6 from both sides.
$x + 7.6 - 7.6 = 14.2 - 7.6$
$x = 6.6$
The solution set is $\{6.6\}$.

5. $0.62 - y = 0.14$
Subtract 0.62 from both sides.
$0.62 - y - 0.62 = 0.14 - 0.62$
$-y = -0.48$
$y = 0.48$
The solution set is $\{0.48\}$.

7. $0.7t = 56$
Multiply both sides by 10.
$10(0.7t) = 10(56)$
$7t = 560$
$t = 80$
The solution set is $\{80\}$.

9. $x = 3.36 - 0.12x$
Multiply both sides by 100.
$100(x) = 100(3.36 - 0.12x)$
$100x = 336 - 12x$
$112x = 336$
$x = 3$
The solution set is $\{3\}$.

11. $s = 35 + 0.3s$
Multiply both sides by 10.
$10(s) = 10(35 + 0.3s)$
$10s = 350 + 3s$
$7s = 350$
$s = 50$
The solution set is $\{50\}$.

13. $s = 42 + 0.4s$
Multiply both sides by 10.
$10(s) = 10(42 + 0.4s)$
$10s = 420 + 4s$
$6s = 420$
$s = 70$
The solution set is $\{70\}$.

15. $0.07x + 0.08(x + 600) = 78$
Multiply both sides by 100.
$100[0.07x + 0.08(x + 600)] = 100(78)$
$7x + 8(x + 600) = 7800$
$7x + 8x + 4800 = 7800$
$15x + 4800 = 7800$
$15x = 3000$
$x = 200$
The solution set is $\{200\}$.

17. $0.09x + 0.1(2x) = 130.5$
Multiply both sides by 100.
$100[0.09x + 0.1(2x)] = 100(130.5)$
$9x + 10(2x) = 13,050$
$9x + 20x = 13,050$
$29x = 13,050$
$x = 450$
The solution set is $\{450\}$.

19. $0.08x + 0.11(500 - x) = 50.5$
Multiply both sides by 100.
$100[0.08x + 0.11(500 - x)] = 100(50.5)$
$8x + 11(500 - x) = 5050$
$8x + 5500 - 11x = 5050$
$-3x + 5500 = 5050$
$-3x = -450$
$x = 150$
The solution set is $\{150\}$.

21. $0.09x = 550 - 0.11(5400 - x)$
Multiply both sides by 100.
$9x = 55,000 - 11(5400 - x)$
$9x = 55,000 - 59,400 + 11x$
$-2x = -4400$
$x = 2200$
The solution set is $\{2200\}$.

23. Let p represent the original price of the trousers.
$(100\%)(p) - (30\%)(p) = 35$
$(70\%)(p) = 35$
$0.7p = 35$
$7p = 350$
$p = 50$
The original price of the trousers was \$50.

25. Let d represent the discount sale price of the sweater. Since the sweater is on sale for 25% off, the discount price is 75% of the original price.
$d = (75\%)(48)$
$d = 0.75(48)$
$d = 36$
The discount sale price is \$36.

27. Let d represent the discount sale price of the putter. Since the putter is on sale for 35% off, the discount price is 65% of the original price.
$d = (65\%)(32)$
$d = 0.65(32)$
$d = 20.8$
The discount sale price is \$20.80.

29. Let r represent the rate of discount.
$180 - r(180) = 126$
$180 - 180r = 126$
$-180r = -54$
$r = 0.3$
The rate of discount is 30%.

31. Let s represent the selling price.
Profit is a percent of cost.
Selling price = Cost + Profit
$s = 5 + (70\%)(5)$
$s = 5 + 0.7(5)$
$s = 5 + 3.5$
$s = 8.5$
The selling price would be \$8.50.

33. Let s represent the selling price.
Profit is a percent of cost.
Selling price = Cost + Profit
$s = 3 + (55\%)(3)$
$s = 3 + 0.55(3)$
$s = 3 + 1.65$
$s = 4.65$
The selling price would be \$4.65.

35. Let s represent the selling price.
Profit is a percent of selling price.
Selling price = Cost + Profit
$s = 400 + (60\%)(s)$
$s = 400 + 0.6s$
$10s = 4000 + 6s$
$4s = 4000$
$s = 1000$
The selling price would be \$1000.

37. Let r represent the rate of profit.
Profit is a percent of cost.
Selling price = Cost + Profit

$$44.8 = 32 + r(32)$$
$$448 = 320 + 320r$$
$$128 = 320r$$
$$0.4 = r$$

The rate of profit would be 40% of the cost.

39.
$$i = Prt$$
$$560 = 3500(r)(2)$$
$$560 = 7000r$$
$$\frac{560}{7000} = r$$
$$0.8 = r$$

The annual interest rate will be 8%.

41.
$$i = Prt$$
$$1000 = P(0.08)(3)$$
$$1000 = 0.24P$$
$$\frac{1000}{0.24} = P$$
$$4166.67 = P$$

The principal will be $4166.67.

43.
$$i = Prt$$
$$i = (5000)(0.068)(10)$$
$$i = 3400$$

The interest earned will be $3400.

45. $i = Prt$

Remember time must be in years.

So 1 month $= \frac{1}{12}$ year.

$$i = 95000(0.08)\left(\frac{1}{12}\right)$$
$$i = 633.33$$

The interest for a month will be $633.33.

PROBLEM SET 3.3 Formulas: Geometric and Others

1. $d = 336,\ r = 48 : d = rt$
$$336 = 48t$$
$$7 = t$$

3. $i = 200,\ r = 0.08,\ t = 5 : i = Prt$
$$200 = P(0.08)(5)$$
$$200 = 0.4P$$
$$500 = P$$

5. $F = 68 : F = \frac{9}{5}C + 32$
$$68 = \frac{9}{5}C + 32$$

Multiply both sides by 5.

$$340 = 9C + 160$$
$$180 = 9C$$
$$20 = C$$

7. $V = 112,\ h = 7 : V = \frac{1}{3}Bh$
$$112 = \frac{1}{3}B(7)$$

Multipy both sides by 3.

$$336 = 7B$$
$$48 = B$$

9. A=652, P=400, r=0.07; $A = P + Prt$
$$652 = 400 + 400(0.07)t$$
$$652 = 400 + 28t$$
$$252 = 28t$$
$$9 = t$$

11. Substitute 14 for l and 9 for w in the formula for finding perimeter of a rectangle.

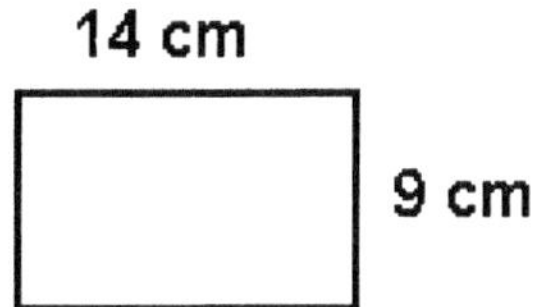

$P = 2l + 2w$
$P = 2(14) + 2(9)$
$P = 28 + 18$
$P = 46$
The perimeter of the rectangle is 46 centimeters.

13. Substitute $3\frac{1}{4}$ feet $= 39$ inches for l and 108 feet for P in the formula for finding perimeter of a rectangle.

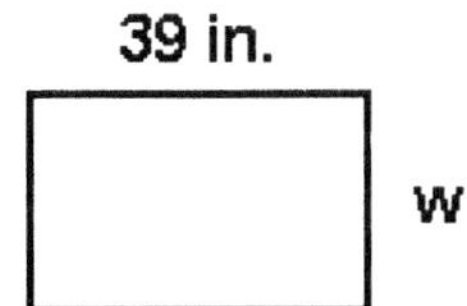

$P = 2l + 2w$
$108 = 2(39) + 2w$
$108 = 78 + 2w$
$30 = 2w$
$15 = w$
The width of the rectangle is 15 inches.

15. Subtract the area of the rectangular garden from the area of the garden plus the dirt path. The width of the larger rectangle is $17 + 4 + 4 = 25$ feet and the length of the larger rectangle is $38 + 4 + 4 = 46$ feet.

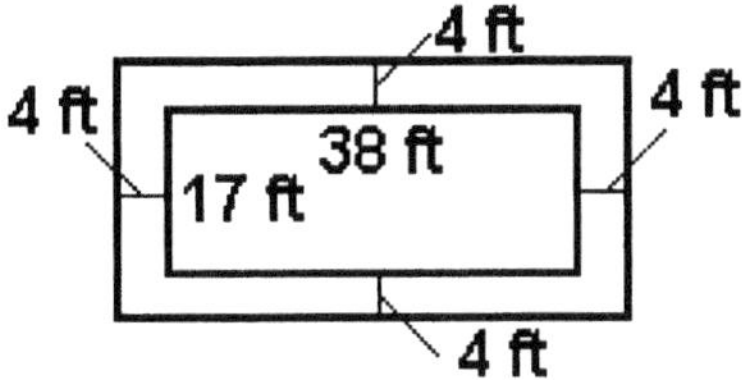

$A =$ large area $-$ garden
$A = (46)(25) - (38)(17)$
$A = 1150 - 646$
$A = 504$
The area of the dirt path is 504 square feet.

17. Since the are is needed in square meters, it is easier to convert the measurements to meters first. Thus, the length of the wood piece is 60 centimeters $= 0.6$ meters and the width is 30 centimeters $= 0.3$ meters. The area of one piece is $(0.3)(0.6) = 0.18$ square meters. There are 50 pieces for a total area of $50(0.18) = 9$ square meters. One liter of paint would cover the total area for a cost of \$2.

19. Let h represent the length of the altitude of the trapezoid.

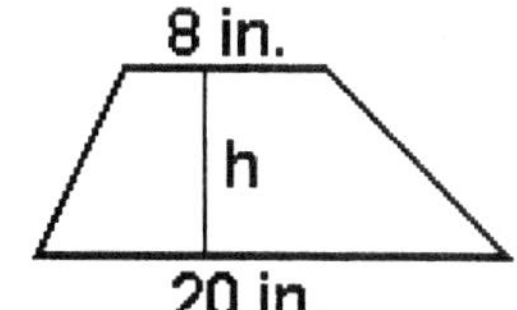

$A = \frac{1}{2}h(b_1 + b_2)$
$98 = \frac{1}{2}h(8 + 20)$
$98 = \frac{1}{2}h(28)$
$98 = 14h$
$7 = h$
The length of the altitude of the trapezoid is 7 inches.

21. The area of one washer can be computed by subtracting the area of the inner circle (diameter $= 2$ cm) from the area of the outer circle (diameter $= 4$ cm). The radius, which is one-half of the diameter, is used in the formula. See Figure 4.19 in the text for a drawing.
$A = \pi(2)^2 - \pi(1)^2 = 4\pi - 1\pi = 3\pi$
The area of one washer is 3π sqare centimeters, so 50 washers would have $50(3\pi) = 150\pi$ square centimeters of metal.

Problem Set 3.3

23. The radius of the circular region is $\frac{1}{2}$ yard.

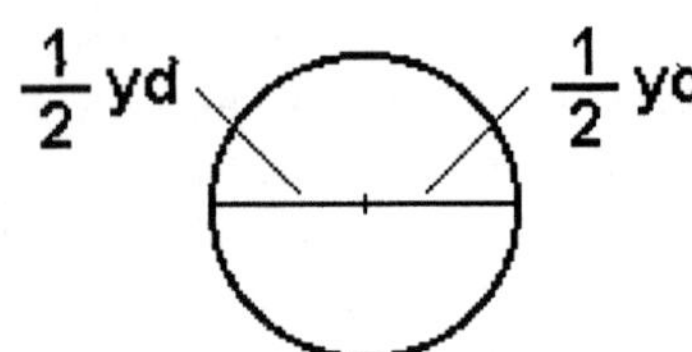

$A = \pi r^2$

$A = \pi\left(\frac{1}{2}\right)^2 = \frac{1}{4}\pi$

The area of the circular region is $\frac{1}{4}\pi$ square yards.

25. Substitute $r = 9$ into both formulas.

$S = 4\pi r^2 = 4\pi(9)^2$

$= 324\pi$ square inches

$V = \frac{4}{3}\pi r^3 = \frac{4}{3}\pi(9)^3$

$= 972\pi$ cubic inches

27. Substitute $r = 8$ and $h = 18$ into both formulas.

$V = \pi r^2 h = \pi(8)^2(18)$

$= 1152\pi$ cubic inches

$S = 2\pi r^2 + 2\pi rh = 2\pi(8)^2 + 2\pi(8)(18)$

$= 128\pi + 288\pi$

$= 416\pi$ square inches

29. Substitute $V = 324\pi$ and $r = 9$ in the formula.

$V = \frac{1}{3}\pi r^2 h$

$324\pi = \frac{1}{3}\pi(9)^2 h$

$324\pi = \frac{81}{3}\pi h$

$324\pi = 27\pi h$

$12 = h$

The height of the cone is 12 inches.

31. Substitute $S = 65\pi$ and $r = 5$ in the formula.

$S = \pi r^2 + \pi rs$

$65\pi = \pi(5)^2 + \pi(5)s$

$65\pi = 25\pi + 5\pi s$

$40\pi = 5\pi s$

$8 = s$

The slant height of the cone is 8 feet.

33. $V = Bh$

Divide both sides by B.

$\frac{V}{B} = \frac{Bh}{B}$

$\frac{V}{B} = h$

35. $V = \frac{1}{3}Bh$

Multiply both sides by 3.

$3V = Bh$

Divide both sides by h.

$\frac{3V}{h} = B$

37. $P = 2l + 2w$

Subtract $2l$ from both sides.

$P - 2l = 2w$

Divide both sides by 2.

$\frac{P - 2l}{2} = w$

39. $V = \frac{1}{3}\pi r^2 h$

Multiply both sides by 3.

$3V = \pi r^2 h$

Divide both sides by πr^2.

$\frac{3V}{\pi r^2} = h$

41. $F = \frac{9}{5}C + 32$

Subtract 32 from both sides.

$F - 32 = \frac{9}{5}C$

Multiply both sides by $\frac{5}{9}$.

$\frac{5}{9}(F - 32) = C$

43. $A = 2\pi r^2 + 2\pi rh$

Subtract $2\pi r^2$ from both sides.

$A - 2\pi r^2 = 2\pi rh$

Divide both sides by $2\pi r$.

$\dfrac{A - 2\pi r^2}{2\pi r} = h$

45. $3x + 7y = 9$

Subtract $7y$ from both sides.

$3x = 9 - 7y$

Divide both sides by 3.

$x = \dfrac{9 - 7y}{3}$

47. $9x - 6y = 13$

Subtracte $9x$ from both sides.

$-6y = 13 - 9x$

Multiply both sides by -1.

$6y = -13 + 9x$

Apply commutative property.

$6y = 9x - 13$

Divide both sides by 6.

$y = \dfrac{9x - 13}{6}$

49. $-2x + 11y = 14$

Subtract $11y$ from both sides.

$-2x = 14 - 11y$

Multiply both sides by -1.

$2x = -14 + 11y$

Apply commutative property.

$2x = 11y - 14$

Divide both sides by 2.

$x = \dfrac{11y - 14}{2}$

51. $y = -3x - 4$

Add 4 to both sides.

$y + 4 = -3x$

Multiply both sides by -1.

$-y - 4 = 3x$

Divide both sides by 3.

$\dfrac{-y - 4}{3} = x$

53. $\dfrac{x - 2}{4} = \dfrac{y - 3}{6}$

Cross products are equal.

$4(y - 3) = 6(x - 2)$

Distributive property.

$4y - 12 = 6x - 12$

Add 12 to both sides..

$4y = 6x$

Divide both sides by 4.

$y = \dfrac{6x}{4} = \dfrac{3x}{2}$

55. $ax - by - c = 0$

Add by to both sides.

$ax - c = by$

Divide both sides by b.

$\dfrac{ax - c}{b} = y$

57. $\dfrac{x + 6}{2} = \dfrac{y + 4}{5}$

Cross products are equal.

$5(x + 6) = 2(y + 4)$

$5x + 30 = 2y + 8$

Subtract 30 from both sides..

$5x = 2y - 22$

Divide both sides by 5.

$x = \dfrac{2y - 22}{5}$

59. $m = \dfrac{y - b}{x}$

Multiply both sides by m.

$mx = y - b$

Add b to both sides.

$mx + b = y$

PROBLEM SET 3.4 Problem Solving

1. $950(0.12)t = 950$
Divide both sides by 950.
$0.12t = 1$
Multiply both sides by 100.
$12t = 100$
Divide both sides by 12 and reduced.
$t = \frac{100}{12} = 8\frac{1}{3}$
The solution set is $\left\{8\frac{1}{3}\right\}$.

3. $l + \frac{1}{4}l - 1 = 19$
Add 1 to both sides.
$l + \frac{1}{4}l = 20$
Multiply both sides by 4.
$4l + l = 80$
$5l = 80$
Divide both sides by 5.
$l = 16$
The solution set is $\{16\}$.

5. $500(0.08)t = 1000$
Simplify the left side.
$40t = 1000$
Divide both sides by 40.
$t = 25$
The solution set is $\{25\}$.

7. $s + (2s - 1) + (3s - 4) = 37$
Combine like terms.
$6s - 5 = 37$
Add 5 to both sides.
$6s = 42$
Divide both sides by 6.
$s = 7$
The solution set is $\{7\}$.

9. $\frac{5}{2}r + \frac{5}{2}(r + 6) = 135$
Multiply both sides by 2.
$2\left[\frac{5}{2}r + \frac{5}{2}(r + 6)\right] = 2(135)$
$5r + 5(r + 6) = 270$
.Apply distributive property.
$5r + 5r + 30 = 270$
Combine like terms.
$10r + 30 = 270$
Subtract 30 from both sides.
$10r = 240$
Divide both sides by 10.
$r = 24$
The solution set is $\{24\}$.

11. $24\left(t - \frac{2}{3}\right) = 18t + 8$
Apply distributive property.
$24t - 16 = 18t + 8$
Subtract $18t$ from both sides.
$6t - 16 = 8$
Add 16 to both sides.
$6t = 24$
Divide both sides by 6.
$t = 4$
The solution set is $\{4\}$.

13. To "double" the interest earned, i would equal 750 when $P = 750$ and $r = 8\%$.
$i = Prt$
$750 = 750(0.08)t$
Divide both sides by 750.
$1 = 0.08t$
Multiply by 100.
$100 = 8t$
Divide both sides by 8.
$12\frac{1}{2} = t$
It would take $12\frac{1}{2}$ years for \$750 to double itself.

15. To "triple itself" the interest earned, i would equal 1600 when $P = 800$ and $r = 10\%$.
$1600 = 800(10\%)t$
Divide both sides by 800.
$2 = 0.1t$
Multiply by 100.
$200 = 10t$
Divide both sides by 10.
$20 = t$
It would take 20 years for \$800 to triple itself.

17. Let w represent the width of the rectangle. Then $3w$ represents the length and the perimeter is 112 inches.

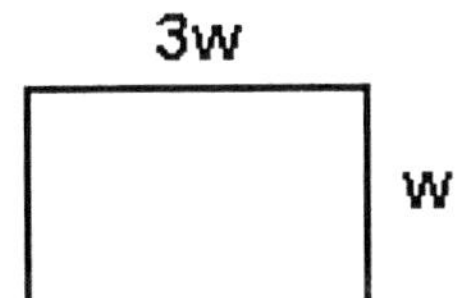

$P = 2l + 2w$
$112 = 2(3w) + 2w$
$112 = 6w + 2w$
$112 = 8w$
$14 = w$
The width is 14 inches and the length is $3(14) = 42$ inches.

19. Let w represent the width of the rectangle. Then $3w - 2$ represents the length and the perimeter is 92 cm.

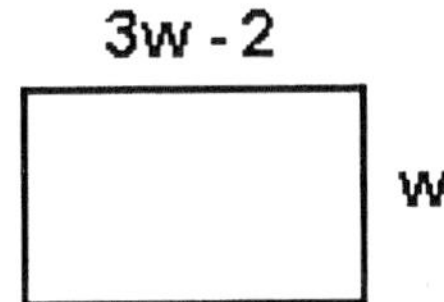

$P = 2l + 2w$
$92 = 2(3w - 2) + 2w$
$92 = 6w - 4 + 2w$
$96 = 8w$
$12 = w$
The width is 12 centimeters and the length is $3(12) - 2 = 34$ centimeters.

21. Let l represent the length of the rectangle; then $\frac{1}{2}l - 3$ represents the width when $P = 42$.

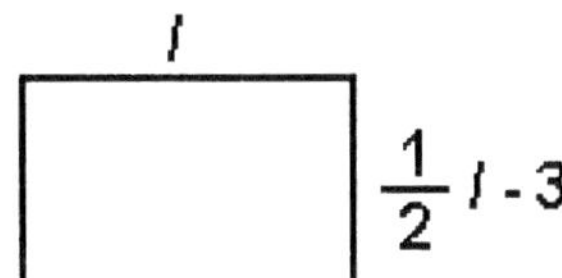

$P = 2l + 2w$
$42 = 2l + 2\left(\frac{1}{2}l - 3\right)$
$42 = 2l + l - 6$
$48 = 3l$
$16 = l$
The length is 16 inches and the width is $\frac{1}{2}(16) - 3 = 5$ inches. Therefore, the area $= lw = (16)(5) = 80$ square inches.

23. Let s represent the shortest side, $2s - 3$ the longest side, and $s + 7$ the third side of a triangle when the perimeter equals 100 feet.

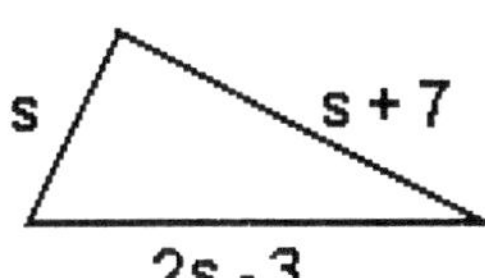

$s + (2s - 3) + (s + 7) = 100$
$4s + 4 = 100$
$4s = 96$
$s = 24$
The lengths of the sides are 24 feet, $2(24) - 3 = 45$ feet, and $24 + 7 = 31$ feet.

25. Let f represent the first side, $3f + 1$ the second side, and $3f + 3$ the third side of a triangle when the perimeter equals 46 centimeters.

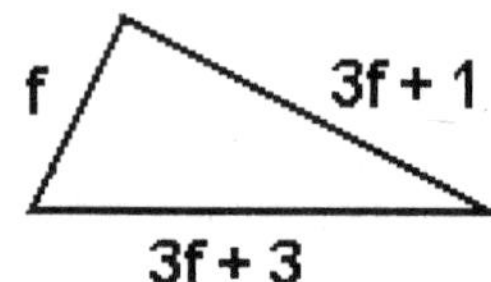

$$f + (3f + 1) + (3f + 3) = 46$$
$$7f + 4 = 46$$
$$7f = 42$$
$$f = 6$$

The lengths of the sides are 6 centimeters, 19 centimeters and 21 centimeters.

27. Let s represent each side of the equilateral triangle; then $s - 4$ represents each side of the square. The perimeter of the triangle is $3s$ and the perimeter of the rectangle is $4(s - 4)$. As stated in the problem, the perimeter of the triangle is 4 centimeters more than the perimeter of the rectangle. This gives the guideline for the equation to be solved.

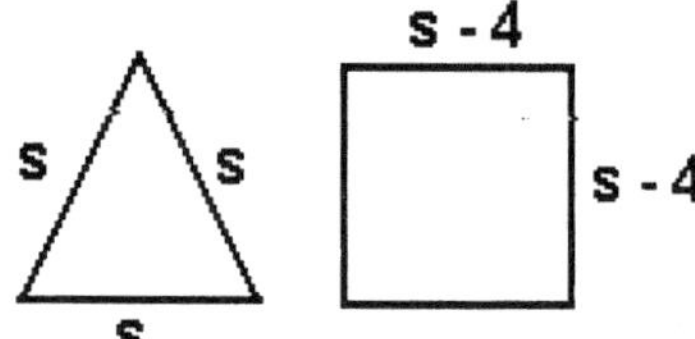

Triangle perimeter = Rectangle perimeter + 4

$$3s = 4(s - 4) + 4$$
$$3s = 4s - 16 + 4$$
$$-s = -12$$
$$s = 12$$

The length of a side of the triangle is 12 centimeters.

29. Let s represent the length of each side of the square; then s also represents the radius of the circle.

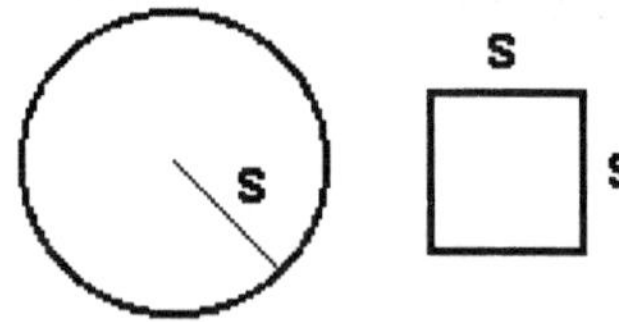

circle circumference = square perimeter + 15.96

$$2\pi r = 2s + 2s + 15.96$$
$$2(3.14)s = 4s + 15.96$$
$$6.28s = 4s + 15.96$$
$$2.28s = 15.96$$
$$s = 7$$

The length of the radius of the circle is 7 centimeters.

31. Let t represent Monica's time; then $t + 1$ represents Sandy's time as she travels one hour longer. A chart of the information would be as follows:

	Rate	Time	Distance ($d = rt$)
Sandy	45	$t + 1$	$45(t + 1)$
Monica	50	t	$50t$

A diagram of the problem would be as follows:

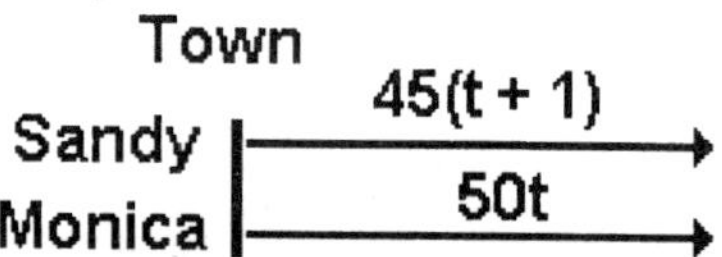

Set the distances equal to each other and solve.

$$45(t + 1) = 50t$$
$$45t + 45 = 50t$$
$$45 = 5t$$
$$9 = t$$

It would take 9 hours for Monica to overtake Sandy.

33. Let t represent the freight train's time; then t also represents the passenger train's time as both trains left at the same time. A chart of the information would be as follows:

	Rate	Time	Distance ($d = rt$)
Freight	40	t	$40t$
Passenger	90	t	$90t$

A diagram of the problem would be as follows:

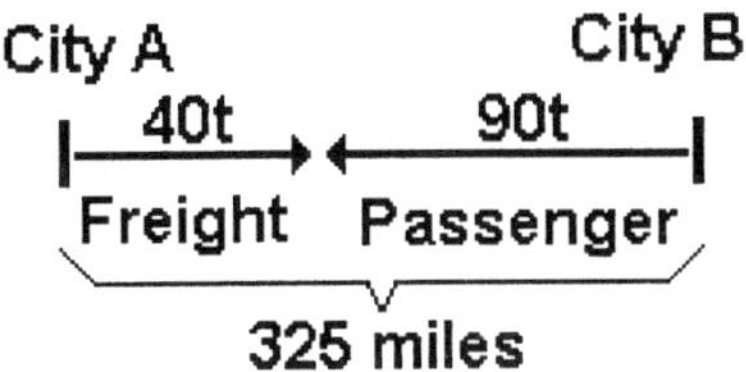

The sum of the two distances would equal the total distance of 325 miles.

$$40t + 90t = 325$$
$$130t = 325$$
$$t = 2\frac{1}{2}$$

It would take $2\frac{1}{2}$ hours for the two trains to meet.

35. Let r represent the second car's rate. Add two hours to the second car's time to find the first car's time. The time also needs to be converted into hours only. A chart of the information would be as follows:

	Rate	Time	Distance $(d = rt)$
First car	40	$7\frac{1}{3} = \frac{22}{3}$	$40\left(\frac{22}{3}\right) = \frac{880}{3}$
Second car	r	$5\frac{1}{3} = \frac{16}{3}$	$\frac{16}{3}r$

A diagram of the problem would be as follows:

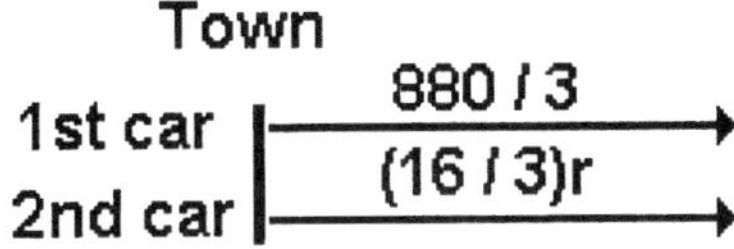

Set the two distances equal and solve.

$$\frac{16}{3}r = \frac{880}{3}$$
$$16r = 880$$
$$r = 55$$

The second car would be traveling 55 mph.

37. Let r represent the rate of the train traveling west; then $r + 8$ represents the rate of the train traveling east The time is the same for both trains. A chart of the information would be as follows:

	Rate	Time	Distance $(d = rt)$
West train	r	$9\frac{1}{2} = \frac{19}{2}$	$\frac{19}{2}r$
East train	$r + 8$	$9\frac{1}{2} = \frac{19}{2}$	$\frac{19}{2}(r + 8)$

A diagram of the problem would be as follows:

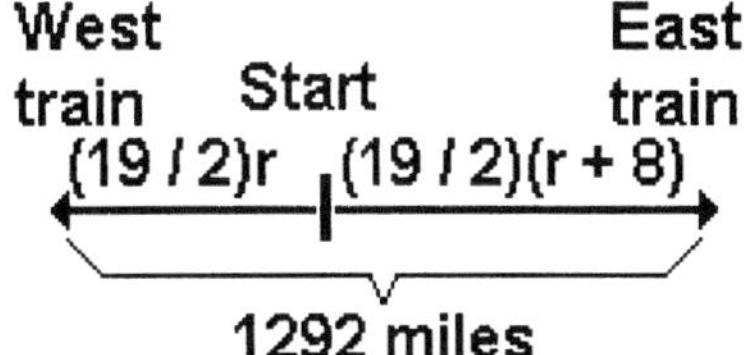

The sum of the two distances would be 1292 miles.

$$\frac{19}{2}r + \frac{19}{2}(r + 8) = 1292$$
$$19r + 19(r + 8) = 2(1292)$$
$$19r + 19r + 152 = 2584$$
$$38r = 2432$$
$$r = 64$$

The train traveling west has a rate of 64 mph and the train traveling east has a rate of 72 mph.

39. Let r represent Jeff's rate when he was leaving; then $r - 2$ represents his rate returning. A chart of the information would be as follows:

	Rate	Time	Distance $(d = rt)$
Leaving	r	3	$3r$
Returning	$r - 2$	$3\frac{3}{4} = \frac{15}{4}$	$\frac{15}{4}(r - 2)$

A diagram of the problem would be as follows:

Home
Leaving
3r
(15 / 4)(r - 2)
Returning
Distances are equal.

Problem Set 3.4

Set the distances equal to each other and solve for r.

$$3r = \frac{15}{4}(r-2)$$
$$12r = 15(r-2)$$
$$12r = 15r - 30$$
$$-3r = -30$$
$$r = 10$$

Jeff's rate leaving was 10 mph. Therefore, he travelled 30 miles leaving and 30 miles returning for a total of 60 miles.

PROBLEM SET 3.5 More about Problem Solving

1. $0.3x + 0.7(20 - x) = 0.4(20)$
$$0.3x + 14 - 0.7x = 8$$
$$14 - 0.4x = 8$$
$$-0.4x = -6$$
$$x = 15$$
The solution set is $\{15\}$.

3. $0.2(20) + x = 0.3(20 + x)$
$$4 + x = 6 + 0.3x$$
$$4 + 0.7x = 6$$
$$0.7x = 2$$
$$7x = 20$$
$$x = \frac{20}{7}$$
The solution set is $\left\{\frac{20}{7}\right\}$.

5. $0.7(15) - x = 0.6(15 - x)$
$$10[0.7(15)] - 10x = 10[0.6(15 - x)]$$
$$7(15) - 10x = 6(15 - x)$$
$$105 - 10x = 90 - 6x$$
$$105 - 4x = 90$$
$$-4x = -15$$
$$x = \frac{15}{4}$$
The solution set is $\left\{\frac{15}{4}\right\}$.

7.
$$\begin{aligned} 0.4(10) - 0.4x + x &= 0.5(10) \\ 4 - 0.4x + x &= 5 \\ 4 + 0.6x &= 5 \\ 0.6x &= 1 \\ 6x &= 10 \\ x &= \frac{10}{6} = \frac{5}{3} \end{aligned}$$

The solution set is $\left\{\frac{5}{3}\right\}$.

9.
$$\begin{aligned} 20x + 12\left(4\frac{1}{2} - x\right) &= 70 \\ 20x + 12\left(\frac{9}{2} - x\right) &= 70 \\ 20x + 54 - 12x &= 70 \\ 8x + 54 &= 70 \\ 8x &= 16 \\ x &= 2 \end{aligned}$$

The solution set is $\{2\}$.

11.
$$\begin{aligned} 3t &= \frac{11}{2}\left(t - \frac{3}{2}\right) \\ 3t &= \frac{11}{2}t - \frac{33}{4} \\ 4(3t) &= 4\left(\frac{11}{2}t - \frac{33}{4}\right) \\ 12t &= 22t - 33 \\ -10t &= -33 \\ t &= \frac{33}{10} \end{aligned}$$

The solution set is $\left\{\frac{33}{10}\right\}$.

13. Let x represent the amount of pure acid to be added; then $100 + x$ represents the amount of final solution.

$$\begin{pmatrix}\text{pure acid in} \\ \text{10\% solution}\end{pmatrix} + \begin{pmatrix}\text{pure acid} \\ \text{to be added}\end{pmatrix} = \begin{pmatrix}\text{pure acid in} \\ \text{final solution}\end{pmatrix}$$

$$\begin{aligned}
(10\%)(100) + x &= (20\%)(100 + x) \\
0.10(100) + x &= 0.20(100 + x) \\
10[0.10(100) + x] &= 10[0.20(100 + x)] \\
1(100) + 10x &= 2(100 + x) \\
100 + 10x &= 200 + 2x \\
100 + 8x &= 200 \\
8x &= 100 \\
x &= 12.5
\end{aligned}$$

We must add 12.5 milliliters of pure acid.

15. Let x represent the amount of distilled water to be added; then $10 + x$ represents the amount of final solution.

$$\begin{pmatrix}\text{pure acid in} \\ \text{50\% solution}\end{pmatrix} + \begin{pmatrix}\text{no acid} \\ \text{to be added}\end{pmatrix} = \begin{pmatrix}\text{pure acid in} \\ \text{final solution}\end{pmatrix}$$

$$\begin{aligned}
(50\%)(10) + 0 &= (20\%)(10 + x) \\
0.50(10) &= 0.20(10 + x) \\
10[0.50(10)] &= 10[0.20(10 + x)] \\
5(10) &= 2(10 + x) \\
50 &= 20 + 2x \\
30 &= 2x \\
15 &= x
\end{aligned}$$

We must add 15 centiliters of distilled water.

17. Let x represent the amount of 30% alcohol to be used; then $10 - x$ represents the amount of 50% solution.

$$\begin{pmatrix}\text{pure alcohol in} \\ \text{30\% solution}\end{pmatrix} + \begin{pmatrix}\text{pure alcohol in} \\ \text{50\% solution}\end{pmatrix} = \begin{pmatrix}\text{pure alcohol in} \\ \text{final 35\% solution}\end{pmatrix}$$

$$\begin{aligned}
(30\%)(x) + (50\%)(10 - x) &= (35\%)(10) \\
0.30(x) + 0.50(10 - x) &= 0.35(10) \\
100[0.30(x) + 0.50(10 - x)] &= 100[0.35(10)] \\
30(x) + 50(10 - x) &= 35(10) \\
30x + 500 - 50x &= 350 \\
500 - 20x &= 350 \\
-20x &= -150 \\
x &= \frac{150}{20} = 7\frac{1}{2}
\end{aligned}$$

We must add $7\frac{1}{2}$ quarts of 30% alcohol and $2\frac{1}{2}$ quarts of 50% alcohol.

19. Let x represent the amount of water to be removed; then $20 - x$ represents the amount of final salt solution.

$$\begin{pmatrix}\text{pure salt in}\\\text{30\% solution}\end{pmatrix} - \begin{pmatrix}\text{no salt to}\\\text{be removed}\end{pmatrix} = \begin{pmatrix}\text{pure salt in}\\\text{final solution}\end{pmatrix}$$

$$\begin{aligned}(30\%)(20) - 0 &= (40\%)(20 - x)\\ 0.30(20) - 0 &= 0.40(20 - x)\\ 6 &= 8 - 0.4x\\ -2 &= -0.4x\\ -20 &= -4x\\ 5 &= x\end{aligned}$$

We must remove 5 gallons of water.

21. Let x represent the amount of pure antifreeze to be added; then x also represents the amount of solution to be drained.

$$\begin{pmatrix}\text{antifreeze in}\\\text{20\% solution}\end{pmatrix} - \begin{pmatrix}\text{antifreeze to}\\\text{be drained}\end{pmatrix} + \begin{pmatrix}\text{antifreeze to}\\\text{be added}\end{pmatrix} = \begin{pmatrix}\text{antifreeze in}\\\text{final solution}\end{pmatrix}$$

$$\begin{aligned}(20\%)(12) - (20\%)(x) + x &= (40\%)(12)\\ 0.20(12) - 0.20x + x &= 0.40(12)\\ 10[0.20(12) - 0.20x + x] &= 10[0.40(12)]\\ 2(12) - 2(x) + 10x &= 4(12)\\ 24 - 2x + 10x &= 48\\ 24 + 8x &= 48\\ 8x &= 24\\ x &= 3\end{aligned}$$

We must drain 3 quarts of the 20% solution and then replace with 3 quarts of pure antifreeze.

23. Let x represent the amount of 15% salt solution to be used; then $8 + x$ represents the amount of final solution.

$$\begin{pmatrix}\text{pure salt in}\\\text{15\% solution}\end{pmatrix} + \begin{pmatrix}\text{pure salt in}\\\text{20\% solution}\end{pmatrix} = \begin{pmatrix}\text{pure salt in}\\\text{final solution}\end{pmatrix}$$

$$\begin{aligned}(15\%)(x) + (20\%)(8) &= (17\%)(8 + x)\\ 0.15(x) + 0.20(8) &= 0.17(8 + x)\\ 15x + 20(8) &= 17(8 + x)\\ 15x + 160 &= 136 + 17x\\ -2x &= -24\\ x &= 12\end{aligned}$$

We must add 12 gallons of the 15% salt solution.

Problem Set 3.5

25. Let p represent the percent of grapefruit juice in the resulting mixture.

$$\begin{pmatrix}\text{pure juice in}\\ \text{10\% solution}\end{pmatrix} + \begin{pmatrix}\text{pure juice in}\\ \text{20\% solution}\end{pmatrix} = \begin{pmatrix}\text{pure juice in}\\ \text{final solution}\end{pmatrix}$$

$$\begin{aligned}(10\%)(30) + (20\%)(50) &= (p)(30+50)\\ 0.10(30) + 0.20(50) &= p(80)\\ 3 + 10 &= 80p\\ 13 &= 80p\\ \frac{13}{80} &= p\\ p = 0.1625 &= 16.25\%\end{aligned}$$

The resulting mixture would be 16.25% grapefruit juice.

27. Let x represent the length of the side of the square; then $2x-9$ represents the width of the rectangle and $2x-3$ represents the length of the rectangle.

2x - 3
2x - 9
x x x x

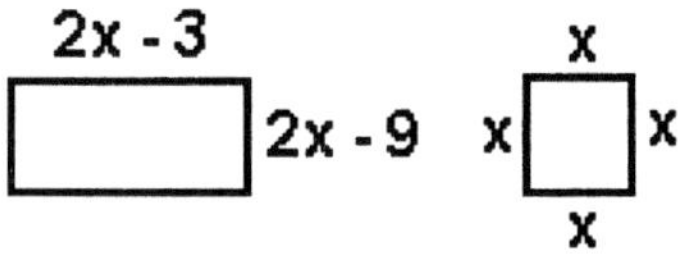

$$\begin{aligned}\text{Rectangular perimeter} &= \text{Square perimeter}\\ 2(2x-9) + 2(2x-3) &= 4x\\ 4x - 18 + 4x - 6 &= 4x\\ 8x - 24 &= 4x\\ -24 &= -4x\\ 6 &= x\end{aligned}$$

Each side of the square is 6 inches, the width of the rectangle is $2(6)-9=3$ inches, and the length of the rectangle is $2(6)-3=9$ inches.

29. Let t represent Dick's time. Add one-half hour to Dick's time to represent Butch's time. A chart of the information would be as follows:

	Rate	Time	Distance ($d = rt$)
Butch	2	$t+\frac{1}{2}$	$2\left(t+\frac{1}{2}\right) = 2t+1$
Dick	$3\frac{1}{2} = \frac{7}{2}$	t	$\frac{7}{2}t$

A diagram of the problem would be as follows:

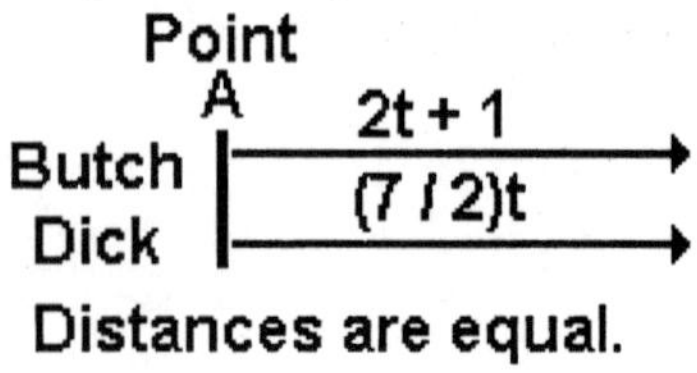

Point
A
2t + 1
Butch
(7 / 2)t
Dick
Distances are equal.

Set the two distances equal and solve.

$$2t + 1 = \frac{7}{2}t$$
$$4t + 2 = 7t$$
$$2 = 3t$$
$$t = \frac{2}{3}\text{hour} = 40 \text{ minutes}$$

Dick would catch up with Butch in 40 minutes.

31. Let the ages be represented as follows:

$2x$: Bill's present age

x : Pam's present age

$2x - 6$: Bill's age 6 years ago

$x - 6$: Pam's age 6 years ago

$$(\text{Bill's age 6 years ago}) = (4)(\text{Pam's age 6 years ago})$$
$$2x - 6 = 4(x - 6)$$
$$2x - 6 = 4x - 24$$
$$-2x - 6 = -24$$
$$-2x = -18$$
$$x = 9$$

Pam is 9 years old and Bill is 18 years old.

33. Let x represent the amount of money invested at 12% interest; then $12,000 - x$ represents the amount of money invested at 14%.

$$\begin{pmatrix}\text{Interest earned}\\ \text{at 12\%}\end{pmatrix} + \begin{pmatrix}\text{Interest earned}\\ \text{at 14\%}\end{pmatrix} = \begin{pmatrix}\text{Total interest}\\ \text{earned}\end{pmatrix}$$

$$(12\%)(x) + (14\%)(12,000 - x) = 1580$$
$$0.12x + 0.14(12,000 - x) = 1580$$
$$12x + 14(12,000 - x) = 158,000$$
$$12x + 168,000 - 14x = 158,000$$
$$-2x = -10,000$$
$$x = 5000$$

She invested \$5000 at 12% and \$7000 at 14%.

35. Let x represent the amount of money invested at 9% interest; then $2x$ represents the amount of money invested at 10% and $3x$ the amount at 11%.

$$\begin{pmatrix}\text{Interest earned}\\ \text{at 9\%}\end{pmatrix} + \begin{pmatrix}\text{Interest earned}\\ \text{at 10\%}\end{pmatrix} + \begin{pmatrix}\text{Interest earned}\\ \text{at 11\%}\end{pmatrix} = \begin{pmatrix}\text{Total interest}\\ \text{earned}\end{pmatrix}$$

$$\begin{aligned}(9\%)(x) + (10\%)(2x) + (11\%)(3x) &= 310\\ 0.09x + 0.10(2x) + 0.11(3x) &= 310\\ 9x + 10(2x) + 11(3x) &= 31,000\\ 9x + 20x + 33x &= 31,000\\ 62x &= 31,000\\ x &= 500\end{aligned}$$

She invested \$500 at 9%, \$1000 at 10%, and \$1500 at 11%.

37. Let x represent the amount of money invested at 10% interest; then $x + 250$ represents the amount invested at 11%.

$$\begin{pmatrix}\text{Interest earned}\\ \text{at 10\%}\end{pmatrix} + \begin{pmatrix}\text{Interest earned}\\ \text{at 11\%}\end{pmatrix} = \begin{pmatrix}\text{Total interest}\\ \text{earned}\end{pmatrix}$$

$$\begin{aligned}(10\%)(x) + (11\%)(x + 250) &= 153.50\\ 0.10(x) + 0.11(x + 250) &= 153.50\\ 10x + 11(x + 250) &= 15,350\\ 10x + 11x + 2750 &= 15,350\\ 21x + 2750 &= 15,350\\ 21x &= 12,600\\ x &= 600\end{aligned}$$

She invested \$600 at 10% and \$850 at 11%.

39. Let x represent the amount of money invested at 12% interest; then $x + 3000$ represents the total amount of money invested at 11%.

$$\begin{pmatrix}\text{Interest earned}\\ \text{at 9\%}\end{pmatrix} + \begin{pmatrix}\text{Interest earned}\\ \text{at 12\%}\end{pmatrix} = \begin{pmatrix}\text{Total interest}\\ \text{at 11\%}\end{pmatrix}$$

$$\begin{aligned}(9\%)(3000) + (12\%)(x) &= (11\%)(x + 3000)\\ 0.09(3000) + 0.12x &= 0.11(x + 3000)\\ 9(3000) + 12x &= 11(x + 3000)\\ 27,000 + 12x &= 11x + 33,000\\ x + 27,000 &= 33,000\\ x &= 6000\end{aligned}$$

The amount invested must be \$6000 at 12%.

41. Let x represent the amount of money invested at 9% interest; then $6000 - x$ represents the amount of money invested at 11%.

$$\begin{pmatrix}\text{Interest earned}\\ \text{at 9\%}\end{pmatrix} = \begin{pmatrix}\text{Interest earned}\\ \text{at 11\%}\end{pmatrix} - 160$$

$$\begin{aligned}(9\%)(x) &= (11\%)(6000 - x) - 160\\ 0.09x &= 0.11(6000 - x) - 160\\ 9x &= 11(6000 - x) - 16,000\\ 9x &= 66,000 - 11x - 16,000\\ 9x &= 50,000 - 11x\\ 20x &= 50,000\\ x &= 2500\end{aligned}$$

The amounts invested are \$2500 at 9% and \$3500 at 11%.

CHAPTER 3 | Review Problem Set

1. $0.5x + 0.7x = 1.7$

Combine like terms.

$$1.2x = 1.7$$

Multiplied both sides by 10.

$$\begin{aligned}12x &= 17\\ x &= \frac{17}{12}\end{aligned}$$

The solution set is $\left\{\frac{17}{12}\right\}$.

2. $0.07t + 0.12(t - 3) = 0.59$

Multiplied by 100.

$$7t + 12(t - 3) = 59$$

Distributive property

$$7t + 12t - 36 = 59$$

Combined like terms.

$$19t - 36 = 59$$

Added 36 to both sides.

$$\begin{aligned}19t &= 95\\ t &= 5\end{aligned}$$

The solution set is $\{5\}$.

3. $0.1x + 0.12(1700 - x) = 188$

Multiplied both sides by 100.

$$\begin{aligned}10x + 12(1700 - x) &= 18,800\\ 10x + 20,400 - 12x &= 18,800\\ -2x + 20,400 &= 18,800\\ -2x &= 1600\\ x &= 800\end{aligned}$$

The solution set is $\{800\}$.

4. $x - 0.25x = 12$

Multiplied both sides by 100.

$$\begin{aligned}100x - 25x &= 1200\\ 75x &= 1200\\ x &= 16\end{aligned}$$

The solution set is $\{16\}$.

5. $0.2(x - 3) = 14$

Multiplied both sides by 10.

$$\begin{aligned}2(x - 3) &= 140\\ 2x - 6 &= 140\\ 2x &= 146\\ x &= 73\end{aligned}$$

The solution set is $\{73\}$.

6. $P = 50,\ l = 19 : P = 2l + 2w$

$$\begin{aligned}50 &= 2(19) + 2w\\ 50 &= 38 + 2w\\ 12 &= 2w\\ 6 &= w\end{aligned}$$

7. $F = 77 : F = \frac{9}{5}C + 32$

$77 = \frac{9}{5}C + 32$

Subtracted 32 from both sides.

$45 = \frac{9}{5}C$

Multiply both sides by $\frac{5}{9}$.

$\frac{5}{9}(45) = \frac{5}{9}\left(\frac{9}{5}\right)C$

$25 = C$

8. $A = P + Prt$

$A - P = Prt$

Divide both sides by Pr.

$\frac{A - P}{Pr} = t$

9. $2x - 3y = 13$

Added $3y$ to both sides.

$2x = 13 + 3y$

Divided both sides by 2.

$x = \frac{13 + 3y}{2}$

10. Substitute $b_1 = 8$, $b_2 = 14$, and $h = 7$ into the formula for the area of a trapezoid.

$A = \frac{1}{2}h(b_1 + b_2) = \frac{1}{2}(7)(8 + 14)$

$= \frac{1}{2}(7)(22) = 77$ square inches

11. Let h represent the altitude when $b = 9$ and $A = 27$.

$A = \frac{1}{2}bh$

$27 = \frac{1}{2}(9)h$

$54 = 9h$

$6 = h$

The altitude of the triangle is 6 centimeters.

12. Let h represent the height when $r = 4$ and $S = 152\pi$.

$S = 2\pi r^2 + 2\pi rh$

$152\pi = 2\pi(4)^2 + 2\pi(4)h$

$152\pi = 32\pi + 8\pi h$

$120\pi = 8\pi h$

$15 = h$

The height of the trapezoid is 15 feet.

13. Let r represent the percent to be found.

$18 = r(30)$

$\frac{18}{30} = r$

$0.6 = r$

Therefore, 18 is 60% of 30.

14. Let n represent one of the numbers, then $96 - n$ represents the other number.

$\frac{5}{7} = \frac{n}{96 - n}$

$7n = 5(96 - n)$

$7n = 480 - 5n$

$12n = 480$

$n = 40$

Therefore, 40 and 56 are the numbers.

15. Let n represent the number.

$(15\%)(n) = 6$

$0.15n = 6$

$15n = 600$

$n = 40$

Therefore, 15% of 40 is 6.

16. Let w represent the width of the rectangle, then $2w + 5$ represents the length when the perimeter is 46 meters.

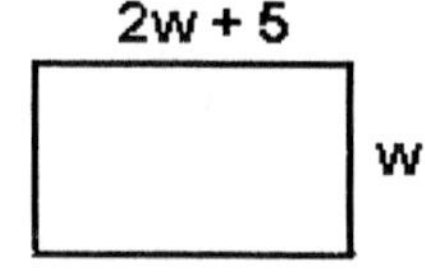

$2(2w + 5) + 2w = 46$

$4w + 10 + 2w = 46$

$6w + 10 = 46$

$6w = 36$

$w = 6$

The dimensions of the rectangle are 6 meters by 17 meters.

17. Let t represent the time for each airplane as both airplanes left at the same time. A chart of the information would be as follows:

	Rate	Time	Distance ($d = rt$)
Plane A	350	t	$350t$
Plane B	400	t	$400t$

A diagram of the problem would be as follows:

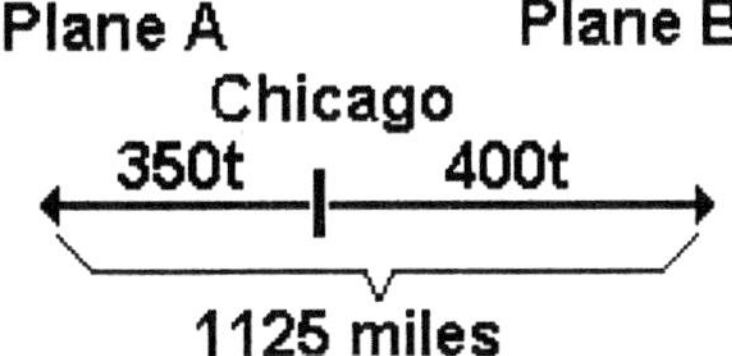

The sum of the two distances would equal the total distance of 1125 miles.

$$350t + 400t = 1125$$
$$750t = 1125$$
$$t = 1\frac{1}{2}$$

It would take $1\frac{1}{2}$ hours for the airplanes to be 1125 miles apart.

18. Let x represent the amount of pure alcohol to be added. Then $10 + x$ represents the amount of final solution.

$$\begin{pmatrix}\text{pure}\\\text{alcohol}\\\text{in 70\%}\\\text{solution}\end{pmatrix} + \begin{pmatrix}\text{pure}\\\text{alcohol}\\\text{to be}\\\text{added}\end{pmatrix} = \begin{pmatrix}\text{pure}\\\text{alcohol}\\\text{in final}\\\text{solution}\end{pmatrix}$$

$$(70\%)(10) + x = (90\%)(10 + x)$$
$$0.70(10) + x = 0.90(10 + x)$$
$$10[0.70(10) + x] = 10[0.90(10 + x)]$$
$$7(10) + 10x = 9(10 + x)$$
$$70 + 10x = 90 + 9x$$
$$70 + x = 90$$
$$x = 20$$

We must add 20 liters of pure alcohol.

19. Let w represent the width of the rectangle, then $2w + 10$ represents the length when the perimeter is 110.

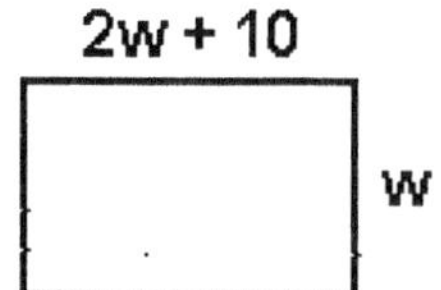

$$2(2w + 10) + 2w = 110$$
$$4w + 20 + 2w = 110$$
$$6w + 20 = 110$$
$$6w = 90$$
$$w = 15$$

The dimensions of the rectangle are 15 centimeters by 40 centimeters.

20. Let w represent the width of the rectangular garden, then $3w - 1$ represents the length when the perimeter is 78 yards.

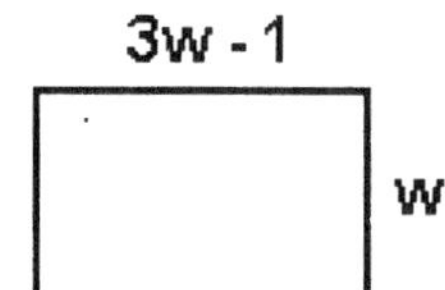

$$2(3w - 1) + 2w = 78$$
$$6w - 2 + 2w = 78$$
$$8w - 2 = 78$$
$$8w = 80$$
$$w = 10$$

The dimensions of the garden are 10 yards by 29 yards.

21. Let a represent the measure of the angle, then $90 - a$ represents the complement of the angle, and $180 - a$ represents the supplement.

$$\frac{\text{complement}}{\text{supplement}} \quad \frac{90 - a}{180 - a} = \frac{7}{16}$$
$$7(180 - a) = 16(90 - a)$$
$$1260 - 7a = 1440 - 16a$$
$$1260 + 9a = 1440$$
$$9a = 180$$
$$a = 20$$

The measure of the angle is $20°$.

22. Let g represent the number of gallons required for the trip.

$$\frac{\text{gallons}}{\text{miles}} \quad \frac{18}{369} = \frac{g}{615}$$

$$369g = 18(615)$$
$$369g = 11{,}070$$
$$g = 30$$

Therefore, 30 gallons would be needed for the 615-mile trip.

23. Let x represent the amount of money invested at 9% interest; then $2100 - x$ represents the amount of money invested at 11%.

$$\begin{pmatrix}\text{Interest earned}\\ \text{at 9\%}\end{pmatrix} + 51 = \begin{pmatrix}\text{Interest earned}\\ \text{at 11\%}\end{pmatrix}$$

$$(9\%)x + 51 = (11\%)(2100 - x)$$
$$0.09x + 51 = 0.11(2100 - x)$$
$$9x + 5100 = 11(2100 - x)$$
$$9x + 5100 = 23{,}100 - 11x$$
$$20x = 18{,}000$$
$$x = 900$$

He invested \$900 at 9% and \$1200 at 11%.

24. Let s represent the selling price. Profit is a percent of selling price.

Selling price = Cost + Profit

$$s = 28 + (30\%)(s)$$
$$s = 28 + 0.3s$$
$$10s = 280 + 3s$$
$$7s = 280$$
$$s = 40$$

The selling price would be \$40.

25. Let r represent the rate of discount.

$$60 - p(60) = 39$$
$$60 - 60p = 39$$
$$-60p = -21$$
$$p = 0.35$$

The percent of discount that she received was 35%.

26. Let a represent the measure of the second angle, then $3a - 3$ represents the third angle. The sum of the measures of the angles of a triangle are 180°.

$$47 + a + (3a - 3) = 180$$
$$44 + 4a = 180$$
$$4a = 136$$
$$a = 34$$

The remaining angles are 34° and $3(34) - 3 = 99°$.

27. Let t represent Zak's time, then $t + 1$ represents Connie's time as she travels one hour longer. A chart of the information would be as follows:

	Rate	Time	Distance ($d = rt$)
Connie	10	$t+1$	$10(t+1)$
Zak	12	t	$12t$

A diagram of the problem would be as follows:

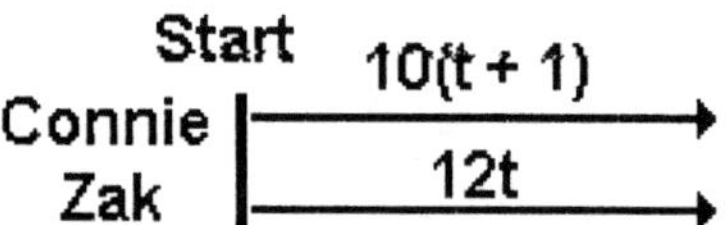

Distances are equal.

Set the distances equal to each other and solve.

$$10(t + 1) = 12t$$
$$10t + 10 = 12t$$
$$10 = 2t$$
$$5 = t$$

It would take 5 hours for Zak to catch up with Connie.

28. Let x represent the amount of 10% salt solution to be used, then $12 + x$ represents the amount of final solution.

$$\begin{pmatrix}\text{pure}\\ \text{salt}\\ \text{in 10\%}\\ \text{solution}\end{pmatrix} + \begin{pmatrix}\text{pure}\\ \text{salt}\\ \text{in 15\%}\\ \text{solution}\end{pmatrix} = \begin{pmatrix}\text{pure}\\ \text{salt}\\ \text{in final}\\ \text{solution}\end{pmatrix}$$

$$(10\%)(x) + (15\%)(12) = (12\%)(12 + x)$$
$$0.10(x) + 0.15(12) = 0.12(12 + x)$$
$$10x + 15(12) = 12(12 + x)$$
$$10x + 180 = 144 + 12x$$
$$-2x = -36$$
$$x = 18$$

We must add 18 gallons of the 10% salt solution.

29. Let p represent the percent of orange juice in the resulting mixture.

$$\begin{pmatrix}\text{pure} \\ \text{juice} \\ \text{in 20\%} \\ \text{solution}\end{pmatrix} + \begin{pmatrix}\text{pure} \\ \text{juice} \\ \text{in 30\%} \\ \text{solution}\end{pmatrix} = \begin{pmatrix}\text{pure} \\ \text{juice} \\ \text{in final} \\ \text{solution}\end{pmatrix}$$

$$(20\%)(20) + (30\%)(30) = (p)(20 + 30)$$
$$0.20(20) + 0.30(30) = p(50)$$
$$4 + 9 = 50p$$
$$13 = 50p$$
$$\frac{13}{50} = p$$
$$p = 0.26 = 26\%$$

The resulting mixture would be 26% orange juice.

30. $i = Prt$

$i = 3500(0.0525)(2)$

$i = 367.5$

The interest is $367.50.

CHAPTER 3 Test

1. $$\frac{x+2}{4} = \frac{x-3}{5}$$

Cross products are equal.

$$5(x + 2) = 4(x - 3)$$
$$5x + 10 = 4x - 12$$
$$x + 10 = -12$$
$$x = -22$$

The solution set is $\{-22\}$.

2. $$\frac{-4}{2x-1} = \frac{3}{3x+5}$$

Cross products are equal.

$$-4(3x + 5) = 3(2x - 1)$$
$$-12x - 20 = 6x - 3$$
$$-18x - 20 = -3$$
$$-18x = 17$$
$$x = -\frac{17}{18}$$

The solution set is $\left\{-\frac{17}{18}\right\}$.

3. $$\frac{x-1}{6} - \frac{x+2}{5} = 2$$

This is NOT a proportion!

Multiply both sides by 30.

$$30\left(\frac{x-1}{6} - \frac{x+2}{5}\right) = 30(2)$$
$$5(x - 1) - 6(x + 2) = 60$$
$$5x - 5 - 6x - 12 = 60$$
$$-x - 17 = 60$$
$$-x = 77$$
$$x = -77$$

The solution set is $\{-77\}$.

4. $$\frac{x+8}{7}-2=\frac{x-4}{4}$$
This is NOT a proportion!
Multiply both sides by 28.
$$28\left(\frac{x+8}{7}-2\right)=28\left(\frac{x-4}{4}\right)$$
$$4(x+8)-28(2)=7(x-4)$$
$$4x+32-56=7x-28$$
$$4x-24=7x-28$$
$$-3x-24=-28$$
$$-3x=-4$$
$$x=\frac{4}{3}$$
The solution set is $\left\{\frac{4}{3}\right\}$.

5. $$\frac{n}{20-n}=\frac{7}{3}$$
Cross products are equal.
$$7(20-n)=3n$$
$$140-7n=3n$$
$$140=10n$$
$$14=n$$
The solution set is $\{14\}$.

6. $$\frac{h}{4}+\frac{h}{6}=1$$
This is NOT a proportion!
Multiply both sides by 12.
$$12\left(\frac{h}{4}+\frac{h}{6}\right)=12(1)$$
$$3h+2h=12$$
$$5h=12$$
$$h=\frac{12}{5}$$
The solution set is $\left\{\frac{12}{5}\right\}$.

7. $0.05n+0.06(400-n)=23$
Multiplied both sides by 100.
$$5n+6(400-n)=2300$$
$$5n+2400-6n=2300$$
$$-n+2400=2300$$
$$-n=-100$$
$$n=100$$
The solution set is $\{100\}$.

8. $s=35+0.5s$
Multiplied both sides by 10.
$$10s=350+5s$$
$$5s=350$$
$$s=70$$
The solution set is $\{70\}$.

9. $0.07n=45.5-0.08(600-n)$
Multiplied both sides by 100.
$$7n=4550-8(600-n)$$
$$7n=4550-4800+8n$$
$$7n=-250+8n$$
$$-n=-250$$
$$n=250$$
The solution set is $\{250\}$.

10. $12t+8\left(\frac{7}{2}-t\right)=50$
Distributive property.
$$12t+28-8t=50$$
$$4t+28=50$$
$$4t=22$$
$$t=\frac{22}{4}=\frac{11}{2}$$
The solution set is $\left\{\frac{11}{2}\right\}$.

11. $$F=\frac{9C+160}{5}$$
Multiplied both sides by 5.
$$5F=9C+160$$
Subtracted 160 from both sides.
$$5F-160=9C$$
Divided both sides by 9.
$$\frac{5F-160}{9}=C$$

12. $y = 2(x-4)$
Distributive Property.
$y = 2x - 8$
Added 8 to both sides.
$y + 8 = 2x$
Divided both sides by 2.
$\frac{y+8}{2} = x$

13. $\frac{x+3}{4} = \frac{y-5}{9}$
Cross products are equal.
$4(y-5) = 9(x+3)$
Distributive property.
$4y - 20 = 9x + 27$
Added 20 to both sides.
$4y = 9x + 47$
Divided both sides by 4.
$y = \frac{9x+47}{4}$

14. Find the radius by solving the formula for circumference for r when $C = 16\pi$.
$C = 2\pi r$
$16\pi = 2\pi r$
$8 = r$
When the radius is 8 centimeters, the area will be $A = \pi(8)^2 = 64\pi$ square centimeters.

15. Let w represent the width of the rectangle when $l = 32$ and $P = 100$.

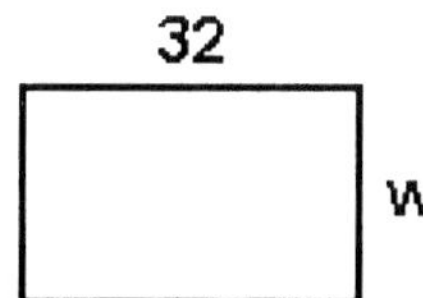

$P = 2l + 2w$
$100 = 2(32) + 2w$
$100 = 64 + 2w$
$36 = 2w$
$18 = w$
The width of the rectangle is 18 inches. Therefore, the area of the rectangle is $(32)(18) = 576$ square inches.

16. Let h represent the altitude of the triangular plot when $A = 133$ and $b = 19$.
$A = \frac{1}{2}bh$
$133 = \frac{1}{2}(19)h$
$266 = 19h$
$14 = h$
The altitude of the triangular plot is 14 yards.

17. $\frac{5}{4} = \frac{n}{100}$
$4n = 500$
$n = 125$
Therefore, $\frac{5}{4} = \frac{125}{100} = 125\%$.

18. Let n represent the number.
$(35\%)n = 24.5$
$0.35n = 24.5$
$35n = 2450$
$n = 70$
Therefore, 35% of 70 is 24.5.

19. Let p represent the original price of the blouse.
$(100\%)(p) - (30\%)(p) = 28$
$(70\%)(p) = 28$
$0.7p = 28$
$7p = 280$
$p = 40$
The original price of the blouse was \$40.

20. Let s represent the selling price of the skirt. Profit is a percent of cost.
Selling price = Cost + Profit
$s = 40 + (30\%)(40)$
$s = 40 + 0.3(40)$
$s = 40 + 12$
$s = 52$
The selling price of the skirt should be \$52.

21. Let r represent the rate of discount he received.

$$80 - r(80) = 48$$
$$80 - 80r = 48$$
$$-80r = -32$$
$$r = 0.40$$

The rate of discount would be 40%.

22. Let f represent the number of females who voted, then $1500 - f$ represents the number of males who voted.

$$\frac{\text{female voters}}{\text{male voters}} \quad \frac{7}{5} = \frac{f}{1500 - f}$$
$$5f = 7(1500 - f)$$
$$5f = 10,500 - 7f$$
$$12f = 10,500$$
$$f = 875$$

There were 875 female voters.

23. Let t represent the time of the second car, then $t + 1$ represents the first car's time as it travels one hour longer. A chart of the information would be as follows:

	Rate	Time	Distance ($d = rt$)
First Car	50	$t+1$	$50(t+1)$
Second Car	55	t	$55t$

A diagram of the problem would be as follows:

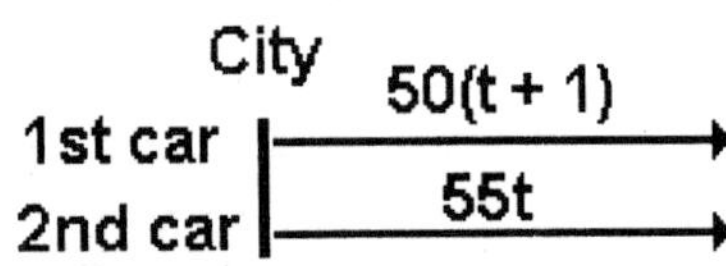

Set the distances equal to each other and solve.

$$50(t + 1) = 55t$$
$$50t + 50 = 55t$$
$$50 = 5t$$
$$10 = t$$

It would take 10 hours for the second car to overtake the first car.

24. Let x represent the amount of pure acid to be added. Then $6 + x$ represents the amount of final solution.

$$\begin{pmatrix}\text{pure}\\ \text{acid}\\ \text{in 50\%}\\ \text{solution}\end{pmatrix} + \begin{pmatrix}\text{pure}\\ \text{acid}\\ \text{to be}\\ \text{added}\end{pmatrix} = \begin{pmatrix}\text{pure}\\ \text{acid}\\ \text{in final}\\ \text{solution}\end{pmatrix}$$

$$(50\%)(6) + x = (70\%)(6 + x)$$
$$0.50(6) + x = 0.70(6 + x)$$
$$10[0.50(6) + x] = 10[0.70(6 + x)]$$
$$5(6) + 10x = 7(6 + x)$$
$$30 + 10x = 42 + 7x$$
$$30 + 3x = 42$$
$$3x = 12$$
$$x = 4$$

We must add 4 centiliters of pure acid.

25.
$$i = Prt$$
$$4000 = 4000(0.09)t$$
$$4000 = 360t$$
$$\frac{4000}{360} = t$$
$$11.1 = t$$

It would take 11.1 years.

CHAPTERS 1-3 Cumulative Review

1. $7x - 9x - 14x =$
$(7 - 9 - 14)x =$
$-16x$

2. $-10a - 4 + 13a + a - 2 = 4a - 6$

3. $5(x - 3) + 7(x + 6) =$
$5x - 15 + 7x + 42 =$
$12x + 27$

4. $3(x - 1) - 4(2x - 1) =$
$3x - 3 - 8x + 4 =$
$-5x + 1$

5. $-3n-2(n-1)+5(3n-2)-n=$
$-3n-2n+2+15n-10-n=$
$9n-8$

6. $6n+3(4n-2)-2(2n-3)-5=$
$6n+12n-6-4n+6-5=$
$14n-5$

7. $\frac{1}{2}x-\frac{3}{4}x+\frac{2}{3}x-\frac{1}{6}x=$
$\frac{6}{12}x-\frac{9}{12}x+\frac{8}{12}x-\frac{2}{12}x=$
$\frac{3}{12}x=\frac{1}{4}x$

8. $\frac{1}{3}n-\frac{4}{15}n+\frac{5}{6}n-n=$
$\frac{10}{30}n-\frac{8}{30}n+\frac{25}{30}n-\frac{30}{30}n=$
$-\frac{3}{30}n=-\frac{1}{10}n$

9. $0.4x+0.7x-0.8x+1.0x=1.3x$

10. $0.5(x-2)+0.4(x+3)-0.2x=$
$0.5x-1.0+0.4x+1.2-0.2x=$
$0.7x+0.2$

11. $x=-2,\ y=5:\ 5x-7y+2xy=$
$5(-2)-7(5)+2(-2)(5)=$
$-10-35-20=-65$

12. $a=3,\ b=-4: 2ab-a+6b=$
$2(3)(-4)-(3)+6(-4)=$
$-24-3-24=-51$

13. $-3(x-1)+2(x+6)=$
$-3x+3+2x+12=-x+15=$
$-(-5)+15=5+15=20,$
when $x=-5$

14. $5(n+3)-(n+4)-n=$
$5n+15-n-4-n=$
$3n+11=3(7)+11$, when $n=7$
$=21+11=32$

15. $x=3,\ y=-6:\ \frac{3x-2y}{2x-3y}=$
$\frac{3(3)-2(-6)}{2(3)-3(-6)}=\frac{9+12}{6+18}=\frac{21}{24}=\frac{7}{8}$

16. $n=-\frac{2}{3}:\ \frac{3}{4}n-\frac{1}{3}n+\frac{5}{6}n=$
$\frac{9}{12}n-\frac{4}{12}n+\frac{10}{12}n=\frac{15}{12}n=\frac{5}{4}n=$
$\frac{5}{4}\left(-\frac{2}{3}\right)=-\frac{5}{6}$

17. $a=0.2,\ b=-0.3:\ 2a^2-4b^2=$
$2(0.2)^2-4(-0.3)^2=$
$2(0.04)-4(0.09)=$
$0.08-0.36=-0.28$

18. $x=\frac{1}{2},\ y=\frac{1}{4}:\ x^2-3xy-2y^2=$
$\left(\frac{1}{2}\right)^2-3\left(\frac{1}{2}\right)\left(\frac{1}{4}\right)-2\left(\frac{1}{4}\right)^2=$
$\frac{1}{4}-\frac{3}{8}-\frac{1}{8}=\frac{2}{8}-\frac{3}{8}-\frac{1}{8}=-\frac{2}{8}=-\frac{1}{4}$

19. $5x-7y-8x+3y=-3x-4y=$
$-3(9)-4(-8)$, when $x=9,\ y=-8$
$=-27+32=5$

20. $a=-1,\ b=3:\ \frac{3a-b-4a+3b}{a-6b-4b-3a}=$
$\frac{-a+2b}{-2a-10b}=\frac{-(-1)+2(3)}{-2(-1)-10(3)}=$
$\frac{1+6}{2-30}=-\frac{7}{28}=-\frac{1}{4}$

21. $3^4=3\cdot3\cdot3\cdot3=81$

22. $-2^6=-(2\cdot2\cdot2\cdot2\cdot2\cdot2)=-64$

23. $\left(\frac{2}{3}\right)^3=\frac{2}{3}\cdot\frac{2}{3}\cdot\frac{2}{3}=\frac{8}{27}$

24. $\left(-\frac{1}{2}\right)^5 =$

$\left(-\frac{1}{2}\right)\left(-\frac{1}{2}\right)\left(-\frac{1}{2}\right)\left(-\frac{1}{2}\right)\left(-\frac{1}{2}\right) =$

$-\frac{1}{32}$

25. $\left(\frac{1}{2}+\frac{1}{3}\right)^2 = \left(\frac{3}{6}+\frac{2}{6}\right)^2 =$

$\left(\frac{5}{6}\right)^2 = \frac{5}{6}\cdot\frac{5}{6} = \frac{25}{36}$

26. $\left(\frac{3}{4}-\frac{7}{8}\right)^3 = \left(\frac{6}{8}-\frac{7}{8}\right)^3 = \left(-\frac{1}{8}\right)^3 =$

$\left(-\frac{1}{8}\right)\left(-\frac{1}{8}\right)\left(-\frac{1}{8}\right) = -\frac{1}{512}$

27. $-5x+2=22$

$-5x=20$

$x=-4$

The solution set is $\{-4\}$.

28. $3x-4=7x+4$

$-4x-4=4$

$-4x=8$

$x=-2$

The solution set is $\{-2\}$.

29. $7(n-3)=5(n+7)$

$7n-21=5n+35$

$2n-21=35$

$2n=56$

$n=28$

The solution set is $\{28\}$.

30. $2(x-1)-3(x-2)=12$

$2x-2-3x+6=12$

$-x+4=12$

$-x=8$

$x=-8$

The solution set is $\{-8\}$.

31. $\frac{2}{5}x-\frac{1}{3}=\frac{1}{3}x+\frac{1}{2}$

Multiply both sides by 30.

$30\left(\frac{2}{5}x-\frac{1}{3}\right)=30\left(\frac{1}{3}x+\frac{1}{2}\right)$

$12x-10=10x+15$

$2x-10=15$

$2x=25$

$x=\frac{25}{2}$

The solution set is $\left\{\frac{25}{2}\right\}$.

32. $\frac{t-2}{4}+\frac{t+3}{3}=\frac{1}{6}$

$12\left(\frac{t-2}{4}+\frac{t+3}{3}\right)=12\left(\frac{1}{6}\right)$

$3(t-2)+4(t+3)=2$

$3t-6+4t+12=2$

$7t+6=2$

$7t=-4$

$t=-\frac{4}{7}$

The solution set is $\left\{-\frac{4}{7}\right\}$.

33. $\frac{2n-1}{5}-\frac{n+2}{4}=1$

Multiply both sides by 20.

$20\left(\frac{2n-1}{5}-\frac{n+2}{4}\right)=20(1)$

$4(2n-1)-5(n+2)=20$

$8n-4-5n-10=20$

$3n-14=20$

$3n=34$

$n=\frac{34}{3}$

The solution set is $\left\{\frac{34}{3}\right\}$.

34. $0.09x + 0.12(500 - x) = 54$

Multiply both sides by 100.

$$100[0.09x + 0.12(500 - x)] = 100(54)$$
$$9x + 12(500 - x) = 5400$$
$$9x + 6000 - 12x = 5400$$
$$6000 - 3x = 5400$$
$$-3x = -600$$
$$x = 200$$

The solution set is $\{200\}$.

35. $-5(n-1) - (n-2) = 3(n-1) - 2n$

Distributive property.

$$-5n + 5 - n + 2 = 3n - 3 - 2n$$

Combined like terms.

$$-6n + 7 = n - 3$$
$$-7n = -10$$
$$n = \frac{10}{7}$$

The solution set is $\left\{\frac{10}{7}\right\}$.

36. $\dfrac{-2}{x-1} = \dfrac{-3}{x+4}$

Cross products are equal.

$$-2(x+4) = -3(x-1)$$
$$-2x - 8 = -3x + 3$$
$$x - 8 = 3$$
$$x = 11$$

The solution set is $\{11\}$.

37. $0.2x + 0.1(x-4) = 0.7x - 1$

Multiplied both sides by 10.

$$2x + 1(x-4) = 7x - 10$$
$$2x + x - 4 = 7x - 10$$
$$3x - 4 = 7x - 10$$
$$-4x - 4 = -10$$
$$-4x = -6$$
$$x = \frac{6}{4} = \frac{3}{2}$$

The solution set is $\left\{\frac{3}{2}\right\}$.

38. $-(t-2) + (t-4) = 2\left(t - \frac{1}{2}\right) - 3\left(t + \frac{1}{3}\right)$

Distributive Property.

$$-t + 2 + t - 4 = 2t - 1 - 3t - 1$$
$$-2 = -t - 2$$
$$0 = -t$$
$$0 = t$$

The solution set is $\{0\}$.

39. $4x - 6 > 3x + 1$

$$x - 6 > 1$$
$$x > 7$$

The solution set is

$\{x|x > 7\}$ or $(7, \infty)$.

40. $-3x - 6 < 12$

$$-3x < 18$$

Reversed the inequality.

$$x > -6$$

The solution set is

$\{x|x > -6\}$ or $(-6, \infty)$.

41. $-2(n-1) \le 3(n-2) + 1$

Distributive property.

$$-2n + 2 \le 3n - 6 + 1$$
$$-2n + 2 \le 3n - 5$$
$$-5n + 2 \le -5$$
$$-5n \le -7$$

Reversed the inequality.

$$n \ge \frac{7}{5}$$

The solution set is

$\left\{n|n \ge \frac{7}{5}\right\}$ or $\left[\frac{7}{5}, \infty\right)$.

42. $\frac{2}{7}x - \frac{1}{4} \geq \frac{1}{4}x + \frac{1}{2}$

Multiply both sides by 28.

$$28\left(\frac{2}{7}x - \frac{1}{4}\right) \geq 28\left(\frac{1}{4}x + \frac{1}{2}\right)$$
$$8x - 7 \geq 7x + 14$$
$$x - 7 \geq 14$$
$$x \geq 21$$

The solution set is

$\{x|x \geq 21\}$ or $[21, \infty)$.

43. $0.08t + 0.1(300 - t) > 28$

Multiplied both sides by 100.

$$8t + 10(300 - t) > 2800$$
$$8t + 3000 - 10t > 2800$$
$$3000 - 2t > 2800$$
$$-2t > -200$$

Reversed the inequality.

$$t < 100$$

The solution set is

$\{t|t < 100\}$ or $(-\infty, 100)$.

44. $-4 > 5x - 2 - 3x$

$-4 > 2x - 2$

$-2 > 2x$

$-1 > x$ means $x < -1$

The solution set is

$\{x|x < -1\}$ or $(-\infty, -1)$.

45. $\frac{2}{3}n - 2 \geq \frac{1}{2}n + 1$

Multiply both sides by 6.

$$6\left(\frac{2}{3}n - 2\right) \geq 6\left(\frac{1}{2}n + 1\right)$$
$$4n - 12 \geq 3n + 6$$
$$n - 12 \geq 6$$
$$n \geq 18$$

The solution set is

$\{n|n \geq 18\}$ or $[18, \infty)$.

46. $-3 < -2(x - 1) - x$

$-3 < -2x + 2 - x$

$-3 < -3x + 2$

$-5 < -3x$

$\frac{5}{3} > x$ means $x < \frac{5}{3}$

The solution set is

$\left\{x|x < \frac{5}{3}\right\}$ or $\left(-\infty, \frac{5}{3}\right)$.

47. Let s represent her salary five years ago.

$$2s + 2000 = 32,000$$
$$2s = 30,000$$
$$s = 15,000$$

Five years ago, her salary was $\$15,000$.

48. Let a represent the measure of one angle; then $180 - a$ represents the supplementary angle.

$$a = 4(180 - a) - 45$$
$$a = 720 - 4a - 45$$
$$a = 675 - 4a$$
$$5a = 675$$
$$a = 135$$

One angle is $135°$ and the other is $180 - (135) = 45°$.

49. Let n represent the number of nickels; then $25 - n$ represents the number of dimes. The value of the nickels in cents is represented by $5n$ and the value of the dimes in cents is $10(25 - n)$.

$$5n + 10(25 - n) = 210$$
$$5n + 250 - 10n = 210$$
$$-5n + 250 = 210$$
$$-5n = -40$$
$$n = 8$$

Jasmal has 8 nickels and $25 - 8 = 17$ dimes.

50. Let x represent what Hana's score in the third game.

$$\frac{144 + 176 + x}{3} \geq 150$$
$$144 + 176 + x \geq 450$$
$$320 + x \geq 450$$
$$x \geq 130$$

Hana must bowl 130 or higher in the third game.

51. Let x represent the shorter piece of the board; then $30 - x$ represents the longer piece.

$$\frac{x}{30 - x} = \frac{2}{3}$$
$$3x = 2(30 - x)$$
$$3x = 60 - 2x$$
$$5x = 60$$
$$x = 12$$

The pieces are 12 feet and $30 - 12 = 18$ feet.

52. Let x represent the selling price of the shoes. Profit is a percent of the selling price.

Selling price = Cost + Profit

$$s = 32 + (20\%)(s)$$
$$s = 32 + 0.2s$$
$$10s = 320 + 2s$$
$$8s = 320$$
$$s = 40$$

The selling price would be \$40.

53. Let r represent the rate of one car; then $r + 5$ represents the other car's rate. Both cars travel for 6 hours. A chart of the information would be as follows:

	Rate	Time	Distance ($d = rt$)
Car A	r	6	$6r$
Car B	$r+5$	6	$6(r+5)$

A diagram of the problem would be as follows:

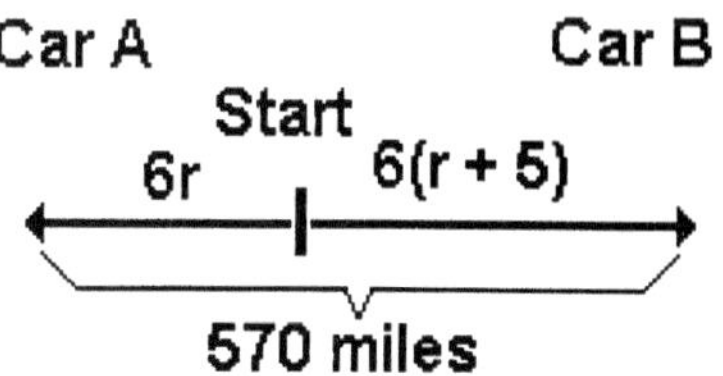

The sum of the two distances would equal the total distance of 570 miles.

$$6r + 6(r + 5) = 570$$
$$6r + 6r + 30 = 570$$
$$12r + 30 = 570$$
$$12r = 540$$
$$r = 45$$

The speeds for the cars were 45 mph and 50 mph.

54. Let x represent the amount of pure alcohol to be added. Then $15 + x$ represents the amount of final solution.

$$\begin{pmatrix}\text{pure}\\\text{alcohol}\\\text{in 20\%}\\\text{solution}\end{pmatrix} + \begin{pmatrix}\text{pure}\\\text{alcohol}\\\text{to be}\\\text{added}\end{pmatrix} = \begin{pmatrix}\text{pure}\\\text{alcohol}\\\text{in final}\\\text{solution}\end{pmatrix}$$

$$(20\%)(15) + x = (40\%)(15 + x)$$
$$0.20(15) + x = 0.40(15 + x)$$
$$10[0.20(15) + x] = 10[0.40(15 + x)]$$
$$2(15) + 10x = 4(15 + x)$$
$$30 + 10x = 60 + 4x$$
$$30 + 6x = 60$$
$$6x = 30$$
$$x = 5$$

We must add 5 liters of pure alcohol.

Chapter 4 Introduction to Graphing and Functions

PROBLEM SET 4.1 Pie, Bar, and Line Graphs

1. Kayaks 8%
Sailboats 6%
Together 14%
14% of the boat rentals were from kayaks or sailboats.

3. Jon Boats were 20%.
20% of 2400 = .20(2400) = 480
Jon Boats were rented 480 times.

5. Sailboats 6%
Ski Boats 28%
Together 34%
The rentals not from Sailboats or Ski Boats are 100% – 34% = 66%.

7. Physics

9. Chemistry 13%
Physics 8%
Together 21%
21% of the students chose chemistry or physics.

11. Oceanography 22%
Astronomy 16%
Together 38%
The percent of students who did not choose oceanography or astronomy is 100% – 38% = 62%.

13. Space Center 1500
Water Park 500
Difference 1000
1000 more people preferred the space center to the water park.

15. Beach 2000
Golf Course 500
Difference 1500
The difference is 1500 people.

17. January and February

19. February and March

21. Bank 8.2%
Credit Union 7.8%
Difference 0.4%
The difference in the interest rates is 0.4%.

23. Friday 120
Saturday 180
Sunday 90
Together 390
$390 in tips were earned for Friday, Saturday, and Sunday.

25. Monday or Wednesday

27. Sunday 90
Monday 40
Tuesday 50
Wednesday 40
Thursday 60
Friday 120
Saturday 180
Together 580
$580 was earned in tips for the week.

29. High-Tech Fund 16%
Utility Fund 14%
Difference 2%
The difference in annual total return is 2%.

31. Year 1997 15%
Year 1996 12%
Difference 3%
The change in the annual total return is 3%.

33. 1997 and 1998

35 a. $\text{Average} = \dfrac{16+17+20+5+4+10}{6}$
$= \dfrac{72}{6} = 12$

The average annual total return is 12%.

b. $\text{Average} = \dfrac{14+12+15+11+9+11}{6}$
$= \dfrac{72}{6} = 12$

The average annual total return is 12%.

c. Neither, the average is the same.

PROBLEM SET 4.2 Cartesian Coordinate System

1. $3x + 7y = 13$
$7y = 13 - 3x$
$y = \dfrac{13 - 3x}{7}$

3. $x - 3y = 9$
$x = 9 + 3y$

5. $-x + 5y = 14$
$5y = x + 14$
$y = \dfrac{x + 14}{5}$

7. $-3x + y = 7$
$-3x = -y + 7$
$x = \dfrac{-y+7}{-3} \cdot \dfrac{-1}{-1} = \dfrac{y-7}{3}$

9. $-2x + 3y = -5$
$3y = 2x - 5$
$y = \dfrac{2x - 5}{3}$

For Proplems 11 – 33, a sample of at least three points is given. For a graph of the equation see answers in the back of the text.

11. $y = x + 1$
$x = -2: \quad y = -2 + 1 = -1$
$x = 0: \quad y = 0 + 1 = 1$
$x = 2: \quad y = 2 + 1 = 3$

Point
$(-2,\ -1)$
$(0,\ 1)$
$(2,\ 3)$

13. $y = x - 2$
$x = -2: \quad y = -2 - 2 = -4$
$x = 0: \quad y = 0 - 2 = -2$
$x = 2: \quad y = 2 - 2 = 0$

Point
$(-2,\ -4)$
$(0,\ -2)$
$(2,\ 0)$

15. $y = (x-2)^2$
$x = -2: \quad y = (-2-2)^2 = (-4)^2 = 16$
$x = 0: \quad y = (0-2)^2 = (-2)^2 = 4$
$x = 2: \quad y = (2-2)^2 = (0)^2 = 0$
$x = 4: \quad y = (4-2)^2 = (2)^2 = 4$
$x = 6: \quad y = (6-2)^2 = (4)^2 = 16$

Problem Set 4.2

Point
$(-2, 16)$
$(0, 4)$
$(2, 0)$
$(4, 4)$
$(6, 16)$

17.

$$y = x^2 - 2$$

$x = -2:$ $y = (-2)^2 - 2 = 4 - 2 = 2$

$x = -1:$ $y = (-1)^2 - 2 = 1 - 2 = -1$

$x = 0:$ $y = (0)^2 - 2 = 0 - 2 = -2$

$x = 1:$ $y = (1)^2 - 2 = 1 - 2 = -1$

$x = 2:$ $y = (2)^2 - 2 = 4 - 2 = 2$

Point
$(-2, 2)$
$(-1, -1)$
$(0, -2)$
$(1, -1)$
$(2, 2)$

19.

$$y = \frac{1}{2}x + 3$$

$x = -2:$ $y = \frac{1}{2}(-2) + 3 = -1 + 3 = 2$

$x = 0:$ $y = \frac{1}{2}(0) + 3 = 0 + 3 = 3$

$x = 2:$ $y = \frac{1}{2}(2) + 3 = 1 + 3 = 4$

Point
$(-2, -2)$
$(0, 3)$
$(2, 4)$

21. (Use values of x that are divisible by 2.)

$$x + 2y = 4, \text{ so } y = \frac{4 - x}{2}$$

$x = -2:$ $y = \frac{4 - (-2)}{2} = \frac{6}{2} = 3$

$x = 0:$ $y = \frac{4 - (0)}{2} = \frac{4}{2} = 2$

$x = 2:$ $y = \frac{4 - (2)}{2} = \frac{2}{2} = 1$

Point
$(-2, 3)$
$(0, 2)$
$(2, 1)$

23. (Use values of x that are divisible by 5.)

$$2x - 5y = 4, \text{ so } y = \frac{2x - 10}{5}$$

$x = -5:$ $y = \frac{2(-5) - 10}{5} = \frac{-10 - 10}{5}$

$= \frac{-20}{5} = -4$

$x = 0:$ $y = \frac{2(0) - 10}{5} = \frac{-10}{5} = -2$

$x = 5:$ $y = \frac{2(5) - 10}{5} = \frac{10 - 10}{5}$

$= \frac{0}{5} = 0$

Point
$(-5, -4)$
$(0, -2)$
$(5, 0)$

25.

$$y = x^3$$

$x = -2:$ $y = (-2)^3 = -8$

$x = -1:$ $y = (-1)^3 = -1$

$x = 0:$ $y = (0)^3 = 0$

$x = 1:$ $y = (1)^3 = 1$

$x = 2:$ $y = (2)^3 = 8$

Point
$(-2, -8)$
$(-1, -1)$
$(0, 0)$
$(1, 1)$
$(2, 8)$

27.

$$y = -x^2$$

$x = -2:$ $y = -(-2)^2 = -4$

$x = -1:$ $y = -(-1)^2 = -1$

$x = 0:$ $y = -(0)^2 = 0$

$x = 1:$ $y = -(1)^2 = -1$

$x = 2:$ $y = -(2)^2 = -4$

Point
$(-2, -4)$
$(-1, -1)$
$(0, 0)$
$(1, -1)$
$(2, -4)$

29.

$$\begin{aligned} & y = x \\ x = -2: \quad & y = -2 \\ x = 0: \quad & y = 0 \\ x = 2: \quad & y = 2 \end{aligned}$$

Point
$(-2, -2)$
$(0, 0)$
$(2, 2)$

31.

$$\begin{aligned} & y = -3x + 2 \\ x = -2: \quad & y = -3(-2) + 2 = 6 + 2 = 8 \\ x = 0: \quad & y = -3(0) + 2 = 0 + 2 = 2 \\ x = 2: \quad & y = -3(2) + 2 = -6 + 2 = -4 \end{aligned}$$

Point
$(-2, 8)$
$(0, 2)$
$(2, -4)$

33.

$$\begin{aligned} & y = 2x^2 \\ x = -2: \quad & y = 2(-2)^2 = 2(4) = 8 \\ x = -1: \quad & y = 2(-1)^2 = 2(1) = 2 \\ x = 0: \quad & y = 2(0)^2 = 2(0) = 0 \\ x = 1: \quad & y = 2(1)^2 = 2(1) = 2 \\ x = 2: \quad & y = 2(2)^2 = 2(4) = 8 \end{aligned}$$

Point
$(-2, -8)$
$(-1, 2)$
$(0, 0)$
$(1, 2)$
$(2, 8)$

PROBLEM SET 4.3 Graphing Linear Equations

For Problems 1 – 35, the y-intercept, the x-intercept and one "check" point is given. If the line contains the origin or is parallel to an axis, two points in addition to the one intercept is given. For a graph of the equation see the answers in the back of the text.

1.

$$\begin{aligned} & x + y = 2 \\ x = 0: \quad & 0 + y = 2 \\ y = 0: \quad & x + 0 = 2 \\ x = 1: \quad & 1 + y = 2 \\ & y = 1 \end{aligned}$$

Point
$(0, 2)$
$(2, 0)$
$(1, 1)$

3.

$$\begin{aligned} & x - y = 3 \\ x = 0: \quad & 0 - y = 3 \\ & -y = 3 \\ & y = -3 \\ y = 0: \quad & x - 0 = 3 \\ y = 1: \quad & x - 1 = 3 \\ & x = 4 \end{aligned}$$

Point
$(0, -3)$
$(3, 0)$
$(4, 1)$

5.

$$\begin{aligned} & x - y = -4 \\ x = 0: \quad & 0 - y = -4 \\ & -y = -4 \\ & y = 4 \\ y = 0: \quad & x - 0 = -4 \\ y = 2: \quad & x - 2 = -4 \\ & x = -2 \end{aligned}$$

Point
$(0, 4)$
$(-4, 0)$
$(-2, 2)$

7.

$$\begin{aligned} & x + 2y = 2 \\ x = 0: \quad & 0 + 2y = 2 \\ & y = 1 \\ y = 0: \quad & x + 0 = 2 \\ y = 2: \quad & x + 4 = 2 \\ & x = -2 \end{aligned}$$

Point
$(0, 1)$
$(2, 0)$
$(-2, 2)$

Problem Set 4.3

9.

$$\begin{aligned} & 3x - y = 6 \\ x = 0: \quad & 0 - y = 6 \\ & -y = 6 \\ & y = -6 \\ y = 0: \quad & 3x - 0 = 6 \\ & x = 2 \\ x = 1: \quad & 3 - y = 6 \\ & -y = 3 \\ & y = -3 \end{aligned}$$

Point
$(0, -6)$
$(2, 0)$
$(1, -3)$

11.

$$\begin{aligned} & 3x - 2y = 6 \\ x = 0: \quad & 0 - 2y = 6 \\ & -2y = 6 \\ & y = -3 \\ y = 0: \quad & 3x - 0 = 6 \\ & x = 2 \\ x = 4: \quad & 12 - 2y = 6 \\ & -2y = -6 \\ & y = 3 \end{aligned}$$

Point
$(0, -3)$
$(2, 0)$
$(4, 3)$

13.

$$\begin{aligned} & x - y = 0 \\ x = 0: \quad & 0 - y = 0 \\ & y = 0 \\ x = 2: \quad & 2 - y = 0 \\ & y = 2 \\ y = 3: \quad & x - 3 = 0 \\ & x = 3 \end{aligned}$$

Point
$(0, 0)$
$(2, 2)$
$(3, 3)$

15.

$$\begin{aligned} & y = 3x \\ x = 0: \quad & y = 3(0) = 0 \\ x = 1: \quad & y = 3(1) = 3 \\ x = -2: \quad & y = 3(-2) = -6 \end{aligned}$$

Point
$(0, 0)$
$(1, 3)$
$(-2, -6)$

17.

$$\begin{aligned} & x = -2 \\ y = 0: \quad & x = -2 \\ y = 2: \quad & x = -2 \\ y = -3: \quad & x = -2 \end{aligned}$$

Point
$(-2, 0)$
$(-2, 2)$
$(-2, -3)$

19.

$$\begin{aligned} & y = 0 \\ x = 0: \quad & y = 0 \\ x = 2: \quad & y = 0 \\ x = -3: \quad & y = 0 \end{aligned}$$

Point
$(0, 0)$
$(2, 0)$
$(-3, 0)$

21.

$$\begin{aligned} & y = -2x - 1 \\ x = 0: \quad & y = 0 - 1 \\ & y = -1 \\ y = 0: \quad & 0 = -2x - 1 \\ & 2x = -1 \\ & x = -\frac{1}{2} \\ x = 2: \quad & y = -2(2) - 1 \\ & y = -4 - 1 \\ & y = -5 \end{aligned}$$

Point
$(0, -1)$
$\left(-\frac{1}{2}, 0\right)$
$(2, -5)$

23.

$$\begin{aligned} & y = \frac{1}{2}x + 1 \\ x = 0: \quad & y = 0 + 1 \\ & y = 1 \\ y = 0: \quad & 0 = \frac{1}{2}x + 1 \\ & 0 = x + 2 \\ & x = -2 \\ x = 2: \quad & y = \frac{1}{2}(2) + 1 \\ & y = 1 + 1 \\ & y = 2 \end{aligned}$$

Point
$(0, 1)$
$(-2, 0)$
$(2, 2)$

25.

$$\begin{aligned} & y = -\frac{1}{3}x - 2 \\ x = 0: \quad & y = 0 - 2 \\ & y = -2 \\ y = 0: \quad & 0 = -\frac{1}{3}x - 2 \\ & 0 = -x - 6 \\ & x = -6 \\ x = 3: \quad & y = -\frac{1}{3}(3) - 2 \\ & y = -1 - 2 \\ & y = -3 \end{aligned}$$

Point
$(0, -2)$
$(-6, 0)$
$(3, -3)$

27.

		Point
	$4x + 5y = -10$	
$x = 0:$	$0 + 5y = -10$	$(0, -2)$
	$y = -2$	$\left(-\frac{5}{2}, 0\right)$
$y = 0:$	$4x + 0 = -10$	$(5, -6)$
	$x = -\frac{10}{4}$	
	$x = -\frac{5}{2}$	
$x = 5:$	$20 + 5y = -10$	
	$5y = -30$	
	$y = -6$	

29.

		Point
	$-2x + y = -4$	
$x = 0:$	$0 + y = -4$	$(0, -4)$
	$y = -4$	$(2, 0)$
$y = 0:$	$-2x + 0 = -4$	$(4, 4)$
	$x = 2$	
$x = 4:$	$-8 + y = -4$	
	$y = 4$	

31.

		Point
	$3x - 4y = 7$	
$x = 0:$	$0 - 4y = 7$	$\left(0, -\frac{7}{4}\right)$
	$y = -\frac{7}{4}$	$\left(\frac{7}{3}, 0\right)$
$y = 0:$	$3x - 0 = 7$	$(-7, -7)$
	$x = \frac{7}{3}$	
$y = -7:$	$3x + 28 = 7$	
	$3x = -21$	
	$x = -7$	

33.

		Point
	$y + 4x = 0$	
$x = 0:$	$y + 0 = 0$	$(0, 0)$
	$y = 0$	$(1, -4)$
$x = 1:$	$y + 4 = 0$	$(-1, 4)$
	$y = -4$	
$x = -1:$	$y - 4 = 0$	
	$y = 4$	

35.

		Point
	$x = 2y$	
$x = 0:$	$0 = 2y$	$(0, 0)$
	$y = 0$	$(2, 1)$
$y = 1:$	$x = 2$	$(4, 2)$
$y = 2:$	$x = 4$	

PROBLEM SET 4.4 Relations and Functions

1. Domain: $\{4, 6, 8, 10\}$
Range: $\{7, 11, 20, 28\}$
It is a function.

3. Domain: $\{-2, -1, 0, 1\}$
Range: $\{1, 2, 3, 4\}$
It is a function.

5. Domain: $\{4, 9\}$
Range: $\{-3, -2, 2, 3\}$
It is not a function.

7. Domain: $\{3, 4, 5, 6\}$
Range: $\{15\}$
It is a function.

9. Domain: {Carol}
Range: $\{22400, 23700, 25200\}$
It is not a function.

11. Domain: $\{-6\}$
Range: $\{1, 2, 3, 4\}$
It is not a function.

13. Domain: $\{-2, -1, 0, 1, 2\}$
Range: $\{0, 1, 4\}$
It is a function.

15. All Reals

17. The denominator can not equal zero.
So set $x + 8 = 0$
$x = -8$
Therefore -8 is excluded from the domain.
The domain is all real numbers except -8.

Problem Set 4.4

19. The denominator can not equal zero.
So set $x - 6 = 0$
$$x = 6$$
Therefore 6 is excluded from the domain.
The domain is all real numbers except 6.

21. The denominator can not equal zero.
So set $2x - 10 = 0$
$$2x = 10$$
$$x = 5$$
Therefore 5 is excluded from the domain.
The domain is all real numbers except 5.

23. The denominator can not equal zero.
So set $5x - 8 = 0$
$$5x = 8$$
$$x = \frac{8}{5}$$
Therefore $\frac{8}{5}$ is excluded from the domain.
The domain is all real numbers except $\frac{8}{5}$.

25. Since the denominator is a constant, it can not eqal zero. The domain is all real numbers.

27. The denominator can not equal zero.
So set $x = 0$
Therefore 0 is excluded from the domain.
The domain is all real numbers except 0.

29. The domain is all real numbers.

31. $f(x) = 3x + 4$

$$\begin{aligned} f(0) &= 3(0) + 4 = 4 \\ f(1) &= 3(1) + 4 = 3 + 4 = 7 \\ f(-1) &= 3(-1) + 4 = -3 + 4 = 1 \\ f(6) &= 3(6) + 4 = 18 + 4 = 22 \end{aligned}$$

33. $f(x) = -5x - 1$

$$\begin{aligned} f(3) &= -5(3) - 1 \\ &= -15 - 1 = -16 \\ f(-4) &= -5(-4) - 1 \\ &= 20 - 1 = 19 \\ f(-5) &= -5(-5) - 1 \\ &= 25 - 1 = 24 \\ f(t) &= -5(t) - 1 = -5t - 1 \end{aligned}$$

35. $g(x) = \frac{2}{3}x + \frac{3}{4}$

$$\begin{aligned} g(3) &= \frac{2}{3}(3) + \frac{3}{4} = 2 + \frac{3}{4} \\ &= \frac{8}{4} + \frac{3}{4} = \frac{11}{4} \\ g\left(\frac{1}{2}\right) &= \frac{2}{3}\left(\frac{1}{2}\right) + \frac{3}{4} = \frac{1}{3} + \frac{3}{4} \\ &= \frac{4}{12} + \frac{9}{12} = \frac{13}{12} \\ g\left(-\frac{1}{3}\right) &= \frac{2}{3}\left(-\frac{1}{3}\right) + \frac{3}{4} = -\frac{2}{9} + \frac{3}{4} \\ &= -\frac{8}{36} + \frac{27}{36} = \frac{19}{36} \\ g(-2) &= \frac{2}{3}(-2) + \frac{3}{4} = -\frac{4}{3} + \frac{3}{4} \\ &= -\frac{16}{12} + \frac{9}{12} = -\frac{7}{12} \end{aligned}$$

37. $f(x) = x^2 - 4$

$$\begin{aligned} f(2) &= (2)^2 - 4 = 4 - 4 = 0 \\ f(-2) &= (-2)^2 - 4 = 4 - 4 = 0 \\ f(7) &= (7)^2 - 4 = 49 - 4 = 45 \\ f(0) &= (0)^2 - 4 = 0 - 4 = -4 \end{aligned}$$

39. $f(x) = -x^2 + 1$

$$\begin{aligned} f(-1) &= -(-1)^2 + 1 \\ &= -1 + 1 = 0 \\ f(2) &= -(2)^2 + 1 \\ &= -4 + 1 = -3 \\ f(-2) &= -(-2)^2 + 1 \\ &= -4 + 1 = -3 \\ f(-3) &= -(-3)^2 + 1 \\ &= -9 + 1 = -8 \end{aligned}$$

$g(x) = x^2 - 2x$

$$\begin{aligned} g(-1) &= (-1)^2 - 2(-1) = 3 \\ g(4) &= (4)^2 - 2(4) = 8 \end{aligned}$$

41. $f(x) = 4x + 3$

$$\begin{aligned} f(5) &= 4(5) + 3 = 23 \\ f(-6) &= 4(-6) + 3 = -21 \end{aligned}$$

43. $f(x) = 3x^2 - x + 4$

$$\begin{aligned} f(-1) &= 3(-1)^2 - (-1) + 4 = 8 \\ f(4) &= 3(4)^2 - (4) + 4 = 48 \end{aligned}$$

$g(x) = -3x + 5$

$$\begin{aligned} g(-1) &= -3(-1) + 5 = 8 \\ g(4) &= -3(4) + 5 = -7 \end{aligned}$$

PROBLEM SET 4.5 Application of Functions

1.
$$\begin{aligned} A(s) &= s^2 \\ A(3) &= 3^2 = 9 \\ A(17) &= 17^2 = 289 \\ A(8.5) &= 8.5^2 = 72.25 \\ A(20.75) &= 20.75^2 = 430.56 \\ A(11.25) &= 11.25^2 = 126.56 \end{aligned}$$

3.
$$\begin{aligned} C(n) &= 12n + 44500 \\ C(35000) &= 12(35000) + 44500 \\ &= 420,000 + 44,500 \\ &= 464,500 \end{aligned}$$
The cost is $\$464,500$.

5.
$$\begin{aligned} s(c) &= 1.5c \\ s(4.50) &= 1.5(4.50) = 6.75 \\ s(6.75) &= 1.5(6.75) = 10.13 \\ s(9.00) &= 1.5(9.00) = 13.50 \\ s(16.40) &= 1.5(16.40) = 24.60 \end{aligned}$$

7.
$$\begin{aligned} f(n) &= 0.75n + 15 \\ f(20) &= 0.75(20) + 15 = 15 + 15 = 30 \\ f(0) &= 0.75(0) + 15 = 15 \\ f(16) &= 0.75(16) + 15 = 12 + 15 = 27 \end{aligned}$$

9. $f(x) = 0$ when $x < 2$
$g(x) = 0.30x$ when $x \geq 2$

Since $x = 8$ which is greater than 2 use $g(x) = 0.30x$ to determine the charge.
$g(8) = 0.30(8) = 2.40$.
The charge would be \$2.40.

Since $x = 3$ which is greater than 2 use $g(x) = 0.30x$ to determine the charge.
$g(3) = 0.30(3) = 0.90$
The charge would be \$0.90.

Since $x = 1$ which is less than 2 use $f(x) = 0$ to determine the charge.
$f(1) = 0$
There would be no charge or the charge equals zero.

Since $x = 12$ which is greater than 2 use $g(x) = 0.30x$ to determine the charge.
$g(12) = 0.30(12) = 3.60$
The charge would be \$3.60.

11. $f(h) = 10.50h$ when $x \leq 40$
$g(h) = 15.75h - 210$ when $x > 40$

Since $x = 35$, which is less than 40, use $f(h) = 10.50h$ to determine his pay.
$f(35) = 10.50(35) = 367.50$
His pay is \$367.50.

Since $x = 40$, which is less than or equal to 40, use $f(h) = 10.50h$ to determine his pay. $f(40) = 10.50(40) = 420$
His pay is \$420.

Since $x = 50$, which is more than 40, use $g(h) = 15.75h - 210$ to determine his pay.
$g(50) = 15.75(50) - 210$
$g(50) = 787.50 - 210$
$g(50) = 577.50$
His pay is \$577.50.

Since $x = 20$, which is less than 40, use $f(h) = 10.50h$ to determine his pay.
$f(20) = 10.50(20) = 210$
His pay is \$210.

13. $f(n) = 48.50 + 10(n - 1)$
$f(2) = 48.50 + 10(2 - 1)$
$f(2) = 48.50 + 10(1)$
$f(2) = 48.50 + 10$
$f(2) = 58.50$

$f(3) = 48.50 + 10(3 - 1)$
$f(3) = 48.50 + 10(2)$
$f(3) = 48.50 + 20$
$f(3) = 68.50$

$f(1) = 48.50 + 10(1 - 1)$
$f(1) = 48.50 + 10(0)$
$f(1) = 48.50 + 0$
$f(1) = 48.50$

$f(4) = 48.50 + 10(4 - 1)$
$f(4) = 48.50 + 10(3)$
$f(4) = 48.50 + 30$
$f(4) = 78.50$

15. $f(x) = \frac{x}{10}$
$x = 0 \quad f(0) = \frac{0}{10} = 0$
$x = 1 \quad f(1) = \frac{1}{10}$
$x = 2 \quad f(2) = \frac{2}{10} = \frac{1}{5}$
$x = 3 \quad f(3) = \frac{3}{10}$
$x = 4 \quad f(4) = \frac{4}{10} = \frac{2}{5}$

x	0	1	2	3	4
$f(x)$	0	$\frac{1}{10}$	$\frac{1}{5}$	$\frac{3}{10}$	$\frac{2}{5}$

17. $f(n) = 0.0003n$

$n = 10,000$
$f(10000) = 0.0003(10,000) = 3$

$n = 15,000$
$f(15000) = 0.0003(15,000) = 4.5$

$n = 20,000$
$f(20000) = 0.0003(20,000) = 6$

$n = 25,000$
$f(25000) = 0.0003(25,000) = 7.5$

$n = 30,000$
$f(30000) = 0.0003(30,000) = 9$

x in gallons, $f(x)$ in pounds

x	10000	15000	20000	25000	30000
$f(x)$	3	4.5	6	7.5	9

19. $P(s) = 45$
a. $P(3) = 4(3) = 12$
b. $P(5) = 4(5) = 20$
c. See back of textbook for graph.
d. $P(4.25) = 4(4.25) = 17$

21. $V(t) = 32500 - 1950t$

a. $V(6) = 32500 - 1950(6)$
$V(6) = 32500 - 11700$
$V(6) = 20800$

b. $V(9) = 32500 - 1950(9)$
$V(9) = 32500 - 17550$
$V(9) = 14950$

c. See back of textbook for graph.

d. $V(10) = 32500 - 1950(10)$
$V(10) = 32500 - 19500$
$V(10) = 13000$
The value after 10 years will be \$13000.

e. answers vary

f.
$$0 = 32500 - 1950t$$
$$-32500 = -1950t$$
$$\frac{-32500}{-1950} = t$$
$$16.7 = t$$
The value will be zero in 16.7 years.

23 a.
$$f(t) = \frac{5}{9}(t - 32)$$
$$f(50) = \frac{5}{9}(50 - 32) = \frac{5}{9}(18) = 10$$
$$f(41) = \frac{5}{9}(41 - 32) = \frac{5}{9}(9) = 5$$
$$f(-4) = \frac{5}{9}(-4 - 32)$$
$$= \frac{5}{9}(-36) = -20$$
$$f(212) = \frac{5}{9}(212 - 32)$$
$$= \frac{5}{9}(180) = 100$$
$$f(95) = \frac{5}{9}(95 - 32) = \frac{5}{9}(63) = 35$$
$$f(77) = \frac{5}{9}(77 - 32) = \frac{5}{9}(45) = 25$$
$$f(59) = \frac{5}{9}(59 - 32) = \frac{5}{9}(27) = 15$$

t	50	41	-4	212	95	77	59
$f(t)$	10	5	-20	100	35	25	15

b. See back of textbook for graph.

c. $f(20) = \frac{5}{9}(20 - 32)$
$f(20) = \frac{5}{9}(-12)$
$f(20) = -6.7$

CHAPTER 4 Review Problem Set

1.

Ankle	8%
Hip	15%
Knee	32%
Together	55%

2.

Hip	15%
Neck	4%
Back	25%
Together	44%

3.

Knee	32%
Elbow	16%
Together	48%

The percent not seen for knee or elbow is 100% – 48% = 52%.

4.

Television	35
Studying	10
Difference	25

There were 25 more students watching television than studying.

Chapter 4 Review Problem Set

5. Shopping 20 student
Workout 20 student

6. Dinner Out, Television, Movie, Shopping, Workout, Sports, Study

7. 1996

8.

Band	6500
Sports	5000
Difference	1500

The largest difference is \$1500.

9. Average =

$$\frac{6500 + 5500 + 6000 + 5000 + 5500 + 6000}{6}$$

$= 5750$

The average over the six years is \$5750.

10. Average =

$$\frac{5000 + 5500 + 5000 + 5500 + 6000 + 6500}{6}$$

$= 5583$

The average over the six years is \$5583.

11. $2x - 5y = 10$

$x = 0$	$y = 0$
$2(0) - 5y = 10$	$2x - 5(0) = 10$
$-5y = 10$	$2x = 10$
$y = -2$	$x = 5$
$(0, -2)$	$(5, 0)$

See back of textbook for graph.

12.

13. $y = 2x^2 + 1$

$x = 1$	$x = 2$	$x = 0$
$y = 2(1)^2 + 1$	$y = 2(2)^2 + 1$	$y = 2(0)^2 + 1$
$y = 2(1) + 1$	$y = 2(4) + 1$	$y = 2(0) + 1$
$y = 2 + 1$	$y = 8 + 1$	$y = 0 + 1$
$y = 3$	$y = 9$	$y = 1$
$(1, 3)$	$(2, 9)$	$(0, 1)$

$x = -1$	$x = -2$
$y = 2(-1)^2 + 1$	$y = 2(-2)^2 + 1$
$y = 2(1) + 1$	$y = 2(4) + 1$
$y = 2 + 1$	$y = 8 + 1$
$y = 3$	$y = 9$
$(-1, 3)$	$(-2, 9)$

See back of textbook for graph.

14.

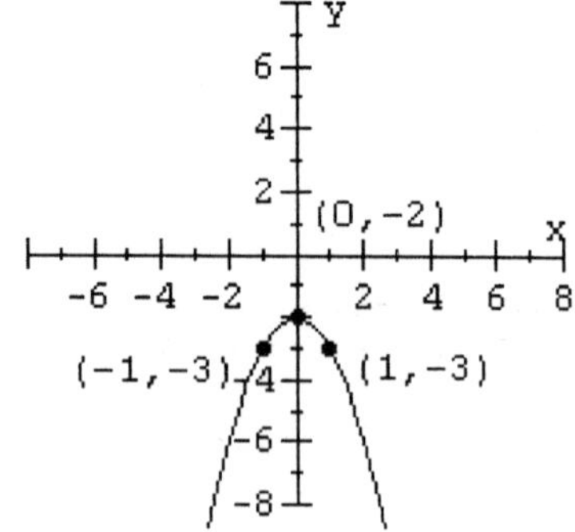

15. $2x - 3y = 0$

$x = 0$	$x = 3$
$2(0) - 3y = 0$	$2(3) - 3y = 0$
$-3y = \dfrac{0}{-3}$	$-3y = -6$
$y = 0$	$x = \dfrac{-6}{-3} = 2$
$(0, 0)$	$(3, 2)$

See back of textbook for graph.

16.

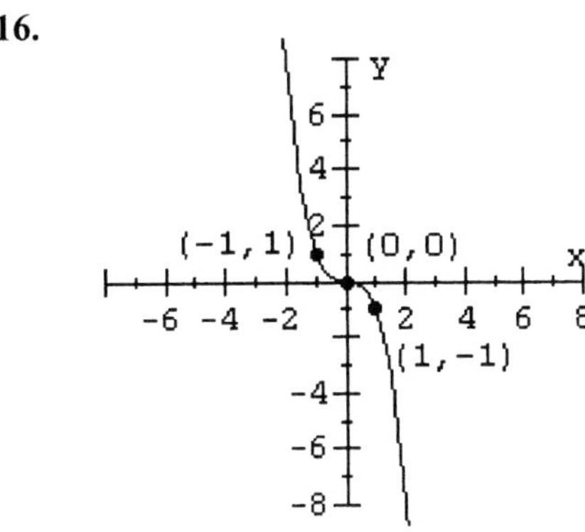

17. $x - y = 4$

$x = 0$	$y = 0$
$0 - y = 4$	$x - 0 = 4$
$-y = 4$	$x = 4$
$y = -4$	
$(0, -4)$	$(4, 0)$

See back of textbook for graph.

18.

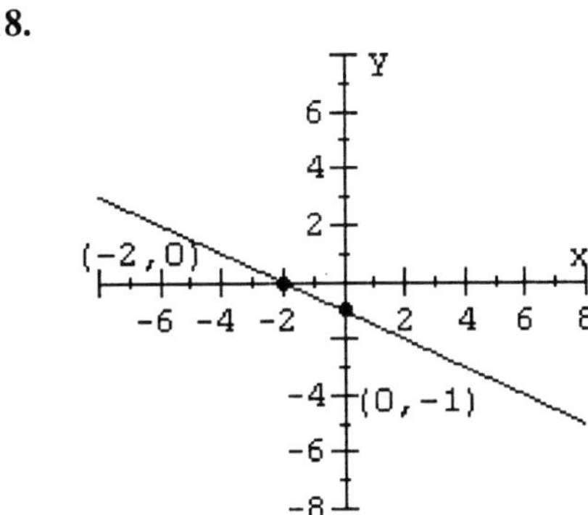

19. $y = \dfrac{2}{3}x - 1$

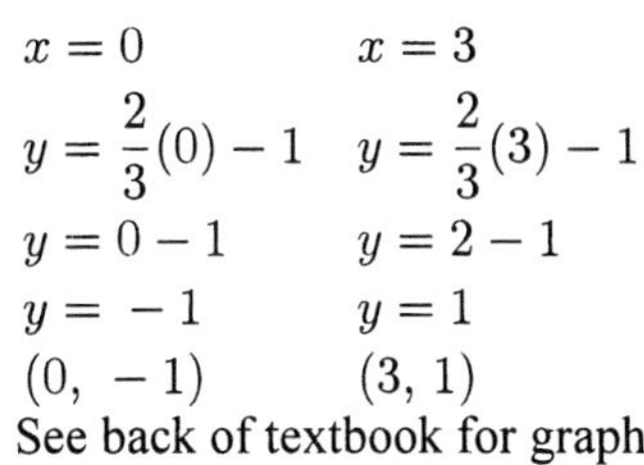

$x = 0$	$x = 3$
$y = \dfrac{2}{3}(0) - 1$	$y = \dfrac{2}{3}(3) - 1$
$y = 0 - 1$	$y = 2 - 1$
$y = -1$	$y = 1$
$(0, -1)$	$(3, 1)$

See back of textbook for graph.

20.

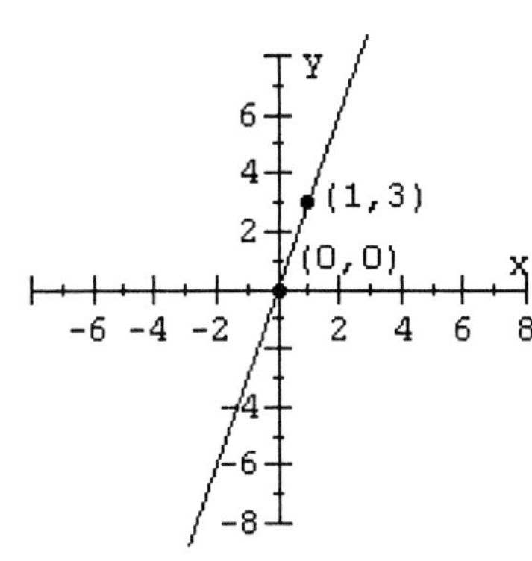

21. $y = -2x$

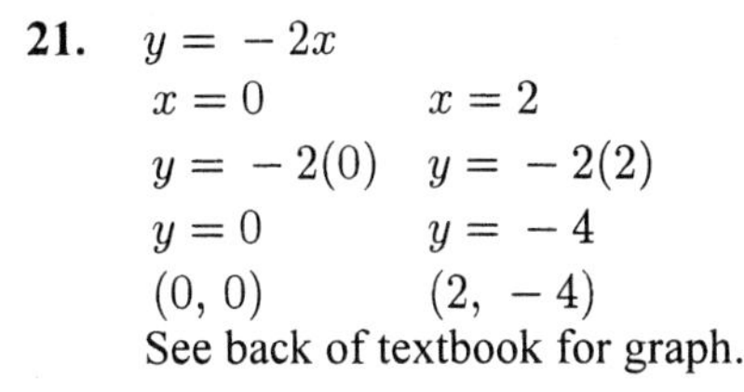

$x = 0$	$x = 2$
$y = -2(0)$	$y = -2(2)$
$y = 0$	$y = -4$
$(0, 0)$	$(2, -4)$

See back of textbook for graph.

22.

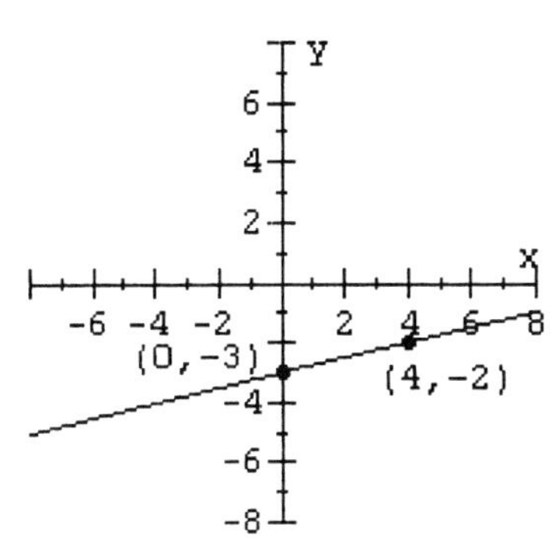

23. $y = x^2 + 2$

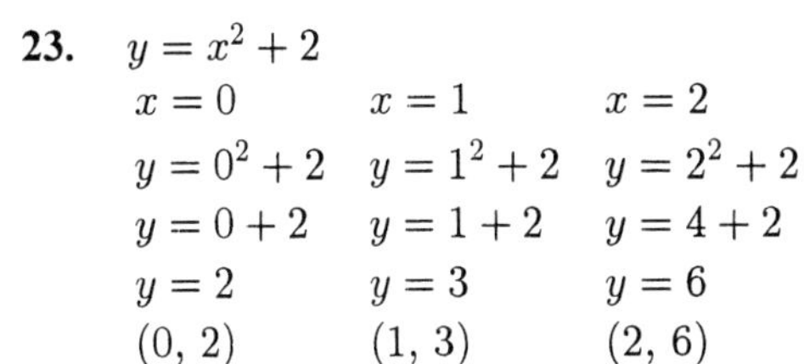

$x = 0$	$x = 1$	$x = 2$
$y = 0^2 + 2$	$y = 1^2 + 2$	$y = 2^2 + 2$
$y = 0 + 2$	$y = 1 + 2$	$y = 4 + 2$
$y = 2$	$y = 3$	$y = 6$
$(0, 2)$	$(1, 3)$	$(2, 6)$

$x = -1$ $\quad$ $x = -2$

$y = (-1)^2 + 2$ $\quad$ $y = (-2)^2 + 2$

$y = 1 + 2$ $\quad$ $y = 4 + 2$

$y = 3$ $\quad$ $y = 6$

$(-1, 3)$ $\quad$ $(-2, 6)$

See back of textbook for graph.

24.

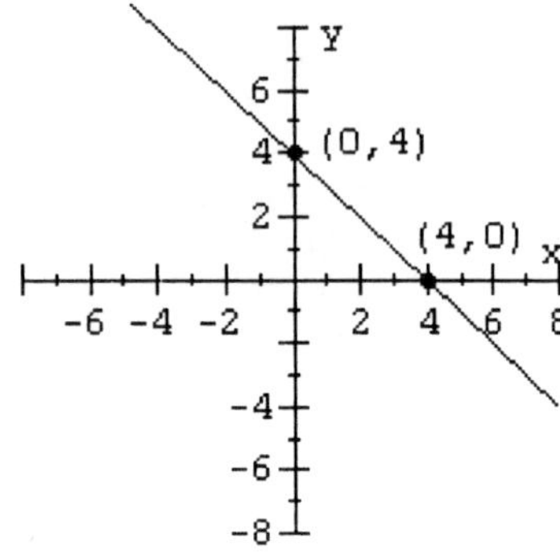

25. Domain {red, blue, green}
Range $\left\{\frac{1}{4}, \frac{1}{8}, \frac{5}{8}\right\}$
It is a function.

26. Domain $\{3, 4, 5\}$
Range $\{5, 7, 9\}$
It is a function.

27. Domain $\{1, 2\}$
Range $\{-16, -8, 8, 16\}$
It is not a function since $(1, 8)$ and $(1, -8)$ has 1 assigned to two different range components.

28. Domain $\{2, 3, 4, 5\}$
Range $\{10\}$
It is a function.

29. Domain $\{-2, -1, 0, 1, 2\}$
Range $\{0, 1, 4\}$
It is a function.

30. Domain $\{1, 2, 3\}$
Range $\{4, 8, 10, 15\}$
It is not a function.

31. The denominator can not equal zero.
So set $x - 6 = 0$
$x = 6$
Therefore 6 is excluded from the domain.
The domain is all real numbers except 6.

32. All reals

33. The domain is all real numbers.

34. The denominator can not equal zero.
So set $x + 4 = 0$
$x = -4$
Therefore -4 is excluded from the domain.
The domain is all real numbers except -4.

35. The denominator can not equal zero.
So set $2x - 1 = 0$
$2x = 1$
$x = \frac{1}{2}$
Therefore $\frac{1}{2}$ is excluded from the domain.
The domain is all real numbers except $\frac{1}{2}$.

36. The denominator can not equal zero.
So set $3x + 1 = 0$
$3x = -1$
$x = -\frac{1}{3}$
Therefore $-\frac{1}{3}$ is excluded from the domain.
The domain is all real numbers except $-\frac{1}{3}$.

37. $f(x) = 3x - 2$

$f(-4) = 3(-4) - 2$ $\quad$ $f(0) = 3(0) - 2$

$f(-4) = -12 - 2$ $\quad$ $f(0) = 0 - 2$

$f(-4) = -14$ $\quad$ $f(0) = -2$

$f(5) = 3(5) - 2$ $\quad$ $f(a) = 3(a) - 2$

$f(5) = 15 - 2$ $\quad$ $f(0) = 3a - 2$

$f(5) = 13$

38. $f(x) = \dfrac{6}{x-4}$

$f(-4) = \dfrac{6}{-4-4} = \dfrac{6}{-8} = -\dfrac{3}{4}$

$f(0) = \dfrac{6}{0-4} = \dfrac{6}{-4} = -\dfrac{3}{2}$

$f(1) = \dfrac{6}{1-4} = \dfrac{6}{-3} = -2$

$f(2) = \dfrac{6}{2-4} = \dfrac{6}{-2} = -3$

39. $f(x) = \dfrac{x}{2x+1}$

$f(-3) = \dfrac{-3}{2(-3)+1} = \dfrac{-3}{-6+1} = \dfrac{-3}{-5} = \dfrac{3}{5}$

$f(0) = \dfrac{0}{2(0)+1} = \dfrac{0}{0+1} = \dfrac{0}{1} = 0$

$f(2) = \dfrac{2}{2(2)+1} = \dfrac{2}{4+1} = \dfrac{2}{5}$

$f(3) = \dfrac{3}{2(3)+1} = \dfrac{3}{6+1} = \dfrac{3}{7}$

40. $f(x) = x^2 + 4x - 3$

$f(-1) = (-1)^2 + 4(-1) - 3$
$= 1 - 4 - 3 = -6$

$f(0) = (0)^2 + 4(0) - 3$
$= 0 + 0 - 3 = -3$

$f(1) = (1)^2 + 4(1) - 3$
$= 1 + 4 - 3 = 2$

$f(2) = (2)^2 + 4(2) - 3$
$= 4 + 8 - 3 = 9$

41 a. $f(2) = 0.20 + 0.30(2)$
$= 0.20 + 0.60 = 0.80$

b. $f(5) = 0.20 + 0.30(5)$
$= 0.20 + 1.50 = 1.70$

c.

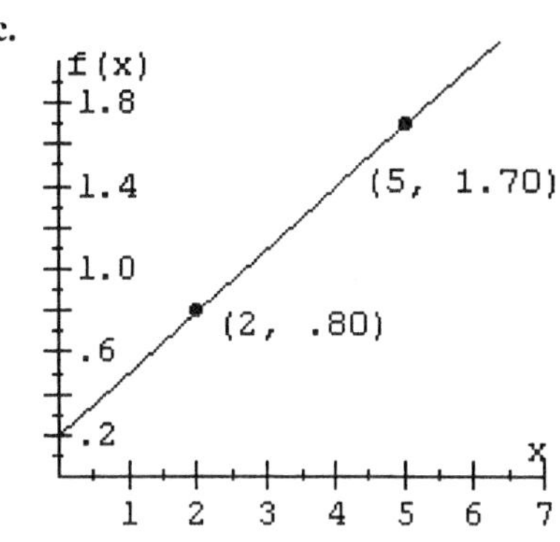

d. $f(4) = 0.20 + 0.30(4)$
$= 0.20 + 1.20 = 1.40$

The charge for 4 ounces will be \$1.40.

42 a. $f(x) = 12 + 0.2x$
$f(50) = 12 + 0.2(50)$
$f(50) = 12 + 10$
$f(50) = 22$

b. $f(100) = 12 + 0.2(100)$
$f(100) = 12 + 20$
$f(100) = 32$

c.

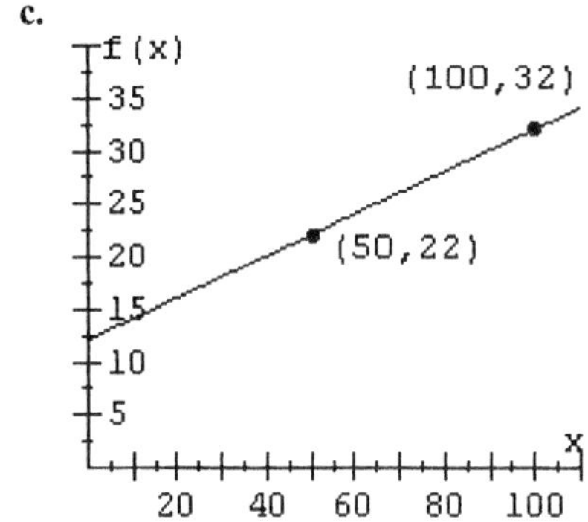

d. $f(80) = 12 + 0.2(80)$
$f(80) = 12 + 16$
$f(80) = 28$

CHAPTER 4 **Test**

1. $5x + 3y = 15$

$x = 0$ | $y = 0$

$5(0) + 3y = 15$ | $5x + 3(0) = 15$

$3y = 15$ | $5x = 15$

$y = 5$ | $x = 3$

$(0, 5)$ | $(3, 0)$

See back of textbook for graph.

2. $-2x + y = -4$

$x = 0$ | $y = 0$

$-2(0) + y = -4$ | $-2x + 0 = -4$

$y = -4$ | $-2x = -4$

$(0, -4)$ | $x = \frac{-4}{-2} = 2$

$(2, 0)$

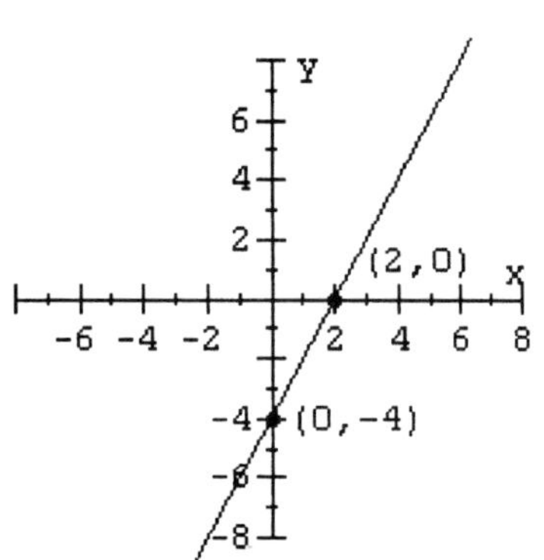

3. $y = -\frac{1}{2}x - 2$

$x = 0$ | $x = 2$

$y = -\frac{1}{2}(0) - 2$ | $y = -\frac{1}{2}(2) - 2$

$y = -2$ | $y = -1 - 2$

$(0, -2)$ | $y = -3$

$(2, -3)$

See back of textbook for graph.

4. $y = 2x^2 - 3$

$x = 0$ | $x = 1$ | $x = 2$

$y = 2(0)^2 - 3$ | $y = 2(1)^2 - 3$ | $y = 2(2)^2 - 3$

$y = 0 + -3$ | $y = 2(1) - 3$ | $y = 2(4) - 3$

$y = -3$ | $y = 2 - 3$ | $y = 8 - 3$

$(0, -3)$ | $y = -1$ | $y = 5$

$(1, -1)$ | $(2, 5)$

$x = -1$ | $x = -2$

$y = 2(-1)^2 - 3$ | $y = 2(-2)^2 - 3$

$y = 2(1) - 3$ | $y = 2(4) - 3$

$y = 2 - 3$ | $y = 8 - 3$

$y = -1$ | $y = 5$

$(-1, -1)$ | $(-2, 5)$

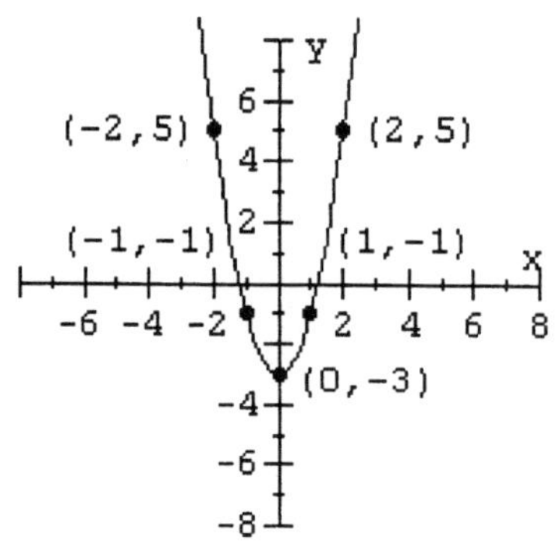

5. $y = \frac{2}{5}x - 3$

$x = 0$ | $x = 5$

$y = \frac{2}{5}(0) - 3$ | $y = \frac{2}{5}(5) - 3$

$y = 0 - 3$ | $y = 2 - 3$

$y = -3$ | $y = -1$

$(0, -3)$ | $(5, -1)$

See back of textbook for graph.

6. $y = 3x$

$x = 0$ | $x = 1$

$y = 3(0)$ | $y = 3(1)$

$y = 0$ | $y = 3$

$(0, 0)$ | $(1, 3)$

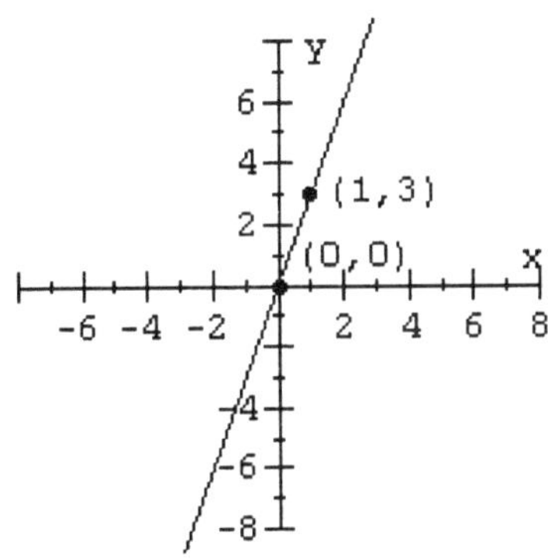

7. $-3x+4y = -12$
let $x=0$
$-3(0)+4y = -12$
$4y = -12$
$y = -3$
The y-intercept is -3.

8. $y=4x+8$
let $y=0$
$0=4x+8$
$-8=4x$
$-2=x$
The x-intercept is -2.

9. Men – Armageddon
Women – Titanic

10.

English Patient	30
Saving Private Ryan	15
Difference	15

15 more women chose English Patient as a favorite instead of Saving Private Ryan.

11.

Armageddon	40
Titanic	25
Difference	15

15 more men chose Armageddon as a favorite instead of Titanic.

12.

Women	50
Men	25
Difference	25

25 more women than men chose Titanic as a favorite.

13. March

14.

February	1700
May	1300
Difference	400

400 more people attended in February than in May.

15. February and June or January and July

16. Attendance decreased from March to May, then increased from May to July.

17. Domain $\{5, 6, 7, 8\}$
Range $\{0, -3, 8\}$
It is a function.

18. Domain $\{1, 2, 3, 4\}$
Range {black, red}
It is a function.

19. Domain $\{1\}$
Range $\{6, 7, 8, 9\}$
It is not a function since $(1, 6)$ and $(1, 7)$ have the same domain element and different range elements.

20. Domain $\{4, 8\}$
Range $\{-2, -1, 1, 2\}$
It is not a function since $(4, 1)$ and $(4, -1)$ have the same domain element with two different range elements.

21. The denominator can not equal zero.
So set $2x-7=0$
$2x=7$
$x=\frac{7}{2}$
Therefore $\frac{7}{2}$ is excluded from the domain.
The domain is all real numbers except $\frac{7}{2}$.

22. All real numbers.

23.
$$f(x) = 5x - 6$$
$$f(2) = 5(2) - 6 = 10 - 6 = 4$$
$$f(0) = 5(0) - 6 = 0 - 6 = -6$$
$$f(-3) = 5(-3) - 6 = -15 - 6 = -21$$
$$f(4) = 5(4) - 6 = 20 - 6 = 14$$

24.
$$P(x) = 20 + 0.10x$$
$$P(4270) = 20 + 0.10(4270)$$
$$P(4270) = 20 + 427$$
$$P(4270) = 447$$

25.
$$f(x) = 0.07x \qquad x \leq 200,000$$
$$g(x) = 0.05x + 4000 \quad x > 200,000$$

Since $165,000$ is less than $200,000$ use $f(x) = 0.07x$ to determine the commission.

$$f(165000) = 0.07(165000) = 11550$$

The commision for a selling price of $\$165,000$ would be $\$11,550$.

Since $245,000$ is more than $165,000$ use $g(x) = 0.05x + 4000$ to determine the commision.

$$g(245000) = 0.05(245000) + 400$$
$$g(245000) = 12250 + 4000$$
$$g(245000) = 16250$$

The commision for a selling price of $\$245,000$ would be $\$16,250$.

Chapter 5 Exponents and Polynomials

PROBLEM SET 5.1 Addition and Subtraction of Polynomials

1. The degree of $7x^2y+6xy$ is 3 because the degree of the term $7x^2y$ is 3.

3. The degree of $5x^2-9$ is 2 because the degree of the term $5x^2$ is 2.

5. The degree of $5x^3-x^2-x+3$ is 3 because the degree of the term $5x^3$ is 3.

7. The degree of $5xy$ is 2 because the degree of the term $5xy$ is 2.

9. $(3x+4)+(5x+7)=$
$(3+5)x+(4+7)=$
$8x+11$

11. $(-5y-3)+(9y+13)=$
$(-5+9)y+(-3+13)=$
$4y+10$

13. $(-2x^2+7x-9)+(4x^2-9x-14)=$
$(-2+4)x^2+(7-9)x+(-9-14)=$
$2x^2-2x-23$

15. $(5x-2)+(3x-7)+(9x-10)=$
$(5+3+9)x+(-2-7-10)=$
$17x-19$

17. $(2x^2-x+4)+(-5x^2-7x-2)$
$+(9x^2+3x-6)=$
$(2-5+9)x^2+(-1-7+3)x$
$+(4-2-6)=$
$6x^2-5x-4$

19. $(-4n^2-n-1)+(4n^2+6n-5)=$
$(-4+4)n^2+(-1+6)n+(-1-5)=$
$5n-6$

21. $(2x^2-7x-10)+(-6x-2)+(-9x^2+5)=$
$(2-9)x^2+(-7-6)x+(-10-2+5)=$
$-7x^2-13x-7$

23. $(12x+6)-(7x+1)=$
$12x+6-7x-1=$
$(12-7)x+(6-1)=$
$5x+5$

25. $(3x-7)-(5x-2)=$
$3x-7-5x+2=$
$(3-5)x+(-7+2)=$
$-2x-5$

27. $(-4x+6)-(-x-1)=$
$-4x+6+x+1=$
$(-4+1)x+(6+1)=$
$-3x+7$

29. $(3x^2+8x-4)-(x^2-7x+2)=$
$3x^2+8x-4-x^2+7x-2=$
$(3-1)x^2+(8+7)x+(-4-2)=$
$2x^2+15x-6$

31. $(3n^2-n+7)-(-2n^2-3n+4)=$
$3n^2-n+7+2n^2+3n-4=$
$(3+2)n^2+(-1+3)n+(7-4)=$
$5n^2+2n+3$

33. $(-7x^3+x^2+6x-12)$
$-(-4x^3-x^2+6x-1)=$
$-7x^3+x^2+6x-12+4x^3$
$+x^2-6x+1=$
$(-7+4)x^3+(1+1)x^2+(6-6)x$
$+(-12+1)=$
$-3x^3+2x^2-11$

35.

$12x-4$	Add	$12x-4$
$3x-2$	the	$-3x+2$
	opposite.	$9x-2$

37. Add the opposite.

$$\begin{array}{r} -3a+9 \\ \underline{-5a-6} \end{array} \qquad \begin{array}{r} -3a+9 \\ \underline{5a+6} \\ 2a+15 \end{array}$$

39. Add the opposite.

$$\begin{array}{r} 6x^2-x+11 \\ \underline{8x^2-x+6} \end{array} \qquad \begin{array}{r} 6x^2-x+11 \\ \underline{-8x^2+x-6} \\ -2x^2 \qquad +5 \end{array}$$

41.

$$\begin{array}{r} 4x^3+6x^2+7x-14 \\ \underline{-2x^3-6x^2+7x-9} \end{array}$$

Add the opposite.

$$\begin{array}{r} 4x^3+6x^2+7x-14 \\ \underline{2x^3+6x^2-7x+9} \\ 6x^3+12x^2 \qquad -5 \end{array}$$

43.

$$\begin{array}{r} 4x^3-6x^2+7x-2 \\ \underline{2x^2-6x-14} \end{array}$$

Add the opposite.

$$\begin{array}{r} 4x^3-6x^2+7x-2 \\ \underline{-2x^2+6x+14} \\ 4x^3-8x^2+13x+12 \end{array}$$

45. $(5x+3)-(7x-2)+(3x+6)=$
$5x+3-7x+2+3x+6=$
$(5-7+3)x+(3+2+6)=$
$x+11$

47. $(-x-1)-(-2x+6)+(-4x-7)=$
$-x-1+2x-6-4x-7=$
$(-1+2-4)x+(-1-6-7)=$
$-3x-14$

49. $(x^2-7x-4)+(2x^2-8x-9)$
$-(4x^2-2x-1)=$
$x^2-7x-4+2x^2-8x-9$
$-4x^2+2x+1=$
$(1+2-4)x^2+(-7-8+2)x$
$+(-4-9+1)=$
$-x^2-13x-12$

51. $(-x^2-3x+4)+(-2x^2-x-2)$
$-(-4x^2+7x+10)=$
$-x^2-3x+4-2x^2-x-2$
$+4x^2-7x-10=$
$(-1-2+4)x^2+(-3-1-7)x$
$+(4-2-10)=$
$x^2-11x-8$

53. $(3a-2b)-(7a+4b)-(6a-3b)=$
$3a-2b-7a-4b-6a+3b=$
$(3-7-6)a+(-2-4+3)b=$
$-10a-3b$

55. $(n-6)-(2n^2-n+4)+(n^2-7)=$
$n-6-2n^2+n-4+n^2-7=$
$(-2+1)n^2+(1+1)n+(-6-4-7)=$
$-n^2+2n-17$

57. $7x+[3x-(2x-1)]=$
$7x+[3x-2x+1]=$
$7x+[x+1]=$
$7x+x+1=$
$(7+1)x+1=$
$8x+1$

59. $-7n-[4n-(6n-1)]=$
$-7n-[4n-6n+1]=$
$-7n-[-2n+1]=$
$-7n+2n-1=$
$(-7+2)n-1=$
$-5n-1$

61. $(5a-1)-[3a+(4a-7)]=$
$(5a-1)-[3a+4a-7]=$
$(5a-1)-[7a-7]=$
$5a-1-7a+7=$
$(5-7)a+(-1+7)=$
$-2a+6$

63. $13x-[5x-[4x-(x-6)]]=$
$13x-[5x-[4x-x+6]]=$
$13x-[5x-[3x+6]]=$
$13x-[5x-3x-6]=$
$13x-[2x-6]=$
$13x-2x+6=$
$11x+6$

65. $[(4x-2)+(7x+6)]-(5x-3)=$
$[4x-2+7x+6]-(5x-3)=$
$[4x+7x-2+6]-(5x-3)=$
$[11x+4]-(5x-3)=$
$11x+4-5x+3=$
$11x-5x+4+3=$
$6x+7$

67. $(-8n+9)-[(-2n-5)+(-n+7)]=$
$(-8n+9)-[-2n-5-n+7]=$
$(-8n+9)-[-2n-n-5+7]=$
$(-8n+9)-[-3n+2]=$
$-8n+9+3n-2=$
$-8n+3n+9-2=$
$-5n+7$

69. Use the formula $P=2l+2w$ with $l=3x+5$ and $w=x-3$.
$P=2(3x+5)+2(x-2)$
$=6x+10+2x-4$
$=6x+2x+10-4$
$=8x+6$

71. Use the formula $A=lw$ for all the figures.
Remember, $x(x)=x^2$
$A=3x(x)+4x(x)+2x(2x)+3x(3x)$
$=3x^2+4x^2+4x^2+9x^2$
$=(3+4+4+9)x^2$
$=20x^2$

PROBLEM SET 5.2 Multiplying Monomials

1. $(5x)(9x)=5\cdot9\cdot x\cdot x=45x^{1+1}=45x^2$

3. $(3x^2)(7x)=3\cdot7\cdot x^2\cdot x=21x^{2+1}=21x^3$

5. $(-3xy)(2xy)=-3\cdot2\cdot x\cdot x\cdot y\cdot y=$
$-6x^{1+1}y^{1+1}=-6x^2y^2$

7. $(-2x^2y)(-7x)=$
$(-2)(-7)(x^2)(x)(y)=$
$14x^{2+1}y^1=14x^3y$

9. $(4a^2b^2)(-12ab)=$
$(4)(-12)(a^2)(a)(b^2)(b)=$
$-48a^{2+1}b^{2+1}=-48a^3b^3$

11. $(-xy)(-5x^3)=$
$(-1)(-5)(x)(x^3)(y)=$
$5x^{1+3}y^1=5x^4y$

13. $(8ab^2c)(13a^2c)=$
$8\cdot13\cdot a\cdot a^2\cdot b^2\cdot c\cdot c=$
$104a^{1+2}b^2c^{1+1}=104a^3b^2c^2$

15. $(5x^2)(2x)(3x^3)=5\cdot2\cdot3\cdot x^2\cdot x\cdot x^3=$
$30x^{2+1+3}=30x^6$

17. $(4xy)(-2x)(7y^2)=$
$(4)(-2)(7)(x)(x)(y)(y^2)=$
$-56x^{1+1}y^{1+2}=-56x^2y^3$

19. $(-2ab)(-ab)(-3b)=$
$(-2)(-1)(-3)(a^{1+1})(b^{1+1+1})=$
$-6a^2b^3$

21. $(6cd)(-3c^2d)(-4d)=$
$(6)(-3)(-4)(c^{1+2}d^{1+1+1})=$
$72c^3d^3$

23. $\left(\frac{2}{3}xy\right)\left(\frac{3}{5}x^2y^4\right)=$
$\left(\frac{2}{3}\right)\left(\frac{3}{5}\right)x^{1+2}y^{1+4}=$
$\frac{2}{5}x^3y^5$

25. $\left(-\frac{7}{12}a^2b\right)\left(\frac{8}{21}b^4\right)=$
$\left(-\frac{\overset{1}{\cancel{7}}}{\underset{3}{\cancel{12}}}\right)\left(\frac{\overset{2}{\cancel{8}}}{\underset{3}{\cancel{21}}}\right)a^2b^{1+4}=-\frac{2}{9}a^2b^5$

Problem Set 5.2

27. $(0.4x^5)(0.7x^3) = (0.4)(0.7)x^{5+3} = 0.28x^8$

29. $(-4ab)(1.6a^3b) =$
$(-4)(1.6)a^{1+3}b^{1+1} = -6.4a^4b^2$

31. $(2x^4)^2 = (2)^2(x^4)^2 = 4x^{4\cdot 2} = 4x^8$

33. $(-3a^2b^3)^2 = (-3)^2(a^2)^2(b^3)^2 = 9a^4b^6$

35. $(3x^2)^3 = (3)^3(x^2)^3 = 27x^6$

37. $(-4x^4)^3 = (-4)^3(x^4)^3 = -64x^{12}$

39. $(9x^4y^5)^2 = (9)^2(x^4)^2(y^5)^2 = 81x^8y^{10}$

41. $(2x^2y)^4 = (2)^4(x^2)^4(y)^4 = 16x^8y^4$

43. $(-3a^3b^2)^4 = (-3)^4(a^3)^4(b^2)^4 =$
$81a^{12}b^8$

45. $(-x^2y)^6 = (-1)^6(x^2)^6(y)^6 =$
$1x^{12}y^6 = x^{12}y^6$

47. $5x(3x+2) = 5x(3x) + 5x(2) =$
$15x^2 + 10x$

49. $3x^2(6x-2) = 3x^2(6x) - 3x^2(2) =$
$18x^3 - 6x^2$

51. $-4x(7x^2-4) =$
$-4x(7x^2) - (-4x)(4) =$
$-28x^3 + 16x$

53. $2x(x^2-4x+6) =$
$2x(x^2) - 2x(4x) + 2x(6) =$
$2x^3 - 8x^2 + 12x$

55. $-6a(3a^2-5a-7) =$
$-6a(3a^2) - (-6a)(5a) - (-6a)(7) =$
$-18a^3 + 30a^2 + 42a$

57. $7xy(4x^2-x+5) =$
$7xy(4x^2) - 7xy(x) + 7xy(5) =$
$28x^3y - 7x^2y + 35xy$

59. $-xy(9x^2-2x-6) =$
$-xy(9x^2) - (-xy)(2x) - (-xy)(6) =$
$-9x^3y + 2x^2y + 6xy$

61. $5(x+2y) + 4(2x+3y) =$
$5x + 10y + 8x + 12y =$
$13x + 22y$

63. $4(x-3y) - 3(2x-y) =$
$4x - 12y - 6x + 3y =$
$-2x - 9y$

65. $2x(x^2-3x-4) + x(2x^2+3x-6) =$
$2x^3 - 6x^2 - 8x + 2x^3 + 3x^2 - 6x =$
$4x^3 - 3x^2 - 14x$

67. $3[2x-(x-2)] - 4(x-2) =$
$3[2x-x+2] - 4(x-2) =$
$3[x+2] - 4(x-2) =$
$3x + 6 - 4x + 8 =$
$-x + 14$

69. $-4(3x+2) - 5[2x-(3x+4)] =$
$-4(3x+2) - 5[2x-3x-4] =$
$-4(3x+2) - 5[-x-4] =$
$-12x - 8 + 5x + 20 =$
$-7x + 12$

71. $(3x)^2(2x)^3 = (3)^2(x)^2(2)(x)^3 =$
$(9)(x^2)(2)(x^3) = 18x^5$

73. $(-3x)^3(-4x)^2 =$
$(-3)^3(x)^3(-4)^2(x)^2 =$
$(-27)(x^3)(16)(x^2) =$
$-432x^5$

75. $(5x^2y)^2(xy^2)^3 =$
$(5)^2(x^2)^2(y)^2(x)^3(y^2)^3 =$
$(25)(x^4)(y^2)(x^3)(y^6) =$
$25x^7y^8$

77. $(-a^2bc^3)^3(a^3b)^2 =$
$(-1)^3(a^2)^3(b)^3(c^3)^3(a^3)^2(b)^2 =$
$-1(a^6)(b^3)(c^9)(a^6)(b^2) =$
$-a^{12}b^5c^9$

79. $(-2x^2y^2)^4(-xy^3)^3 =$
$(-2)^4(x^2)^4(y^2)^4(-1)^3(x)^3(y^3)^3 =$
$(16)(x^8)(y^8)(-1)(x^3)(y^9) =$
$-16x^{11}y^{17}$

81. Use the formula $A = lw$ for both rectangles and then add.
Area = Left Area + Right Area
$A = 3(x-1) + 4(x+2)$
$= 3x - 3 + 4x + 8$
$= 7x + 5$

83. Use the formula $A = \pi r^2$ for the both circles, then subtract the area of the smaller circle from the area of the larger circle.
$$\text{Area} = \begin{array}{c}\text{Area of}\\ \text{Larger Circle}\end{array} - \begin{array}{c}\text{Area of}\\ \text{Smaller Circle}\end{array}$$
$A = \pi(2x)^2 - \pi(x)^2$
$= \pi(4x^2) - \pi(x^2)$
$= 4\pi x^2 - \pi x^2$
$= 3\pi x^2$
The area of the shaded region can be represented by $3\pi x^2$.

PROBLEM SET 5.3 Multiplying Polynomials

1. $(x+2)(y+3) =$
$x(y) + x(3) + 2(y) + 2(3) =$
$xy + 3x + 2y + 6$

3. $(x-4)(y+1) =$
$x(y) + x(1) - 4(y) - 4(1) =$
$xy + x - 4y - 4$

5. $(x-5)(y-6) =$
$x(y) + x(-6) - 5(y) - 5(-6) =$
$xy - 6x - 5y + 30$

7. $(x+2)(y+z+1) =$
$x(y) + x(z) + x(1) + 2(y) + 2(z) + 2(1) =$
$xy + xz + x + 2y + 2z + 2$

9. $(2x+3)(3y+1) =$
$2x(3y) + 2x(1) + 3(3y) + 3(1) =$
$6xy + 2x + 9y + 3$

11. $(x+3)(x+7) =$
$x(x) + x(7) + 3(x) + 3(7) =$
$x^2 + 7x + 3x + 21 =$
$x^2 + 10x + 21$

13. $(x+8)(x-3) =$
$x(x) + x(-3) + 8(x) + 8(-3) =$
$x^2 - 3x + 8x - 24 =$
$x^2 + 5x - 24$

15. $(x-7)(x+1) =$
$x(x) + x(1) - 7(x) - 7(1) =$
$x^2 + 1x - 7x - 7 =$
$x^2 - 6x - 7$

17. $(n-4)(n-6) =$
$n(n) + n(-6) - 4(n) - 4(-6) =$
$n^2 - 6n - 4n + 24 =$
$n^2 - 10n + 24$

19. $(3n+1)(n+6) =$
$3n(n) + 3n(6) + 1(n) + 1(6) =$
$3n^2 + 18n + 1n + 6 =$
$3n^2 + 19n + 6$

21. $(5x-2)(3x+7) =$
$5x(3x) + 5x(7) - 2(3x) - 2(7) =$
$15x^2 + 35x - 6x - 14 =$
$15x^2 + 29x - 14$

23. $(x+3)(x^2+4x+9) =$
$x(x^2) + x(4x) + x(9) + 3(x^2)$
$+ 3(4x) + 3(9) =$
$x^3 + 4x^2 + 9x + 3x^2 + 12x + 27 =$
$x^3 + 7x^2 + 21x + 27$

Problem Set 5.3

25. $(x+4)(x^2-x-6)=$
$x(x^2)+x(-x)+x(-6)+4(x^2)$
$+4(-x)+4(-6)=$
$x^3-x^2-6x+4x^2-4x-24=$
$x^3+3x^2-10x-24$

27. $(x-5)(2x^2+3x-7)=$
$x(2x^2)+x(3x)+x(-7)-5(2x^2)$
$-5(3x)-5(-7)=$
$2x^3+3x^2-7x-10x^2-15x+35=$
$2x^3-7x^2-22x+35$

29. $(2a-1)(4a^2-5a+9)=$
$2a(4a^2)+2a(-5a)+2a(9)-1(4a^2)$
$-1(-5a)-1(9)=$
$8a^3-10a^2+18a-4a^2+5a-9=$
$8a^3-14a^2+23a-9$

31. $(3a+5)(a^2-a-1)=$
$3a(a^2)+3a(-a)+3a(-1)+5(a^2)$
$+5(-a)+5(-1)=$
$3a^3-3a^2-3a+5a^2-5a-5=$
$3a^3+2a^2-8a-5$

33. $(x^2+2x+3)(x^2+5x+4)=$
$x^2(x^2+5x+4)+2x(x^2+5x+4)$
$+3(x^2+5x+4)=$
$x^4+5x^3+4x^2+2x^3+10x^2+8x$
$+3x^2+15x+12=$
$x^4+7x^3+17x^2+23x+12$

35. $(x^2-6x-7)(x^2+3x-9)=$
$x^2(x^2+3x-9)-6x(x^2+3x-9)$
$-7(x^2+3x-9)=$
$x^4+3x^3-9x^2-6x^3-18x^2+54x$
$-7x^2-21x+63=$
$x^4-3x^3-34x^2+33x+63$

37. $(x+2)(x+9)=$
$x^2+(2+9)x+18=$
$x^2+11x+18$

39. $(x+6)(x-2)=$
$x^2+(6-2)x-12=$
$x^2+4x-12$

41. $(x+3)(x-11)=$
$x^2+(3-11)x-33=$
$x^2-8x-33$

43. $(n-4)(n-3)=$
$n^2+(-4-3)n+12=$
$n^2-7n+12$

45. $(n+6)(n+12)=$
$n^2+(6+12)n+72=$
$n^2+18n+72$

47. $(y+3)(y-7)=$
$y^2+(3-7)y-21=$
$y^2-4y-21$

49. $(y-7)(y-12)=$
$y^2+(-7-12)y+84=$
$y^2-19y+84$

51. $(x-5)(x+7)=$
$x^2+(-5+7)x-35=$
$x^2+2x-35$

53. $(x-14)(x+8)=$
$x^2+(-14+8)x-112=$
$x^2-6x-112$

55. $(a+10)(a-9)=$
$a^2+(10-9)a-90=$
a^2+a-90

57. $(2a+1)(a+6)=$
$2a^2+(12+1)a+6=$
$2a^2+13a+6$

59. $(5x-2)(x+7)=$
$5x^2+(35-2)x-14=$
$5x^2+33x-14$

61. $(3x-7)(2x+1)=$
$6x^2+(3-14)x-7=$
$6x^2-11x-7$

63. $(4a+3)(3a-4)=$
$12a^2+(-16+9)a-12=$
$12a^2-7a-12$

65. $(6n-5)(2n-3)=$
$12n^2+(-18-10)n+15=$
$12n^2-28n+15$

67. $(7x-4)(2x+3)=$
$14x^2+(21-8)x-12=$
$14x^2+13x-12$

69. $(5-x)(9-2x)=$
$45+(-10-9)x+2x^2=$
$45-19x+2x^2$

71. $(-2x+3)(4x-5)=$
$-8x^2+(10+12)x-15=$
$-8x^2+22x-15$

73. $(-3x-1)(3x-4)=$
$-9x^2+(12-3)x+4=$
$-9x^2+9x+4$

75. $(8n+3)(9n-4)=$
$72n^2+(-32+27)n-12=$
$72n^2-5n-12$

77. $(3-2x)(9-x)=$
$27+(-3-18)x+2x^2=$
$27-21x+2x^2$

79. $(-4x+3)(-5x-2)=$
$20x^2+(8-15)x-6=$
$20x^2-7x-6$

81. Use the pattern
$(a+b)^2=a^2+2ab+b^2$
$(x+7)^2=x^2+2(x)(7)+(7)^2=$
$x^2+14x+49$

83. Use the pattern
$(a+b)(a-b)=a^2-b^2$
$(5x-2)(5x+2)=(5x)^2-(2)^2=$
$25x^2-4$

85. Use the pattern
$(a-b)^2=a^2-2ab+b^2$
$(x-1)^2=x^2-2(x)(1)+(1)^2=$
x^2-2x+1

87. Use the pattern
$(a+b)^2=a^2+2ab+b^2$
$(3x+7)^2=(3x)^2+2(3x)(7)+(7)^2=$
$9x^2+42x+49$

89. Use the pattern
$(a-b)^2=a^2-2ab+b^2$
$(2x-3)^2=(2x)^2-2(2x)(3)+(3)^2=$
$4x^2-12x+9$

91. Use the pattern
$(a+b)(a-b)=a^2-b^2$
$(2x+3y)(2x-3y)=(2x)^2-(3y)^2=$
$4x^2-9y^2$

93. Use the pattern
$(a-b)^2=a^2-2ab+b^2$
$(1-5n)^2=1^2-2(1)(5n)+(5n)^2=$
$1-10n+25n^2$

95. Use the pattern
$(a+b)^2=a^2+2ab+b^2$
$(3x+4y)^2=(3x)^2+2(3x)(4y)+(4y)^2=$
$9x^2+24xy+16y^2$

97. Use the pattern
$(a+b)^2=a^2+2ab+b^2$
$(3+4y)^2=3^2+2(3)(4y)+(4y)^2=$
$9+24y+16y^2$

99. Use the pattern
$(a+b)(a-b)=a^2-b^2$
$(1+7n)(1-7n)=(1)^2-(7n)^2=$
$1-49n^2$

101. Use the pattern
$(a-b)^2=a^2-2ab+b^2$
$(4a-7b)^2=(4a)^2-2(4a)(7b)+(7b)^2=$
$16a^2-56ab+49b^2$

103. Use the pattern
$(a+b)^2=a^2+2ab+b^2$
$(x+8y)^2=x^2+2(x)(8y)+(8y)^2=$
$x^2+16xy+64y^2$

105. Use the pattern
$(a+b)(a-b)=a^2-b^2$
$(5x-11y)(5x+11y)=(5x)^2-(11y)^2=$
$25x^2-121y^2$

107. Use the pattern
$(a+b)(a-b)=a^2-b^2$
$x(8x+1)(8x-1)=x\left[(8x)^2-(1)^2\right]$
$x[64x^2-1]=64x^3-x$

109. Use the pattern
$(a+b)(a-b)=a^2-b^2$
$-2x(4x+y)(4x-y)=-2x\left[(4x)^2-(y)^2\right]$
$-2x[16x^2-y^2]=-32x^3+2xy^2$

111. $(x+2)^3=$
$(x+2)(x+2)(x+2)=$
$(x+2)(x^2+4x+4)=$
$x(x^2+4x+4)+2(x^2+4x+4)=$
$x^3+4x^2+4x+2x^2+8x+8=$
$x^3+6x^2+12x+8$

113. $(x-3)^3=$
$(x-3)(x-3)(x-3)=$
$(x-3)(x^2-6x+9)=$
$x(x^2-6x+9)-3(x^2-6x+9)=$
$x^3-6x^2+9x-3x^2+18x-27=$
$x^3-9x^2+27x-27$

115. $(2n+1)^3=$
$(2n+1)(2n+1)(2n+1)=$
$(2n+1)(4n^2+4n+1)=$
$2n(4n^2+4n+1)+1(4n^2+4n+1)=$
$8n^3+8n^2+2n+4n^2+4n+1=$
$8n^3+12n^2+6n+1$

117. $(3n-2)^3=$
$(3n-2)(3n-2)(3n-2)=$
$(3n-2)(9n^2-12n+4)=$
$3n(9n^2-12n+4)-2(9n^2-12n+4)=$
$27n^3-36n^2+12n-18n^2+24n-8=$
$27n^3-54n^2+36n-8$

119. Let $x+3$ represent the width of the rectangle, then $x+5$ represents the length. Therefore, the area of the figure is represented by

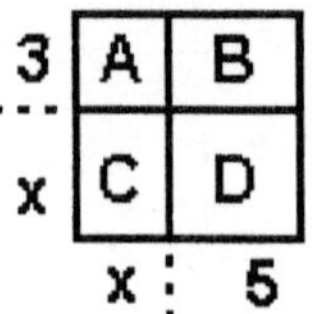

$$\begin{aligned}\text{Area} &= (\text{length})(\text{width})\\ &= (x+3)(x+5)\\ &= x^2+8x+15\end{aligned}$$

Geometrically, the sum of the area of each section would be

$$\text{Area} = \begin{matrix}\text{Area}\\ \text{of A}\end{matrix} + \begin{matrix}\text{Area}\\ \text{of B}\end{matrix} + \begin{matrix}\text{Area}\\ \text{of C}\end{matrix} + \begin{matrix}\text{Area}\\ \text{of D}\end{matrix}$$
$$A = 3x + 15 + x^2 + 5x$$
$$= x^2+8x+15$$

121. Each side of the box will be $14-2x$ and the height will be x.

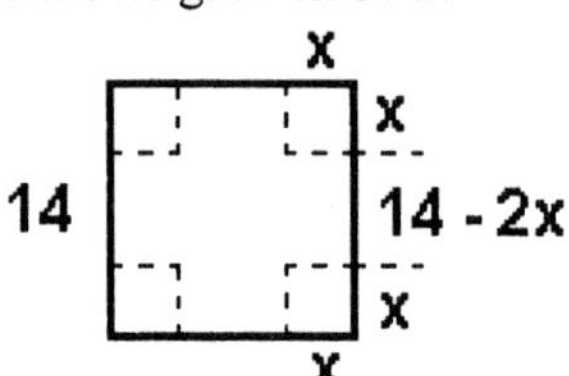

$$\begin{aligned}\text{The volume is } V &= lwh\\ &= (14-2x)(14-2x)x\\ &= \left(196-56x+4x^2\right)x\\ &= 196x-56x^2+4x^3\end{aligned}$$

$$\begin{matrix}\text{The outside}\\ \text{surface area is}\end{matrix}\ S = \begin{matrix}\text{Original}\\ \text{Area}\end{matrix} - 4\text{ Corners}$$
$$= (14)^2-4(x)^2$$
$$= 196-4x^2$$

PROBLEM SET 5.4 Dividing by Monomials

1. $\frac{x^{10}}{x^2} = x^{10-2} = x^8$

3. $\frac{4x^3}{2x} = 2x^{3-1} = 2x^2$

5. $\frac{-16n^6}{2n^2} = -8n^{6-2} = -8n^4$

7. $\frac{72x^3}{-9x^3} = -8(1) = -8$

9. $\frac{65x^2y^3}{5xy} = 13x^{2-1}y^{3-1} = 13xy^2$

11. $\frac{-91a^4b^6}{-13a^3b^4} = 7a^{4-3}b^{6-4} = 7ab^2$

13. $\frac{18x^2y^6}{xy^2} = 18x^{2-1}y^{6-2} = 18xy^4$

15. $\frac{32x^6y^2}{-x} = -32^{6-1}y^2 = -32x^5y^2$

17. $\frac{-96x^5y^7}{12y^3} = -8x^5y^{7-3} = -8x^5y^4$

19. $\frac{-ab}{ab} = -1(1)(1) = -1$

21. $\frac{56a^2b^3c^5}{4abc} = 14a^{2-1}b^{3-1}c^{5-1} = 14ab^2c^4$

23. $\frac{-80xy^2z^6}{-5xyz^2} = 16(1)y^{2-1}z^{6-2} = 16yz^4$

25. $\frac{8x^4+12x^5}{2x^2} = \frac{8x^4}{2x^2} + \frac{12x^5}{2x^2} = 4x^2 + 6x^3$

27. $\frac{9x^6-24x^4}{3x^3} = \frac{9x^6}{3x^3} - \frac{24x^4}{3x^3} = 3x^3 - 8x$

29. $\frac{-28n^5+36n^2}{4n^2} = \frac{-28n^5}{4n^2} + \frac{36n^2}{4n^2} =$

$-7n^3 + 9$

31. $\frac{35x^6-56x^5-84x^3}{7x^2} =$

$\frac{35x^6}{7x^2} - \frac{56x^5}{7x^2} - \frac{84x^3}{7x^2} =$

$5x^4 - 8x^3 - 12x$

33. $\frac{-24n^8+48n^5-78n^3}{-6n^3} =$

$\frac{-24n^8}{-6n^3} + \frac{48n^5}{-6n^3} - \frac{78n^3}{-6n^3} =$

$4n^5 - 8n^2 + 13$

35. $\frac{-60a^7-96a^3}{-12a} = \frac{-60a^7}{-12a} - \frac{96a^3}{-12a} =$

$5a^6 + 8a^2$

37. $\dfrac{27x^2y^4 - 45xy^4}{-9xy^3} =$

$\dfrac{27x^2y^4}{-9xy^3} - \dfrac{45xy^4}{-9xy^3} =$

$-3xy + 5y$

39. $\dfrac{48a^2b^2 + 60a^3b^4}{-6ab} =$

$\dfrac{48a^2b^2}{-6ab} + \dfrac{60a^3b^4}{-6ab} =$

$-8ab - 10a^2b^3$

41. $\dfrac{12a^2b^2c^2 - 52a^2b^3c^5}{-4a^2bc} =$

$\dfrac{12a^2b^2c^2}{-4a^2bc} - \dfrac{52a^2b^3c^5}{-4a^2bc} =$

$-3bc + 13b^2c^4$

43. $\dfrac{9x^2y^3 - 12x^3y^4}{-xy} = \dfrac{9x^2y^3}{-xy} - \dfrac{12x^3y^4}{-xy} =$

$-9xy^2 + 12x^2y^3$

45. $\dfrac{-42x^6 - 70x^4 + 98x^2}{14x^2} =$

$\dfrac{-42x^6}{14x^2} - \dfrac{70x^4}{14x^2} + \dfrac{98x^2}{14x^2} =$

$-3x^4 - 5x^2 + 7$

47. $\dfrac{15a^3b - 35a^2b - 65ab^2}{-5ab} =$

$\dfrac{15a^3b}{-5ab} - \dfrac{35a^2b}{-5ab} - \dfrac{65ab^2}{-5ab} =$

$-3a^2 + 7a + 13b$

49. $\dfrac{-xy + 5x^2y^3 - 7x^2y^6}{xy} =$

$\dfrac{-xy}{xy} + \dfrac{5x^2y^3}{xy} - \dfrac{7x^2y^6}{xy} =$

$-1 + 5xy^2 - 7xy^5$

PROBLEM SET 5.5 Dividing by Binomials

1.
$$\begin{array}{r}
x + 12 \\
x + 4\overline{)x^2 + 16x + 48} \\
\underline{x^2 + 4x} \\
12x + 48 \\
\underline{12x + 48}
\end{array}$$

3.
$$\begin{array}{r}
x + 2 \\
x - 7\overline{)x^2 - 5x - 14} \\
\underline{x^2 - 7x} \\
2x - 14 \\
\underline{2x - 14}
\end{array}$$

5.
$$\begin{array}{r}
x + 8 \\
x + 3\overline{)x^2 + 11x + 28} \\
\underline{x^2 + 3x} \\
8x + 28 \\
\underline{8x + 24} \\
4
\end{array}$$

7.
$$\begin{array}{r} x+4 \\ x-8\,\overline{)\,x^2-4x-39} \\ \underline{x^2-8x\qquad} \\ 4x-39 \\ \underline{4x-32} \\ -7 \end{array}$$

9.
$$\begin{array}{r} 5n+4 \\ n-1\,\overline{)\,5n^2-\ n-4} \\ \underline{5n^2-5n\qquad} \\ 4n-4 \\ \underline{4n-4} \end{array}$$

11.
$$\begin{array}{r} 8y-3 \\ y+7\,\overline{)\,8y^2+53y-19} \\ \underline{8y^2+56y\qquad} \\ -3y-19 \\ \underline{-3y-21} \\ 2 \end{array}$$

13.
$$\begin{array}{r} 4x-7 \\ 5x+1\,\overline{)\,20x^2-31x-7} \\ \underline{20x^2+\ 4x\qquad} \\ -35x-7 \\ \underline{-35x-7} \end{array}$$

15.
$$\begin{array}{r} 3x+2 \\ 2x+7\,\overline{)\,6x^2+25x+\ 8} \\ \underline{6x^2+21x\qquad} \\ 4x+\ 8 \\ \underline{4x+14} \\ -6 \end{array}$$

17.
$$\begin{array}{r} 2x^2+3x+4 \\ x-2\,\overline{)\,2x^3-\ x^2-2x-8} \\ \underline{2x^3-4x^2\qquad\qquad} \\ 3x^2-2x-8 \\ \underline{3x^2-6x\qquad} \\ 4x-8 \\ \underline{4x-8} \end{array}$$

19.
$$\begin{array}{r} 5n^2-\ 4n-3 \\ n+3\,\overline{)\,5n^3+11n^2-15n-9} \\ \underline{5n^3+15n^2\qquad\qquad} \\ -4n^2-15n-9 \\ \underline{-4n^2-12n\qquad} \\ -3n-9 \\ \underline{-3n-9} \end{array}$$

21.
$$\begin{array}{r} n^2+\ 6n-4 \\ n-6\,\overline{)\,n^3+0n^2-40n+24} \\ \underline{n^3-6n^2\qquad\qquad} \\ 6n^2-40n+24 \\ \underline{6n^2-36n\qquad} \\ -4n+24 \\ \underline{-4n+24} \end{array}$$

23.
$$\begin{array}{r} x^2+3x+9 \\ x-3\,\overline{)\,x^3+0x^2+0x-27} \\ \underline{x^3-3x^2\qquad\qquad} \\ 3x^2-0x-27 \\ \underline{3x^2-9x\qquad} \\ 9x-27 \\ \underline{9x-27} \end{array}$$

25.
$$\begin{array}{r} 9x^2+12x+16 \\ 3x-4\,\overline{)\,27x^3+\ 0x^2+\ 0x-64} \\ \underline{27x^3-36x^2\qquad\qquad} \\ 36x^2+\ 0x-64 \\ \underline{36x^2-48x\qquad} \\ 48x-64 \\ \underline{48x-64} \end{array}$$

27.
$$\begin{array}{r} 3n-\ 8 \\ n+2\,\overline{)\,3n^2-2n+\ 1} \\ \underline{3n^2+6n\qquad} \\ -8n+\ 1 \\ \underline{-8n-16} \\ 17 \end{array}$$

29.

$$\begin{array}{r} 3t+2 \\ 3t-1\overline{)9t^2+3t+4} \\ \underline{9t^2-3t} \\ 6t+4 \\ \underline{6t-2} \\ 6 \end{array}$$

31.

$$\begin{array}{r} 3n^2-\ \ n-4 \\ 2n-1\overline{)6n^3-5n^2-7n+4} \\ \underline{6n^3-3n^2} \\ -2n^2-7n+4 \\ \underline{-2n^2+1n} \\ -8n+4 \\ \underline{-8n+4} \end{array}$$

33.

$$\begin{array}{r} 4x^2-\ \ 5x+5 \\ x+7\overline{)4x^3+23x^2-30x+32} \\ \underline{4x^3+28x^2} \\ -5x^2-30x+32 \\ \underline{-5x^2-35x} \\ 5x+32 \\ \underline{5x+35} \\ -3 \end{array}$$

35.

$$\begin{array}{r} x+4 \\ x^2-2x\overline{)x^3+2x^2-3x-1} \\ \underline{x^3-2x^2} \\ 4x^2-3x-1 \\ \underline{4x^2-8x} \\ 5x-1 \end{array}$$

37.

$$\begin{array}{r} 2x-12 \\ x^2+4x\overline{)2x^3-4x^2+\ \ x-\ 5} \\ \underline{2x^3+8x^2} \\ -12x^2+\ \ x-\ 5 \\ \underline{-12x^2-48x} \\ 49x-\ 5 \end{array}$$

39.

$$\begin{array}{r} x^3-2x^2+4x-8 \\ x+2\overline{)x^4+0x^3+0x^2+0x-16} \\ \underline{x^4+2x^3} \\ -2x^3+0x^2+0x-16 \\ \underline{-2x^3-4x^2} \\ 4x^2+0x-16 \\ \underline{4x^2+8x} \\ -8x-16 \\ \underline{-8x-16} \end{array}$$

PROBLEM SET 5.6 Zero and Negative Integers as Exponents

1. $3^{-2}=\frac{1}{3^2}=\frac{1}{9}$

3. $4^{-3}=\frac{1}{4^3}=\frac{1}{64}$

5. $\left(\frac{3}{2}\right)^{-1}=\left(\frac{2}{3}\right)^{1}=\frac{2}{3}$

7. $\frac{1}{2^{-4}}=\frac{1}{\frac{1}{2^4}}=2^4=16$

9. $\left(-\frac{4}{3}\right)^{0}=1$

11. $\left(-\frac{2}{3}\right)^{-3}=\left(-\frac{3}{2}\right)^{3}=-\frac{27}{8}$

13. $(-2)^{-2}=\frac{1}{(-2)^2}=\frac{1}{4}$

15. $-\left(3^{-2}\right)=-\frac{1}{3^2}=-\frac{1}{9}$

17. $\frac{1}{\left(\frac{3}{4}\right)^{-3}}=\left(\frac{3}{4}\right)^{3}=\frac{27}{64}$

19. $2^6 \cdot 2^{-9} = 2^{6-9} = 2^{-3} = \frac{1}{2^3} = \frac{1}{8}$

21. $3^6 \cdot 3^{-3} = 3^{6-3} = 3^3 = 27$

23. $\frac{10^2}{10^{-1}} = 10^{2-(-1)} = 10^{2+1} = 10^3 = 1000$

25. $\frac{10^{-1}}{10^2} = 10^{-1-2} = 10^{-3} = \frac{1}{10^3} = \frac{1}{1000}$

27. $(2^{-1} \cdot 3^{-2})^{-1} = (2^{-1})^{-1} \cdot (3^{-2})^{-1} =$
$2 \cdot 3^2 = 2 \cdot 9 = 18$

29. $\left(\frac{4^{-1}}{3}\right)^{-2} = \frac{(4^{-1})^{-2}}{3^{-2}} = \frac{4^2}{3^{-2}} =$

$4^2 \cdot 3^2 = 16 \cdot 9 = 144$

31. $x^6x^{-1} = x^{6-1} = x^5$

33. $n^{-4}n^2 = n^{-4+2} = n^{-2} = \frac{1}{n^2}$

35. $a^{-2}a^{-3} = a^{-2-3} = a^{-5} = \frac{1}{a^5}$

37. $(2x^3)(4x^{-2}) = 2 \cdot 4 \cdot x^{3-2} = 8x$

39. $\left(3x^{-6}\right)\left(9x^2\right) = 3 \cdot 9 \cdot x^{-6+2} = 27x^{-4} = \frac{27}{x^4}$

41. $(5y^{-1})(-3y^{-2}) = (5)(-3)y^{-1-2} =$

$-15y^{-3} = -\frac{15}{y^3}$

43. $(8x^{-4})(12x^4) = 8 \cdot 12 \cdot x^{-4+4} =$
$96x^0 = 96(1) = 96$

45. $\frac{x^7}{x^{-3}} = x^{7-(-3)} = x^{7+3} = x^{10}$

47. $\frac{n^{-1}}{n^3} = n^{-1-3} = n^{-4} = \frac{1}{n^4}$

49. $\frac{4n^{-1}}{2n^3} = 2n^{-1-(-3)} = 2n^{-1+3} = 2n^2$

51. $\frac{-24x^{-6}}{8x^{-2}} = -3x^{-6-(-2)} =$
$-3x^{-6+2} = -3x^{-4} = -\frac{3}{x^4}$

53. $\frac{-52y^{-2}}{-13y^{-2}} = 4y^{-2-(-2)} = 4y^{-2+2} =$
$4y^0 = 4(1) = 4$

55. $\left(x^{-3}\right)^{-2} = x^{-3(-2)} = x^6$

57. $\left(x^2\right)^{-2} = x^{2(-2)} = x^{-4} = \frac{1}{x^4}$

59. $\left(x^3y^4\right)^{-1} = \left(x^3\right)^{-1}\left(y^4\right)^{-1} =$

$x^{-3}y^{-4} = \frac{1}{x^3x^4}$

61. $\left(x^{-2}y^{-1}\right)^3 = \left(x^{-2}\right)^3\left(y^{-1}\right)^3 =$

$x^{-6}y^{-3} = \frac{1}{x^6y^3}$

63. $\left(2n^{-2}\right)^3 = (2)^3\left(n^{-2}\right)^3 = 8n^{-6} = \frac{8}{n^6}$

65. $\left(4n^3\right)^{-2} = (4)^{-2}\left(n^3\right)^{-2} = 4^{-2}n^{-6} =$

$\frac{1}{4^2n^6} = \frac{1}{16n^6}$

67. $\left(3a^{-2}\right)^4 = 3^4\left(a^{-2}\right)^4 = 81a^{-8} = \frac{81}{a^8}$

69. $\left(5x^{-1}\right)^{-2} = 5^{-2}\left(x^{-1}\right)^{-2} = 5^{-2}x^2 =$

$\frac{x^2}{5^2} = \frac{x^2}{25}$

71. $\left(2x^{-2}y^{-1}\right)^{-1} = 2^{-1}\left(x^{-2}\right)^{-1}\left(y^{-1}\right)^{-1} =$

$2^{-1}x^2y^1 = \frac{x^2y}{2}$

Problem Set 5.6

73. $\left(\frac{x^2}{y}\right)^{-1} = \frac{(x^2)^{-1}}{y^{-1}} = \frac{x^{-2}}{y^{-1}} = \frac{y}{x^2}$

75. $\left(\frac{a^{-1}}{b^2}\right)^{-4} = \frac{a^4}{b^{-8}} = a^4b^8$

77. $\left(\frac{x^{-1}}{y^{-3}}\right)^{-2} = \frac{x^2}{y^6}$

79. $\left(\frac{x^2}{x^3}\right)^{-1} = \left(x^{2-3}\right)^{-1} = \left(x^{-1}\right)^{-1} = x$

81. $\left(\frac{2x^{-1}}{x^{-2}}\right)^{-3} = \left(2x^{-1-(-2)}\right)^{-3} =$

$\left(2x^{-1+2}\right)^{-3} = (2x)^{-3} = \frac{1}{(2x)^3} = \frac{1}{8x^3}$

83. $\left(\frac{18x^{-1}}{9x}\right)^{-2} = \left(2x^{-1-1}\right)^{-2} =$

$\left(2x^{-2}\right)^{-2} = 2^{-2}x^4 = \frac{x^4}{2^2} = \frac{x^4}{4}$

85. $321 = (3.21)(10^2)$
Number > 10, positive exponent.

87. $8000 = (8)(10^3)$
Number > 10, positive exponent.

89. $0.00246 = (2.46)(10^{-3})$
Number < 1, negative exponent.

91. $0.0000179 = (1.79)(10^{-5})$
Number < 1, negative exponent.

93. $87,000,000 = (8.7)(10^7)$
Number > 10, positive exponent.

95. $(8)(10^3) = 8000$
Positive exponent, move decimal right.

97. $(5.21)(10^4) = 52,100$
Positive exponent, move decimal right.

99. $(1.14)(10^7) = 11,400,000$
Positive exponent, move decimal right.

101. $(7)(10^{-2}) = 0.07$
Negative exponent, move decimal left.

103. $(9.87)(10^{-4}) = 0.000987$
Negative exponent, move decimal left.

105. $(8.64)(10^{-6}) = 0.00000864$
Negative exponent, move decimal left.

107. $(0.007)(120) =$
$(7)(10^{-3})(1.2)(10^2) =$
$(7)(1.2)(10^{-3})(10^2) =$
$(8.4)(10^{-1}) = 0.84$

109. $(5,000,000)(0.00009) =$
$(5)(10^6)(9)(10^{-5}) =$
$(5)(9)(10^6)(10^{-5}) =$
$(45)(10^1) = 450$

111. $\frac{6000}{0.0015} = \frac{(6)(10^3)}{(1.5)(10^{-3})} =$
$(4)\left(10^{3-(-3)}\right) = (4)(10^6) =$
$4,000,000$

113. $\frac{0.00086}{4300} = \frac{(8.6)(10^{-4})}{(4.3)(10^3)} =$
$(2)(10^{-4-3}) = (2)(10^{-7}) =$
0.0000002

115. $\frac{0.00039}{0.0013} = \frac{(3.9)(10^{-4})}{(1.3)(10^{-3})} =$
$(3)\left(10^{-4-(-3)}\right) = (3)(10^{-1}) = 0.3$

117. $\frac{(0.0008)(0.07)}{(20,000)(0.0004)} =$

$\frac{(8)(10^{-4})(7)(10^{-2})}{(2)(10^4)(4)(10^{-4})} =$

$\frac{(56)(10^{-6})}{(8)(10^0)} =$

$(7)(10^{-6}) = 0.000007$

CHAPTER 5 Review Problem Set

1. $(5x^2 - 6x + 4) + (3x^2 - 7x - 2) =$
$(5 + 3)x^2 + (-6 - 7)x + (4 - 2) =$
$8x^2 - 13x + 2$

2. $(7y^2 + 9y - 3) - (4y^2 - 2y + 6) =$
$7y^2 + 9y - 3 - 4y^2 + 2y - 6 =$
$3y^2 + 11y - 9$

3. $(2x^2 + 3x - 4) + (4x^2 - 3x - 6)$
$- (3x^2 - 2x - 1) =$
$2x^2 + 3x - 4 + 4x^2 - 3x - 6 - 3x^2$
$+ 2x + 1 =$
$3x^2 + 2x - 9$

4. $(-3x^2 - 2x + 4) - (x^2 - 5x - 6)$
$- (4x^2 + 3x - 8) =$
$-3x^2 - 2x + 4 - x^2 + 5x + 6$
$- 4x^2 - 3x + 8 =$
$-8x^2 + 18$

5. $5(2x - 1) + 7(x + 3) - 2(3x + 4) =$
$10x - 5 + 7x + 21 - 6x - 8 =$
$11x + 8$

6. $3(2x^2 - 4x - 5) - 5(3x^2 - 4x + 1) =$
$6x^2 - 12x - 15 - 15x^2 + 20x - 5 =$
$-9x^2 + 8x - 20$

7. $6(y^2 - 7y - 3) - 4(y^2 + 3y - 9) =$
$6y^2 - 42y - 18 - 4y^2 - 12y + 36 =$
$2y^2 - 54y + 18$

8. $3(a - 1) - 2(3a - 4) - 5(2a + 7) =$
$3a - 3 - 6a + 8 - 10a - 35 =$
$-13a - 30$

9. $-(a + 4) + 5(-a - 2) - 7(3a - 1) =$
$-a - 4 - 5a - 10 - 21a + 7 =$
$-27a - 7$

10. $-2(3n - 1) - 4(2n + 6) + 5(3n + 4) =$
$-6n + 2 - 8n - 24 + 15n + 20 =$
$n - 2$

11. $3(n^2 - 2n - 4) - 4(2n^2 - n - 3) =$
$3n^2 - 6n - 12 - 8n^2 + 4n + 12 =$
$-5n^2 - 2n$

12. $-5(-n^2 + n - 1) + 3(4n^2 - 3n - 7) =$
$5n^2 - 5n + 5 + 12n^2 - 9n - 21 =$
$17n^2 - 14n - 16$

13. $(5x^2)(7x^4) = 5 \cdot 7 \cdot x^2 \cdot x^4$
$= 35x^{2+4} = 35x^6$

14. $(-6x^3)(9x^5) = -6 \cdot 9 \cdot x^3 \cdot x^5 =$
$-54x^{3+5} = -54x^8$

15. $(-4xy^2)(-6x^2y^3) =$
$(-4)(-6)(x^{1+2})(y^{2+3}) =$
$24x^3y^5$

16. $(2a^3b^4)(-3ab^5) =$
$(2)(-3)(a^{3+1})(b^{4+5}) =$
$-6a^4b^9$

17. $(2a^2b^3)^3 = (2)^3(a^2)^3(b^3)^3 = 8a^6b^9$

18. $(-3xy^2)^2 = (-3)^2(x)^2(y^2)^2 =$
$9x^2y^4$

19. $5x(7x + 3) = 35x^2 + 15x$

20. $(-3x^2)(8x - 1) = -24x^3 + 3x^2$

21. $(x + 9)(x + 8) =$
$x^2 + (9 + 8)x + 72 =$
$x^2 + 17x + 72$

22. $(3x + 7)(x + 1) =$
$3x^2 + (3 + 7)x + 7 =$
$3x^2 + 10x + 7$

Chapter 5 Review Problem Set

23. $(x-5)(x+2) =$
$x^2 + (-5+2)x - 10 =$
$x^2 - 3x - 10$

24. $(y-4)(y-9) =$
$y^2 + (-9-4)y + 36 =$
$y^2 - 13y + 36$

25. $(2x-1)(7x+3) =$
$14x^2 + (6-7)x - 3 =$
$14x^2 - x - 3$

26. $(4a-7)(5a+8) =$
$20a^2 + (32-35)a - 56 =$
$20a^2 - 3a - 56$

27. $(3a-5)^2 =$
$(3a)^2 - 2(3a)(5) + (5)^2 =$
$9a^2 - 30a + 25$

28. $(x+6)(2x^2+5x-4) =$
$x(2x^2+5x-4) + 6(2x^2+5x-4) =$
$2x^3 + 5x^2 - 4x + 12x^2 + 30x - 24 =$
$2x^3 + 17x^2 + 26x - 24$

29. $(5n-1)(6n+5) =$
$30n^2 + (25-6)n - 5 =$
$30n^2 + 19n - 5$

30. $(3n+4)(4n-1) =$
$12n^2 + (-3+16)n - 4 =$
$12n^2 + 13n - 4$

31. $(2n+1)(2n-1) =$
$(2n)^2 - (1)^2 =$
$4n^2 - 1$

32. $(4n-5)(4n+5) =$
$(4n)^2 - (5)^2 =$
$16n^2 - 25$

33. $(2a+7)^2 =$
$(2a)^2 + 2(2a)(7) + (7)^2 =$
$4a^2 + 28a + 49$

34. $(3a+5)^2 =$
$(3a)^2 + 2(3a)(5) + (5)^2 =$
$9a^2 + 30a + 25$

35. $(x-2)(x^2-x+6) =$
$x(x^2-x+6) - 2(x^2-x+6) =$
$x^3 - x^2 + 6x - 2x^2 + 2x - 12 =$
$x^3 - 3x^2 + 8x - 12$

36. $(2x-1)(x^2+4x+7) =$
$2x(x^2+4x+7) - 1(x^2+4x+7) =$
$2x^3 + 8x^2 + 14x - x^2 - 4x - 7 =$
$2x^3 + 7x^2 + 10x - 7$

37. $(a+5)^3 = (a+5)(a+5)(a+5) =$
$(a+5)(a^2+10a+25) =$
$a(a^2+10a+25) + 5(a^2+10a+25) =$
$a^3 + 10a^2 + 25a + 5a^2 + 50a + 125 =$
$a^3 + 15a^2 + 75a + 125$

38. $(a-6)^3 = (a-6)(a-6)(a-6) =$
$(a-6)(a^2-12a+36) =$
$a(a^2-12a+36) - 6(a^2-12a+36) =$
$a^3 - 12a^2 + 36a - 6a^2 + 72a - 216 =$
$a^3 - 18a^2 + 108a - 216$

39. $(x^2-x-1)(x^2+2x+5) =$
$x^2(x^2+2x+5) - x(x^2+2x+5)$
$\quad - 1(x^2+2x+5) =$
$x^4 + 2x^3 + 5x^2 - x^3 - 2x^2 - 5x$
$\quad - x^2 - 2x - 5 =$
$x^4 + x^3 + 2x^2 - 7x - 5$

40. $(n^2+2n+4)(n^2-7n-1) =$
$n^2(n^2-7n-1) + 2n(n^2-7n-1)$
$\quad + 4(n^2-7n-1) =$
$n^4 - 7n^3 - n^2 + 2n^3 - 14n^2 - 2n$
$\quad + 4n^2 - 28n - 4 =$
$n^4 - 5n^3 - 11n^2 - 30n - 4$

41. $\dfrac{36x^4y^5}{-3xy^2} = -12x^{4-1}y^{5-2} = -12x^3y^3$

42. $\dfrac{-56a^5b^7}{-8a^2b^3} = 7a^{5-2}b^{7-3} = 7a^3b^4$

43. $\dfrac{-18x^4y^3 - 54x^6y^2}{6x^2y^2} =$

$\dfrac{-18x^4y^3}{6x^2y^2} - \dfrac{54x^6y^2}{6x^2y^2} =$

$-3x^2y - 9x^4$

44. $\dfrac{-30a^5b^{10} + 39a^4b^8}{-3ab} =$

$\dfrac{-30a^5b^{10}}{-3ab} + \dfrac{39a^4b^8}{-3ab} =$

$10a^4b^9 - 13a^3b^7$

45. $\dfrac{56x^4 - 40x^3 - 32x^2}{4x^2} =$

$\dfrac{56x^4}{4x^2} - \dfrac{40x^3}{4x^2} - \dfrac{32x^2}{4x^2} =$

$14x^2 - 10x - 8$

46.

$$\begin{array}{r|l} & x + 4 \\ x+5 & x^2 + 9x - 1 \\ & \underline{x^2 + 5x} \\ & 4x - 1 \\ & \underline{4x + 20} \\ & -21 \end{array}$$

47.

$$\begin{array}{r|l} & 7x - 6 \\ 3x+2 & 21x^2 - 4x - 12 \\ & \underline{21x^2 + 14x} \\ & -18x - 12 \\ & \underline{-18x - 12} \end{array}$$

48.

$$\begin{array}{r|l} & 2x^2 + x + 4 \\ x-2 & 2x^3 - 3x^2 + 2x - 4 \\ & \underline{2x^3 - 4x^2} \\ & x^2 + 2x - 4 \\ & \underline{x^2 - 2x} \\ & 4x - 4 \\ & \underline{4x - 8} \\ & 4 \end{array}$$

49. $3^2 + 2^2 = 9 + 4 = 13$

50. $(3+2)^2 = (5)^2 = 25$

51. $2^{-4} = \dfrac{1}{2^4} = \dfrac{1}{16}$

52. $(-5)^0 = 1$

53. $-5^0 = -1$

54. $\dfrac{1}{3^{-2}} = 3^2 = 9$

55. $\left(\dfrac{3}{4}\right)^{-2} = \left(\dfrac{4}{3}\right)^2 = \dfrac{16}{9}$

56. $\dfrac{1}{\left(\dfrac{1}{4}\right)^{-1}} = \left(\dfrac{1}{4}\right)^1 = \dfrac{1}{4}$

57. $\dfrac{1}{(-2)^{-3}} = (-2)^3 = -8$

58. $2^{-1} + 3^{-2} = \dfrac{1}{2} + \dfrac{1}{3^2} = \dfrac{1}{2} + \dfrac{1}{9} =$

$\dfrac{9}{18} + \dfrac{2}{18} = \dfrac{11}{18}$

59. $3^0 + 2^{-2} = 1 + \dfrac{1}{2^2} =$

$1 + \dfrac{1}{4} = \dfrac{4}{4} + \dfrac{1}{4} = \dfrac{5}{4}$

60. $(2+3)^{-2} = (5)^{-2} = \dfrac{1}{5^2} = \dfrac{1}{25}$

61. $x^5x^{-8} = x^{5-8} = x^{-3} = \dfrac{1}{x^3}$

62. $(3x^5)(4x^{-2}) = 12x^{5-2} = 12x^3$

63. $\dfrac{x^{-4}}{x^{-6}} = x^{-4-(-6)} = x^{-4+6} = x^2$

64. $\dfrac{x^{-6}}{x^{-4}} = x^{-6-(-4)} = x^{-6+4} = x^{-2} = \dfrac{1}{x^2}$

Chapter 5 Review Problem Set

65. $\frac{24a^5}{3a^{-1}} = 8a^{5-(-1)} = 8a^6$

66. $\frac{48n^{-2}}{12n^{-1}} = 4n^{-2-(-1)} =$

$4n^{-2+1} = 4n^{-1} = \frac{4}{n}$

67. $(x^{-2}y)^{-1} = (x^{-2})^{-1}(y)^{-1} = x^2y^{-1} = \frac{x^2}{y}$

68. $(a^2b^{-3})^{-2} = (a^2)^{-2}(b^{-3})^{-2}$

$= a^{-4}b^6 = \frac{b^6}{a^4}$

69. $(2x)^{-1} = \frac{1}{2x}$

70. $(3n^2)^{-2} = 3^{-2}(n^2)^{-2} = 3^{-2}n^{-4} =$

$\frac{1}{3^2n^4} = \frac{1}{9n^4}$

71. $(2n^{-1})^{-3} = (2^{-3})(n^{-1})^{-3} =$

$2^{-3}n^3 = \frac{n^3}{2^3} = \frac{n^3}{8}$

72. $(4ab^{-1})(-3a^{-1}b^2) = -12a^{1-1}b^{-1+2} =$

$-12a^0b^1 = -12b$

73. $(6.1)(10^2) = 610$

74. $(5.6)(10^4) = 56,000$

75. $(8)(10^{-2}) = 0.08$

76. $(9.2)(10^{-4}) = 0.00092$

77. $9000 = (9)(10^3)$

78. $47 = (4.7)(10^1)$

79. $0.047 = (4.7)(10^{-2})$

80. $0.00021 = (2.1)(10^{-4})$

81. $(0.00004)(12,000) =$

$(4)(10^{-5})(1.2)(10^4) =$

$(4)(1.2)(10^{-5})(10^4) =$

$(4.8)(10^{-1}) = 0.48$

82. $(0.0021)(2000) =$

$(2.1)(10^{-3})(2)(10^3) =$

$(2.1)(2)(10^{-3})(10^3) =$

$(4.2)(10^0) = 4.2$

83. $\frac{0.0056}{0.0000028} = \frac{(5.6)(10^{-3})}{(2.8)(10^{-6})} =$

$(2)(10^{-3-(-6)}) = (2)(10^3) = 2000$

84. $\frac{0.00078}{39,000} = \frac{(7.8)(10^{-4})}{(3.9)(10^4)} =$

$(2)(10^{-4-4}) = (2)(10^{-8}) = 0.00000002$

CHAPTER 5 Test

1. $(-7x^2 + 6x - 2) + (5x^2 - 8x + 7) =$

$-2x^2 - 2x + 5$

2. $(-4x^2 + 3x + 6) - (-x^2 + 9x - 14) =$

$-4x^2 + 3x + 6 + x^2 - 9x + 14 =$

$-3x^2 - 6x + 20$

3. $3(2x - 1) - 6(3x - 2) - (x + 7) =$

$6x - 3 - 18x + 12 - x - 7 =$

$-13x + 2$

4. $(-4xy^2)(7x^2y^3) =$

$(-4)(7)(x^{1+2})(y^{2+3}) =$

$-28x^3y^5$

5. $(2x^2y)^2(3xy^3) =$

$(4x^4y^2)(3xy^3) =$

$12x^5y^5$

6. $(x-9)(x+2) =$
$x^2 + (2-9)x - 18 =$
$x^2 - 7x - 18$

7. $(n+14)(n-7) =$
$n^2 + (14-7)n - 98 =$
$n^2 + 7n - 98$

8. $(5a+3)(8a+7) =$
$40a^2 + (35+24)a + 21 =$
$40a^2 + 59a + 21$

9. $(3x-7y)^2 =$
$9x^2 - 2(21xy) + 49y^2 =$
$9x^2 - 42xy + 49y^2$

10. $(x+3)(2x^2-4x-7) =$
$x(2x^2-4x-7) + 3(2x^2-4x-7) =$
$2x^3 - 4x^2 - 7x + 6x^2 - 12x - 21 =$
$2x^3 + 2x^2 - 19x - 21$

11. $(9x-5y)(9x+5y) = 81x^2 - 25y^2$

12. $(3x-7)(5x-11) =$
$15x^2 + (-33-35)x + 77 =$
$15x^2 - 68x + 77$

13. $\frac{-96x^4y^5}{-12x^2y} = 8x^{4-2}y^{5-1} = 8x^2y^4$

14. $\frac{56x^2y - 7xy^2}{-8xy} = \frac{56x^2y}{-8xy} - \frac{72xy^2}{-8xy} =$

$-7x + 9y$

15.

$$\begin{array}{r} x^2 + \ \ 4x - 5 \\ 2x - 3\overline{)2x^3 + 5x^2 - 22x + 15} \\ \underline{2x^3 - 3x^2 \qquad\qquad} \\ 8x^2 - 22x + 15 \\ \underline{8x^2 - 12x \qquad} \\ -10x + 15 \\ \underline{-10x + 15} \end{array}$$

16.

$$\begin{array}{r} 4x^2 - \ \ x + 6 \\ x + 6\overline{)4x^3 + 23x^2 + 0x + 36} \\ \underline{4x^3 + 24x^2 \qquad\qquad} \\ -1x^2 - 0x + 36 \\ \underline{-1x^2 - 6x \qquad} \\ 6x + 36 \\ \underline{6x + 36} \end{array}$$

17. $\left(\frac{2}{3}\right)^{-3} = \left(\frac{3}{2}\right)^{3} = \frac{27}{8}$

18. $4^{-2} + 4^{-1} + 4^0 = \frac{1}{4^2} + \frac{1}{4} + 1 =$

$\frac{1}{16} + \frac{1}{4} + 1 = \frac{1}{16} + \frac{4}{16} + \frac{16}{16} = \frac{21}{16}$

19. $\frac{1}{2^{-4}} = 2^4 = 16$

20. $(-6x^{-4})(4x^2) = -24x^{-2} = -\frac{24}{x^2}$

21. $\left(\frac{8x^{-1}}{2x^2}\right)^{-1} = (4x^{-1-2})^{-1} =$

$(4x^{-3})^{-1} = 4^{-1}x^3 = \frac{x^3}{4}$

22. $(x^{-3}y^5)^{-2} = (x^{-3})^{-2}(y^5)^{-2} =$

$x^6y^{-10} = \frac{x^6}{y^{10}}$

23. $0.00027 = (2.7)(10^{-4})$

24. $(9.2)(10^6) = 9,200,000$

25. $(0.000002)(3000) =$
$(2)(10^{-6})(3)(10^3) =$
$(6)(10^{-3}) = 0.006$

CHAPTERS 1-5 **Cumulative Review**

1. $5+3(2-7)^2 \div 3 \cdot 5$
$5+3(-5)^2 \div 3 \cdot 5$
$5+3(25) \div 3 \cdot 5$
$5+75 \div 3 \cdot 5$
$5+25 \cdot 5$
$5+125$
130

2. $8 \div 2 \cdot (-1)+3$
$4 \cdot (-1)+3$
$-4+3$
-1

3. $7-2^2 \cdot 5 \div (-1)$
$7-4 \cdot 5 \div (-1)$
$7-20 \div (-1)$
$7+20$
27

4. $4+(-2)-3(6)$
$4+(-2)-18$
$2-18$
-16

5. $(-3)^4=(-3)(-3)(-3)(-3)=81$

6. $-2^5=-(2 \cdot 2 \cdot 2 \cdot 2 \cdot 2)=-32$

7. $\left(\frac{2}{3}\right)^{-1}=\frac{2^{-1}}{3^{-1}}=\frac{3^1}{2^1}=\frac{3}{2}$

8. $\frac{1}{4^{-2}}=4^2=16$

9. $\left(\frac{1}{2}-\frac{1}{3}\right)^{-2}=\left(\frac{3}{6}-\frac{2}{6}\right)^{-2}=\left(\frac{1}{6}\right)^{-2}$
$\frac{1^{-2}}{6^{-2}}=\frac{6^2}{1^2}=\frac{36}{1}=36$

10. $2^0+2^{-1}+2^{-2}=1+\frac{1}{2}+\frac{1}{2^2}=$
$1+\frac{1}{2}+\frac{1}{4}=\frac{4}{4}+\frac{2}{4}+\frac{1}{4}=\frac{7}{4}=1\frac{3}{4}$

11. $x=\frac{1}{2},\ y=-\frac{1}{3}:\ \frac{2x+3y}{x-y}=$
$\frac{2\left(\frac{1}{2}\right)+3\left(-\frac{1}{3}\right)}{\frac{1}{2}-\left(-\frac{1}{3}\right)}=\frac{1-1}{\frac{3}{6}+\frac{2}{6}}=$
$\frac{0}{\frac{5}{6}}=0$

12. $n=-\frac{3}{4}:\ \frac{2}{5}n-\frac{1}{3}n-n+\frac{1}{2}n=$
$\frac{12}{30}n-\frac{10}{30}n-\frac{30}{30}n+\frac{15}{30}n=$
$\frac{-13}{30}n=-\frac{13}{30}\left(-\frac{3}{4}\right)=\frac{13}{40}$

13. $a=-1,\ b=-\frac{1}{3}:\ \frac{3a-2b-4a+7b}{-a-3a+b-2b}=$
$\frac{-a+5b}{-4a-b}=\frac{-(-1)+5\left(-\frac{1}{3}\right)}{-4(-1)-\left(-\frac{1}{3}\right)}=$
$\frac{1-\frac{5}{3}}{4+\frac{1}{3}}=\frac{\frac{3}{3}-\frac{5}{3}}{\frac{12}{3}+\frac{1}{3}}=\frac{-\frac{2}{3}}{\frac{13}{3}}=-\frac{2}{13}$

14. $x=-2:\ -2(x-4)+3(2x-1)-(3x-2)=$
$-2x+8+6x-3-3x+2=$
$x+7=(-2)+7=5$

15. $x=-1:\ (x^2+2x-4)-(x^2-x-2)$
$+(2x^2-3x-1)=$
$x^2+2x-4-x^2+x+2+2x^2-3x-1=$
$2x^2-3=$
$2(-1)^2-3=2(1)-3=2-3=-1$

16. $n = 3:\ 2(n^2 - 3n - 1) - (n^2 + n + 4)$
$- 3(2n - 1) =$
$2n^2 - 6n - 2 - n^2 - n - 4 - 6n + 3 =$
$n^2 - 13n - 3 =$
$(3)^2 - 13(3) - 3 = 9 - 39 - 3 = -33$

17. $(3x^2y^3)(-5xy^4) =$
$(3)(-5)(x^{2+1})(y^{3+4}) =$
$-15x^3y^7$

18. $(-6ab^4)(-2b^3) =$
$(-6)(-2)(a^1)(b^{4+3}) =$
$12ab^7$

19. $(-2x^2y^5)^3 =$
$-(2)^3(x^2)^3(y^5)^3 =$
$-8x^6y^{15}$

20. $-3xy(2x - 5y) = -6x^2y + 15xy^2$

21. $(5x - 2)(3x - 1) =$
$15x^2 + (-5 - 6)x + 2 =$
$15x^2 - 11x + 2$

22. $(7x - 1)(3x + 4) =$
$21x^2 + (28 - 3)x - 4 =$
$21x^2 + 25x - 4$

23. $(-x - 2)(2x + 3) =$
$-2x^2 + (-3 - 4)x - 6 =$
$-2x^2 - 7x - 6$

24. $(7 - 2y)(7 + 2y) =$
$49 + (14 - 14)y - 4y^2 =$
$49 - 4y^2$

25. $(x - 2)(3x^2 - x - 4) =$
$x(3x^2 - x - 4) - 2(3x^2 - x - 4) =$
$3x^3 - x^2 - 4x - 6x^2 + 2x + 8 =$
$3x^3 - 7x^2 - 2x + 8$

26. $(2x - 5)(x^2 + x - 4) =$
$2x(x^2 + x - 4) - 5(x^2 + x - 4) =$
$2x^3 + 2x^2 - 8x - 5x^2 - 5x + 20 =$
$2x^3 - 3x^2 - 13x + 20$

27. $(2n + 3)^3 = (2n + 3)(2n + 3)(2n + 3) =$
$(2n + 3)(4n^2 + 12n + 9) =$
$2n(4n^2 + 12n + 9) + 3(4n^2 + 12n + 9) =$
$8n^3 + 24n^2 + 18n + 12n^2 + 36n + 27 =$
$8n^3 + 36n^2 + 54n + 27$

28. $(1 - 2n)^3 = (1 - 2n)(1 - 2n)(1 - 2n) =$
$(1 - 2n)(1 - 4n + 4n^2) =$
$1(1 - 4n + 4n^2) - 2n(1 - 4n + 4n^2) =$
$1 - 4n + 4n^2 - 2n + 8n^2 - 8n^3 =$
$1 - 6n + 12n^2 - 8n^3$

29. $(x^2 - 2x + 6)(2x^2 + 5x - 6) =$
$x^2(2x^2 + 5x - 6) - 2x(2x^2 + 5x - 6)$
$+ 6(2x^2 + 5x - 6) =$
$2x^4 + 5x^3 - 6x^2 - 4x^3 - 10x^2 + 12x$
$+ 12x^2 + 30x - 36 =$
$2x^4 + x^3 - 4x^2 + 42x - 36$

30. $\dfrac{-52x^3y^4}{13xy^2} = -4x^{3-1}y^{4-2} = -4x^2y^2$

31. $\dfrac{-126a^3b^5}{-9a^2b^3} = 14a^{3-2}b^{5-3}$

$= 14a^1b^2 = 14ab^2$

32. $\dfrac{56xy^2 - 64x^3y - 72x^4y^4}{8xy} =$

$\dfrac{56xy^2}{8xy} - \dfrac{64x^3y}{8xy} - \dfrac{72x^4y^4}{8xy} =$

$7y - 8x^2 - 9x^3y^3$

33.

$$\begin{array}{r l}
 & 2x^2 - 4x - 7 \\
x + 3 \overline{)} & 2x^3 + 2x^2 - 19x - 21 \\
 & \underline{2x^3 + 6x^2} \\
 & -4x^2 - 19x - 21 \\
 & \underline{-4x^2 - 12x} \\
 & -7x - 21 \\
 & \underline{-7x - 21}
\end{array}$$

34.

$$\begin{array}{r} x^2 + 6x + 4 \\ 3x - 1 \overline{)3x^3 + 17x^2 + 6x - 4} \\ \underline{3x^3 - x^2 \qquad\qquad} \\ 18x^2 + 6x - 4 \\ \underline{18x^2 - 6x \quad\;} \\ 12x - 4 \\ \underline{12x - 4} \end{array}$$

35. $(-2x^3)(3x^{-4}) = -6x^{3-4} =$

$-6x^{-1} = -\dfrac{6}{x}$

36. $\dfrac{4x^{-2}}{2x^{-1}} = 2x^{-2-(-1)} = 2x^{-2+1} =$

$2x^{-1} = \dfrac{2}{x}$

37. $(3x^{-1}y^{-2})^{-1} = 3^{-1}(x^{-1})^{-1}(y^{-2})^{-1} =$

$3^{-1}x^1y^2 = \dfrac{xy^2}{3}$

38. $(xy^2z^{-1})^{-2} = x^{-2}(y^2)^{-2}(z^{-1})^{-2} =$

$x^{-2}y^{-4}z^2 = \dfrac{z^2}{x^2y^4}$

39. $(0.00003)(4000) =$
$(3)(10^{-5})(4)(10^3) =$
$(12)(10^{-5+3}) =$
$(12)(10^{-2}) = 0.12$

40. $(0.0002)(0.003)^2 =$
$(2)(10^{-4})(3)^2(10^{-3})^2 =$
$(2)(9)(10^{-4})(10^{-6}) =$
$(18)(10^{-4-6}) =$
$(18)(10^{-10}) = 0.0000000018$

41. $\dfrac{0.00034}{0.0000017} = \dfrac{(34)(10^{-4})}{(17)(10^{-6})} =$

$(2)\left(10^{-4-(-6)}\right) = (2)(10^2) = 200$

42.

$$\begin{aligned} 5x + 8 &= 6x - 3 \\ 5x - 6x + 8 &= -3 \\ -x + 8 &= -3 \\ -x &= -3 - 8 \\ -x &= -11 \\ x &= 11 \end{aligned}$$

The solution set is $\{11\}$.

43.

$$\begin{aligned} -2(4x - 1) &= -5x + 3 - 2x \\ -8x + 2 &= -5x + 3 - 2x \\ -8x + 2 &= -7x + 3 \\ -8x + 7x + 2 &= 3 \\ -x + 2 &= 3 \\ -x &= 3 - 2 \\ -x &= 1 \\ x &= -1 \end{aligned}$$

The solution set is $\{-1\}$.

44.

$$\begin{aligned} \frac{y}{2} - \frac{y}{3} &= 8 \\ 6\left(\frac{y}{2} - \frac{y}{3}\right) &= 6(8) \\ 6\left(\frac{y}{2}\right) - 6\left(\frac{y}{3}\right) &= 48 \\ 3y - 2y &= 48 \\ y &= 48 \end{aligned}$$

The solution set is $\{48\}$.

45.

$$\begin{aligned} 6x + 8 - 4x &= 10(3x + 2) \\ 6x + 8 - 4x &= 30x + 20 \\ 2x + 8 &= 30x + 20 \\ 2x - 30x + 8 &= 20 \\ -28x + 8 &= 20 \\ -28x &= 20 - 8 \\ -28x &= 12 \\ x &= \frac{12}{-28} = -\frac{3}{7} \end{aligned}$$

The solution set is $\left\{-\dfrac{3}{7}\right\}$.

46.
$$\begin{aligned} 1.6 - 2.4x &= 5x - 65 \\ 1.6 - 2.4x - 5x &= -65 \\ 1.6 - 7.4x &= -65 \\ -7.4x &= -65 - 1.6 \\ -7.4x &= -66.6 \\ x &= \frac{-66.6}{-7.4} = 9 \end{aligned}$$
The solution set is $\{9\}$.

47.
$$\begin{aligned} -3(x-1) + 2(x+3) &= -4 \\ -3x + 3 + 2x + 6 &= -4 \\ -x + 9 &= -4 \\ -x &= -4 - 9 \\ -x &= -13 \\ x &= 13 \end{aligned}$$
The solution set is $\{13\}$.

48.
$$\begin{aligned} \frac{3n+1}{5} + \frac{n-2}{3} &= \frac{2}{15} \\ 15\left(\frac{3n+1}{5} + \frac{n-2}{3}\right) &= 15\left(\frac{2}{15}\right) \\ 15\left(\frac{3n+1}{5}\right) + 15\left(\frac{n-2}{3}\right) &= 2 \\ 3(3n+1) + 5(n-2) &= 2 \\ 9n + 3 + 5n - 10 &= 2 \\ 14n - 7 &= 2 \\ 14n &= 9 \\ n &= \frac{9}{14} \end{aligned}$$
The solution set is $\left\{\frac{9}{14}\right\}$.

49.
$$\begin{aligned} 0.06x + 0.08(1500 - x) &= 110 \\ 0.06x + 120 - 0.08x &= 110 \\ -0.02x + 120 &= 110 \\ -0.02x &= 110 - 120 \\ -0.02x &= -10 \\ x &= \frac{-10}{-0.02} = 500 \end{aligned}$$
The solution set is $\{500\}$.

50.
$$\begin{aligned} 2x - 7 &\le -3(x+4) \\ 2x - 7 &\le -3x - 12 \\ 2x + 3x - 7 &\le -12 \\ 5x - 7 &\le -12 \\ 5x &\le -12 + 7 \\ 5x &\le -5 \\ x &\le -1 \end{aligned}$$
The solution set is
$\{x|x \le -1\}$ or $(-\infty, -1]$.

51.
$$\begin{aligned} 6x + 5 - 3x &> 5 \\ 3x + 5 &> 5 \\ 3x &> 5 - 5 \\ 3x &> 0 \\ x &> \frac{0}{3} \\ x &> 0 \end{aligned}$$
The solution set is
$\{x|x > 0\}$ or $(0, \infty)$.

52.
$$\begin{aligned} 4(x-5) + 2(3x+6) &< 0 \\ 4x - 20 + 6x + 12 &< 0 \\ 10x - 8 &< 0 \\ 10x &< 8 \\ x &< \frac{8}{10} \\ x &< \frac{4}{5} \end{aligned}$$
The solution set is
$\left\{x|x < \frac{4}{5}\right\}$ or $\left(-\infty, \frac{4}{5}\right)$.

53.
$$\begin{aligned} -5x + 3 &> -4x + 5 \\ -5x + 4x + 3 &> 5 \\ -x + 3 &> 5 \\ -x &> 2 \\ \frac{-x}{-1} &< \frac{2}{-1} \\ x &< -2 \end{aligned}$$
The solution set is
$\{x|x < -2\}$ or $(-\infty, -2)$.

54.
$$\frac{3x}{4}-\frac{x}{2}\le\frac{5x}{6}-1$$
$$12\left(\frac{3x}{4}-\frac{x}{2}\right)\le 12\left(\frac{5x}{6}-1\right)$$
$$12\left(\frac{3x}{4}\right)-12\left(\frac{x}{2}\right)\le 12\left(\frac{5x}{6}\right)-12(1)$$
$$3(3x)-6(x)\le 2(5x)-12$$
$$9x-6x\le 10x-12$$
$$3x\le 10x-12$$
$$3x-10x\le -12$$
$$-7x\le -12$$
$$\frac{-7x}{-7}\ge\frac{-12}{-7}$$
$$x\ge\frac{12}{7}$$

The solution set is

$\left\{x|x\ge\frac{12}{7}\right\}$ or $\left[\frac{12}{7},\infty\right)$.

55. $0.08(700-x)+0.11x\ge 65$
$$56-0.08x+0.11x\ge 65$$
$$56+0.03x\ge 65$$
$$0.03\ge 9$$
$$x\ge\frac{9}{0.03}$$
$$x\ge 300$$

The solution set is

$\{x|x\ge 300\}$ or $[300,\infty)$.

56. $y=x^2-1$

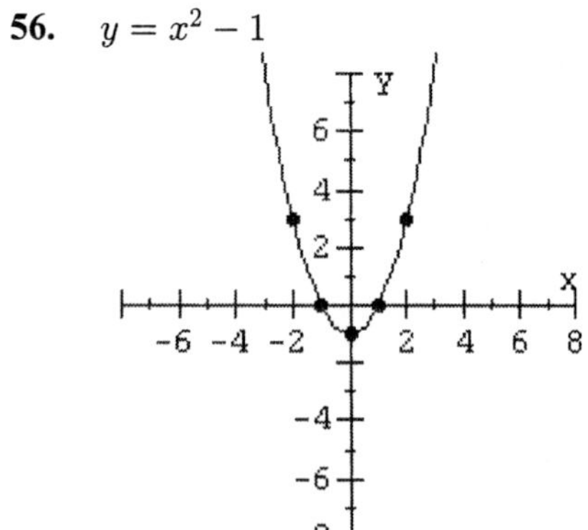

$x=0$	$x=1$	$x=2$
$y=0^2-1$	$y=1^2-1$	$y=2^2-1$
$y=-1$	$y=1-1$	$y=4-1$
$(0,-1)$	$y=0$	$y=3$
	$(1,0)$	$(2,3)$

$x=-1$	$x=-2$
$y=(-1)^2-1$	$y=(-2)^2-1$
$y=1-1$	$y=4-1$
$y=0$	$y=3$
$(-1,0)$	$(-2,3)$

57. $y=2x+3$

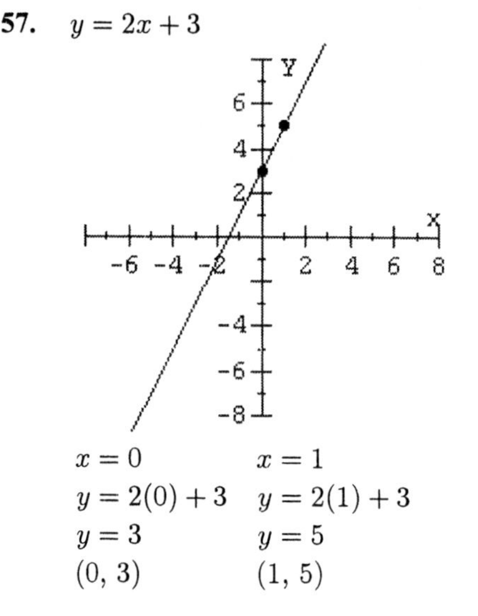

$x=0$	$x=1$
$y=2(0)+3$	$y=2(1)+3$
$y=3$	$y=5$
$(0,3)$	$(1,5)$

58. $y=-5x$

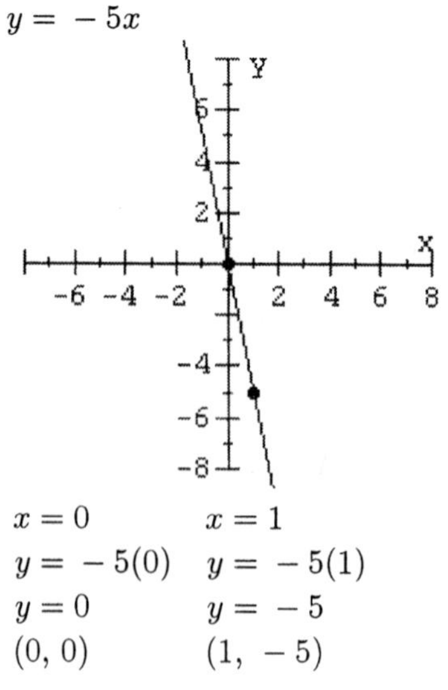

$x=0$	$x=1$
$y=-5(0)$	$y=-5(1)$
$y=0$	$y=-5$
$(0,0)$	$(1,-5)$

59. $x - 2y = 6$

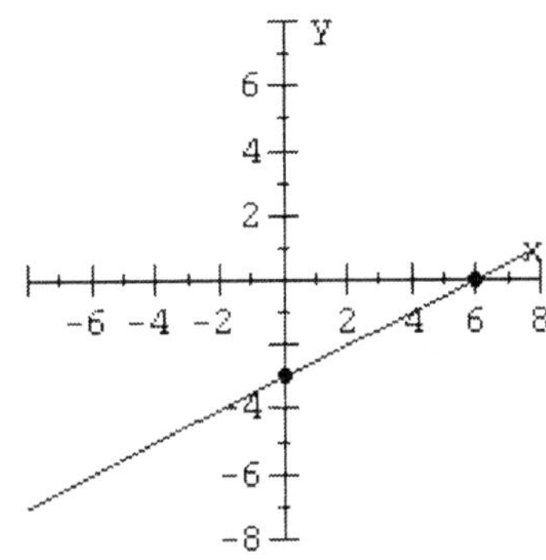

$x = 0$	$y = 0$
$0 - 2y = 6$	$x - 2(0) = 6$
$-2y = 6$	$x - 0 = 6$
$y = -3$	$x = 6$
$(0, -3)$	$(6, 0)$

60. $y = -\frac{1}{2}x + 2$

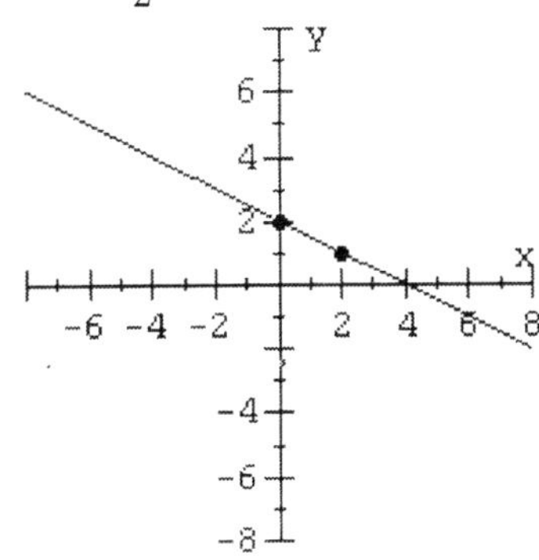

$x = 0$	$x = 2$
$y = -\frac{1}{2}(0) + 2$	$y = -\frac{1}{2}(2) + 2$
$y = 2$	$y = -1 + 2$
$(0, 2)$	$y = 1$
	$(2, 1)$

61. $f(x) = \frac{4x}{x - 8}$

$f(-2) = \frac{4(-2)}{-2-8} = \frac{-8}{-10} = \frac{4}{5}$

62. $f(x) = x^2 - 4x + 3$
$f(7) = (7)^2 - 4(7) + 3$
$f(7) = 49 - 28 + 3$
$f(7) = 24$

63. The denominator can not equal zero.
So set $2x - 5 = 0$
$2x = 5$
$x = \frac{5}{2}$
Therefore $\frac{5}{2}$ is excluded from the domain.
The domain is all real numbers except $\frac{5}{2}$.

64. All real numbers

65. Domain $\{-3, -2, 2, 3\}$
Range $\{4, 6\}$
It is a function.

66. Domain $\{0, 1, 2, 3\}$
Range $\{4, 6\}$
It is a function.

67. $f(x) = 0.20 + 0.03x$
$f(75) = 0.20 + 0.03(75)$
$f(75) = 0.20 + 2.25$
$f(75) = 2.40$
The change is \$2.40.

68. $g(x) = 20 + 0.5x$
$g(90) = 20 + 0.5(90)$
$g(90) = 20 + 45$
$g(90) = 65$
The fee is \$65.

69.

Water	21%
Beer	6%
Difference	15%

Water was favored over beer by 15 percentage points.

70.

Water	21%
Tea	15%
Coffee	12%
Total	48%

48% chose water, tea, or coffee.

71. Let n represent the number.

$$\begin{aligned} 4+3n &= n+10 \\ 4+2n &= 10 \\ 2n &= 6 \\ n &= 3 \end{aligned}$$

The number is 3.

72. Let n represent the number.

$$\begin{aligned} (15\%)(n) &= 6 \\ 0.15n &= 6 \\ 15n &= 600 \\ n &= 40 \end{aligned}$$

Therefore, 15% of $40 = 6$.

73. Let d represent the number of dimes, then $18-d$ represents the number of quarters. The value of the dimes in cents is 10d and the value of the quarters in cents is $25(18-d)$.

$$\begin{aligned} 10d+25(18-d) &= 330 \\ 10d+450-25d &= 330 \\ -15d+450 &= 330 \\ -15d &= 120 \\ d &= 8 \end{aligned}$$

He would have 8 dimes and 10 quarters.

74. Let x represent the amount of money invested at 8% interest; then $1500-x$ represents the amount of money invested at 9%.

$$\begin{pmatrix}\text{Interest}\\ \text{earned}\\ \text{at}\\ 8\%\end{pmatrix}+\begin{pmatrix}\text{Interest}\\ \text{earned}\\ \text{at}\\ 9\%\end{pmatrix}=\begin{pmatrix}\text{Total}\\ \text{interest}\\ \text{earned}\end{pmatrix}$$

$$\begin{aligned} (8\%)(x)+(9\%)(1500-x) &= 128 \\ 0.08x+0.09(1500-x) &= 128 \\ 8x+9(1500-x) &= 128 \\ 8x+13,500-9x &= 12,800 \\ 13,500-x &= 12,800 \\ -x &= -700 \\ x &= 700 \end{aligned}$$

The amounts invested are \$700 at 8% and \$800 at 9%.

75. Let x represent the amount of water to be added, then $15+x$ represents the amount of final salt solution.

$$\begin{pmatrix}\text{Pure}\\ \text{salt in}\\ \text{water}\\ \text{added}\end{pmatrix}+\begin{pmatrix}\text{Pure}\\ \text{salt}\\ \text{in 12\%}\\ \text{solution}\end{pmatrix}=\begin{pmatrix}\text{Pure}\\ \text{salt}\\ \text{in final}\\ \text{solution}\end{pmatrix}$$

$$\begin{aligned} 0+(12\%)(15) &= (10\%)(15+x) \\ 0+0.12(15) &= 0.10(15+x) \\ 12(15) &= 10(15+x) \\ 180 &= 150+10x \\ 30 &= 10x \\ 3 &= x \end{aligned}$$

The amount of water to be added would be 3 gallons.

76. Let t represent the time for both airplanes as they left Atlanta at the same time. A chart of the information would be as follows:

	Rate	Time	Distance ($d=rt$)
Plane A	400	t	$400t$
Plane B	450	t	$450t$

A diagram of the problem would be as follows:

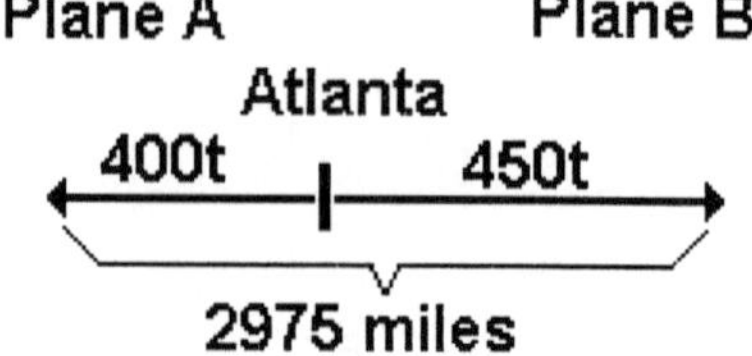

The sum of the two distances would be 2975 miles.

$$\begin{aligned} 400t + 450t &= 2975 \\ 850t &= 2975 \\ t &= \frac{2975}{850} = 3\frac{1}{2} \text{ hours} \end{aligned}$$

It would take $3\frac{1}{2}$ hours for the airplanes to be 2975 miles apart.

77. Let w represent the width of the rectangle, then $2w + 1$ represents the length of the rectangle.

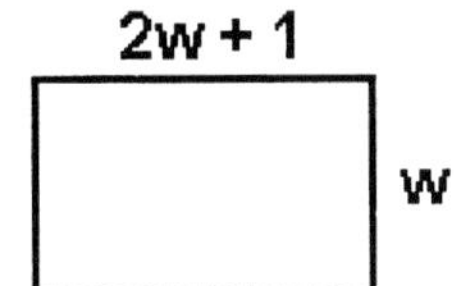

$$\begin{aligned} 2(w) + 2(2w + 1) &= 44 \\ 2w + 4w + 2 &= 44 \\ 6w + 2 &= 44 \\ 6w &= 42 \\ w &= 7 \end{aligned}$$

The width of the rectangle is 7 meters and the length is 15 meters.

Chapter 6 Factoring and Solving Equations

PROBLEM SET 6.1 Factoring by Using the Distributive Property

1. $\left.\begin{aligned} 24y &= 2 \cdot 2 \cdot 2 \cdot 3 \cdot y \\ 30xy &= 2 \cdot 3 \cdot 5 \cdot x \cdot y \end{aligned}\right)$
The greatest common factor is
$2 \cdot 3 \cdot y = 6y$

3. $\left.\begin{aligned} 60x^2y &= 2 \cdot 2 \cdot 3 \cdot 5 \cdot x \cdot x \cdot y \\ 84xy^2 &= 2 \cdot 2 \cdot 3 \cdot 7 \cdot x \cdot y \cdot y \end{aligned}\right)$
The greatest common factor is
$2 \cdot 2 \cdot 3 \cdot x \cdot y = 12xy$

5. $\left.\begin{aligned} 42ab^3 &= 2 \cdot 3 \cdot 7 \cdot a \cdot b \cdot b \cdot b \\ 70a^2b^2 &= 2 \cdot 5 \cdot 7 \cdot a \cdot a \cdot b \cdot b \end{aligned}\right)$
The greatest common factor is
$2 \cdot 7 \cdot a \cdot b \cdot b = 14ab^2$

7. $\left.\begin{aligned} 6x^3 &= 2 \cdot 3 \cdot x \cdot x \cdot x \\ 8x &= 2 \cdot 2 \cdot 2 \cdot x \\ 24x^2 &= 2 \cdot 2 \cdot 2 \cdot 3 \cdot x \cdot x \end{aligned}\right)$
The greatest common factor is
$2 \cdot x = 2x$

9. $\left.\begin{aligned} 16a^2b^2 &= 2 \cdot 2 \cdot 2 \cdot 2 \cdot a \cdot a \cdot b \cdot b \\ 40a^2b^3 &= 2 \cdot 2 \cdot 2 \cdot 5 \cdot a \cdot a \cdot b \cdot b \cdot b \\ 56a^3b^4 &= 2 \cdot 2 \cdot 2 \cdot 7 \cdot a \cdot a \cdot a \cdot b \cdot b \cdot b \cdot b \end{aligned}\right)$
The greatest common factor is
$2 \cdot 2 \cdot 2 \cdot a \cdot a \cdot b \cdot b = 8a^2b^2$

11. $8x + 12y =$
$4(2x) + 4(3y) =$
$4(2x + 3y)$

13. $14xy - 21y =$
$7y(2x) - 7y(3) =$
$7y(2x - 3)$

15. $18x^2 + 45x =$
$9x(2x) + 9x(5) =$
$9x(2x + 5)$

17. $12xy^2 - 30x^2y =$
$6xy(2y) - 6xy(5x) =$
$6xy(2y - 5x)$

19. $36a^2b - 60a^3b^4 =$
$12a^2b(3) - 12a^2b(5ab^3) =$
$12a^2b(3 - 5ab^3)$

21. $16xy^3 + 25x^2y^2 =$
$xy^2(16y) + xy^2(25x) =$
$xy^2(16y + 25x)$

23. $64ab - 72cd =$
$8(8ab) - 8(9cd) =$
$8(8ab - 9cd)$

25. $9a^2b^4 - 27a^2b =$
$9a^2b(b^3) - 9a^2b(3) =$
$9a^2b(b^3 - 3)$

27. $52x^4y^2 + 60x^6y =$
$4x^4y(13y) + 4x^4y(15x^2) =$
$4x^4y(13y + 15x^2)$

29. $40x^2y^2 + 8x^2y =$
$8x^2y(5y) + 8x^2y(1) =$
$8x^2y(5y + 1)$

31. $12x + 15xy + 21x^2 =$
$3x(4) + 3x(5y) + 3x(7x) =$
$3x(4 + 5y + 7x)$

33. $2x^3 - 3x^2 + 4x =$
$x(2x^2) - x(3x) + x(4) =$
$x(2x^2 - 3x + 4)$

35. $44y^5 - 24y^3 - 20y^2 =$
$4y^2(11y^3) - 4y^2(6y) - 4y^2(5) =$
$4y^2(11y^3 - 6y - 5)$

37. $14a^2b^3 + 35ab^2 - 49a^3b =$
$7ab(2ab^2) + 7ab(5b) - 7ab(7a^2) =$
$7ab(2ab^2 + 5b - 7a^2)$

39. $x(y+1) + z(y+1) =$
$(y+1)(x+z)$

41. $a(b-4) - c(b-4) =$
$(b-4)(a-c)$

43. $x(x+3) + 6(x+3) =$
$(x+3)(x+6)$

45. $2x(x+1) - 3(x+1) =$
$(x+1)(2x-3)$

47. $5x + 5y + bx + by =$
$5(x+y) + b(x+y) =$
$(x+y)(5+b)$

49. $bx - by - cx + cy =$
$b(x-y) - c(x-y) =$
$(x-y)(b-c)$

51. $ac + bc + a + b =$
$c(a+b) + 1(a+b) =$
$(a+b)(c+1)$

53. $x^2 + 5x + 12x + 60 =$
$x(x+5) + 12(x+5) =$
$(x+5)(x+12)$

55. $x^2 - 2x - 8x + 16 =$
$x(x-2) - 8(x-2) =$
$(x-2)(x-8)$

57. $2x^2 + x - 10x - 5 =$
$x(2x+1) - 5(2x+1) =$
$(2x+1)(x-5)$

59. $6n^2 - 3n - 8n + 4 =$
$3n(2n-1) - 4(2n-1) =$
$(2n-1)(3n-4)$

61. $x^2 - 8x = 0$
$x(x-8) = 0$
$x = 0$ or $x - 8 = 0$
$x = 0$ or $x = 8$
The solution set is $\{0, 8\}$.

63. $x^2 + x = 0$
$x(x+1) = 0$
$x = 0$ or $x + 1 = 0$
$x = 0$ or $x = -1$
The solution set is $\{-1, 0\}$.

65. $n^2 = 5n$
$n^2 - 5n = 0$
$n(n-5) = 0$
$n = 0$ or $n - 5 = 0$
$n = 0$ or $n = 5$
The solution set is $\{0, 5\}$.

67. $2y^2 - 3y = 0$
$y(2y-3) = 0$
$y = 0$ or $2y - 3 = 0$
$y = 0$ or $2y = 3$
$y = 0$ or $y = \frac{3}{2}$
The solution set is $\left\{0, \frac{3}{2}\right\}$.

69. $7x^2 = -3x$
$7x^2 + 3x = 0$
$x(7x+3) = 0$
$x = 0$ or $7x + 3 = 0$
$x = 0$ or $7x = -3$
$x = 0$ or $x = -\frac{3}{7}$
The solution set is $\left\{-\frac{3}{7}, 0\right\}$.

71. $3n^2 + 15n = 0$
$3n(n+5) = 0$
$3n = 0$ or $n + 5 = 0$
$n = 0$ or $n = -5$
The solution set is $\{-5, 0\}$.

73. $4x^2 = 6x$

$2x^2 = 3x$

$2x^2 - 3x = 0$

$x(2x - 3) = 0$

$x = 0$ or $2x - 3 = 0$

$x = 0$ or $2x = 3$

$x = 0$ or $x = \frac{3}{2}$

The solution set is $\left\{0, \frac{3}{2}\right\}$.

75. $7x - x^2 = 0$

$x(7 - x) = 0$

$x = 0$ or $7 - x = 0$

$x = 0$ or $7 = x$

The solution set is $\{0, 7\}$.

77. $13x = x^2$

$13x - x^2 = 0$

$x(13 - x) = 0$

$x = 0$ or $13 - x = 0$

$x = 0$ or $13 = x$

The solution set is $\{0, 13\}$.

79. $5x = -2x^2$

$2x^2 + 5x = 0$

$x(2x + 5) = 0$

$x = 0$ or $2x + 5 = 0$

$x = 0$ or $2x = -5$

$x = 0$ or $x = -\frac{5}{2}$

The solution set is $\left\{-\frac{5}{2}, 0\right\}$.

81. $x(x + 5) - 4(x + 5) = 0$

$(x + 5)(x - 4) = 0$

$x + 5 = 0$ or $x - 4 = 0$

$x = -5$ or $x = 4$

The solution set is $\{-5, 4\}$.

83. $4(x - 6) - x(x - 6) = 0$

$(x - 6)(4 - x) = 0$

$x - 6 = 0$ or $4 - x = 0$

$x = 6$ or $4 = x$

The solution set is $\{4, 6\}$.

85. Let n represent the number, then n^2 represents the square of the number.

$n^2 = 9n$

$n^2 - 9n = 0$

$n(n - 9) = 0$

$n = 0$ or $n - 9 = 0$

$n = 0$ or $n = 9$

The number is 0 or 9.

87. Let s represent the length of a side of the square.

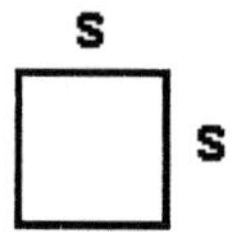

Area = (5)(Perimeter)

$s^2 = 5(4s)$

$s^2 = 20s$

$s^2 - 20s = 0$

$s(s - 20) = 0$

$s = 0$ or $s - 20 = 0$

$s = 0$ or $s = 20$

Since 0 is not a reasonable answer to the problem, the length of the side of the square must be 20 units.

89. Let s represent the length of a side of the square, then s also represents the length of a radius of the circle.

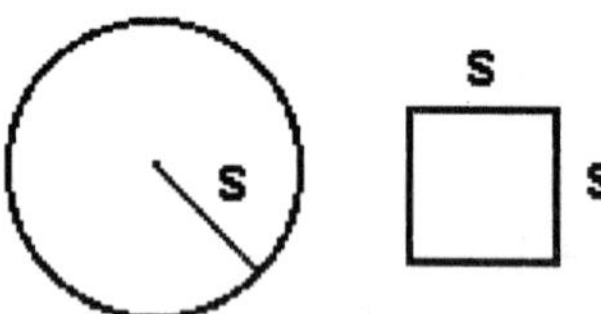

Circle Area = Square Perimeter

$\pi s^2 = 4s$

$\pi s^2 - 4s = 0$

$s(\pi s - 4) = 0$

$s = 0$ or $\pi s - 4 = 0$
$s = 0$ or $\pi s = 4$
$s = 0$ or $s = \frac{4}{\pi}$
The answer of 0 must be discarded, so the length of a side of the square is $\frac{4}{\pi}$ units.

91. Let w represent the width of the rectangle, then w also represents the length of a side of the square.

6 w
w w

Rectangle Area = (2)(Square Area)
$6w = 2(w^2)$
$6w - 2w^2 = 0$
$2w(3 - w) = 0$
$2w = 0$ or $3 - w = 0$
$w = 0$ or $3 = w$
The answer of 0 must be discarded, so the dimensions of the rectangle are 3 inches by 6 inches and the dimensions of the square are 3 inches by 3 inches.

PROBLEM SET 6.2 Factoring the Difference of Two Squares

1. $x^2 - 1 =$
$(x)^2 - (1)^2 =$
$(x - 1)(x + 1)$

3. $x^2 - 100 =$
$(x)^2 - (10)^2 =$
$(x - 10)(x + 10)$

5. $x^2 - 4y^2 =$
$(x)^2 - (2y)^2 =$
$(x - 2y)(x + 2y)$

7. $9x^2 - y^2 =$
$(3x)^2 - (y)^2 =$
$(3x - y)(3x + y)$

9. $36a^2 - 25b^2 =$
$(6a)^2 - (5b)^2 =$
$(6a - 5b)(6a + 5b)$

11. $1 - 4n^2 =$
$(1)^2 - (2n)^2 =$
$(1 - 2n)(1 + 2n)$

13. $5x^2 - 20 =$
$5(x^2 - 4) =$
$5(x - 2)(x + 2)$

15. $8x^2 + 32 = 8(x^2 + 4)$

17. $2x^2 - 18y^2 =$
$2(x^2 - 9y^2) =$
$2(x - 3y)(x + 3y)$

19. $x^3 - 25x =$
$x(x^2 - 25) =$
$x(x - 5)(x + 5)$

21. $x^2 + 9y^2$ is not factorable.

23. $45x^2 - 36xy = 9x(5x - 4y)$

25. $36 - 4x^2 =$
$4(9 - x^2) =$
$4(3 - x)(3 + x)$

27. $4a^4 + 16a^2 = 4a^2(a^2 + 4)$

29. $x^4 - 81 =$
$(x^2 + 9)(x^2 - 9) =$
$(x^2 + 9)(x + 3)(x - 3)$

31. $x^4 + x^2 = x^2(x^2 + 1)$

33. $3x^3 + 48x = 3x(x^2 + 16)$

35. $5x - 20x^3 =$
$5x(1 - 4x^2) =$
$5x(1 - 2x)(1 + 2x)$

37. $4x^2 - 64 =$
$4(x^2 - 16) =$
$4(x - 4)(x + 4)$

39. $75x^3y - 12xy^3 =$
$3xy(25x^2 - 4y^2) =$
$3xy(5x + 2y)(5x - 2y)$

41.
$$x^2 = 9$$
$$x^2 - 9 = 0$$
$$(x - 3)(x + 3) = 0$$
$$x - 3 = 0 \quad \text{or} \quad x + 3 = 0$$
$$x = 3 \quad \text{or} \quad x = -3$$
The solution set is $\{-3, 3\}$.

43.
$$4 = n^2$$
$$4 - n^2 = 0$$
$$(2 - n)(2 + n) = 0$$
$$2 - n = 0 \quad \text{or} \quad 2 + n = 0$$
$$2 = n \quad \text{or} \quad n = -2$$
The solution set is $\{-2, 2\}$.

45.
$$9x^2 = 16$$
$$9x^2 - 16 = 0$$
$$(3x - 4)(3x + 4) = 0$$
$$3x - 4 = 0 \quad \text{or} \quad 3x + 4 = 0$$
$$3x = 4 \quad \text{or} \quad 3x = -4$$
$$x = \frac{4}{3} \quad \text{or} \quad x = -\frac{4}{3}$$
The solution set is $\left\{-\frac{4}{3}, \frac{4}{3}\right\}$.

47.
$$n^2 - 121 = 0$$
$$(n - 11)(n + 11) = 0$$
$$n - 11 = 0 \quad \text{or} \quad n + 11 = 0$$
$$n = 11 \quad \text{or} \quad n = -11$$
The solution set is $\{-11, 11\}$.

49.
$$25x^2 = 4$$
$$25x^2 - 4 = 0$$
$$(5x - 2)(5x + 2) = 0$$
$$5x - 2 = 0 \quad \text{or} \quad 5x + 2 = 0$$
$$5x = 2 \quad \text{or} \quad 5x = -2$$
$$x = \frac{2}{5} \quad \text{or} \quad x = -\frac{2}{5}$$
The solution set is $\left\{-\frac{2}{5}, \frac{2}{5}\right\}$.

51.
$$3x^2 = 75$$
Divided both sides by 3.
$$x^2 = 25$$
$$x^2 - 25 = 0$$
$$(x - 5)(x + 5) = 0$$
$$x - 5 = 0 \quad \text{or} \quad x + 5 = 0$$
$$x = 5 \quad \text{or} \quad x = -5$$
The solution set is $\{-5, 5\}$.

53.
$$3x^3 - 48x = 0$$
Divided both sides by 3.
$$x^3 - 16x = 0$$
$$x(x^2 - 16) = 0$$
$$x(x - 4)(x + 4) = 0$$
$$x = 0 \quad \text{or} \quad x - 4 = 0 \quad \text{or} \quad x + 4 = 0$$
$$x = 0 \quad \text{or} \quad x = 4 \quad \text{or} \quad x = -4$$
The solution set is $\{-4, 0, 4\}$.

55.
$$n^3 = 16n$$
$$n^3 - 16n = 0$$
$$n(n^2 - 16) = 0$$
$$n(n - 4)(n + 4) = 0$$
$$n = 0 \quad \text{or} \quad n - 4 = 0 \quad \text{or} \quad n + 4 = 0$$
$$n = 0 \quad \text{or} \quad n = 4 \quad \text{or} \quad n = -4$$
The solution set is $\{-4, 0, 4\}$.

57.
$$5 - 45x^2 = 0$$
Divided by 5.
$$1 - 9x^2 = 0$$
$$(1 - 3x)(1 + 3x) = 0$$

$1 - 3x = 0$ or $1 + 3x = 0$

$1 = 3x$ or $3x = -1$

$x = \frac{1}{3}$ or $x = -\frac{1}{3}$

The solution set is $\left\{-\frac{1}{3}, \frac{1}{3}\right\}$.

59. $4x^3 - 400x = 0$

Divided by 4.

$x^3 - 100x = 0$

$x(x^2 - 100) = 0$

$x(x-10)(x+10) = 0$

$x = 0$ or $x - 10 = 0$ or $x + 10 = 0$

$x = 0$ or $x = 10$ or $x = -10$

The solution set is $\{-10, 0, 10\}$.

61. $64x^2 = 81$

$64x^2 - 81 = 0$

$(8x-9)(8x+9) = 0$

$8x - 9 = 0$ or $8x + 9 = 0$

$8x = 9$ or $8x = -9$

$x = \frac{9}{8}$ or $x = -\frac{9}{8}$

The solution set is $\left\{-\frac{9}{8}, \frac{9}{8}\right\}$.

63. $36x^3 = 9x$

Divided both sides by 9.

$4x^3 = x$

$4x^3 - x = 0$

$x(4x^2 - 1) = 0$

$x(2x-1)(2x+1) = 0$

$x = 0$ or $2x - 1 = 0$ or $2x + 1 = 0$

$x = 0$ or $2x = 1$ or $2x = -1$

$x = 0$ or $x = \frac{1}{2}$ or $x = -\frac{1}{2}$

The solution set is $\left\{-\frac{1}{2}, 0, \frac{1}{2}\right\}$.

65. Let n represent the number.

$n^2 - 49 = 0$

$(n-7)(n+7) = 0$

$n - 7 = 0$ or $n + 7 = 0$

$n = 7$ or $n = -7$

The number could be either -7 or 7.

67. Let n represent the number.

$5n^3 = 80n$

Divided both sides by 5.

$n^3 = 16n$

$n^3 - 16n = 0$

$n(n^2 - 16) = 0$

$n(n-4)(n+4) = 0$

$n = 0$ or $n - 4 = 0$ or $n + 4 = 0$

$n = 0$ or $n = 4$ or $n = -4$

The number could be -4, 0 or 4.

69. Let x represent the length of a side of the smaller square, then $5x$ represents the length of a side of the larger square.

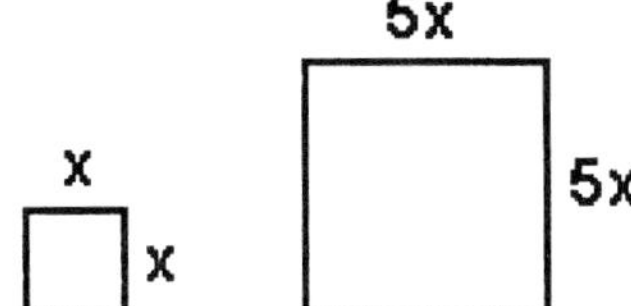

Larger Area + Smaller Area = 234

$(5x)^2 + x^2 = 234$

$25x^2 + x^2 = 234$

$26x^2 = 234$

$x^2 = 9$

$x^2 - 9 = 0$

$(x-3)(x+3) = 0$

$x - 3 = 0$ or $x + 3 = 0$

$x = 3$ or $x = -3$

The solution -3 must be discarded. Thus, the length of a side of the smaller square is 3 inches and the length of a side of the larger square is 15 inches.

71. Let w represent the width of the rectangle, then $2\frac{1}{2}w = \frac{5}{2}w$ represents the length of the rectangle.

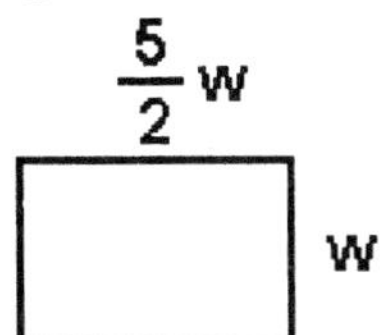

Rectangular Area $= 160$

$$w\left(\frac{5}{2}w\right) = 160$$

$$\frac{5}{2}w^2 = 160$$

Multiplied both sides by $\frac{2}{5}$.

$$w^2 = 64$$
$$w^2 - 64 = 0$$
$$(w-8)(w+8) = 0$$
$$w - 8 = 0 \quad \text{or} \quad w + 8 = 0$$
$$w = 8 \quad \text{or} \quad w = -8$$

The solution of -8 must be discarded. Thus, the width of the rectangle is 8 centimeters and the length is $\frac{5}{2}(8) = 20$ centimeters.

73. Let r represent the length of a radius of the smaller circle, then $2r$ represents the length of a radius of the larger circle.

Smaller Area + Larger Area $= 80\pi$

$$\pi r^2 + \pi(2r)^2 = 80\pi$$
$$\pi r^2 + 4\pi r^2 = 80\pi$$
$$5\pi r^2 = 80\pi$$
$$r^2 = 16$$
$$r^2 - 16 = 0$$
$$(r-4)(r+4) = 0$$
$$r - 4 = 0 \quad \text{or} \quad r + 4 = 0$$
$$r = 4 \quad \text{or} \quad r = -4 \left(\begin{array}{c}\text{Discard this}\\ \text{solution}\end{array}\right)$$

The length of a radius of the smaller circle is 4 meters and the length of a radius of the larger circle is 8 meters.

75. Let x represent a radius of the base, then x also represents the altitude of the cylinder.

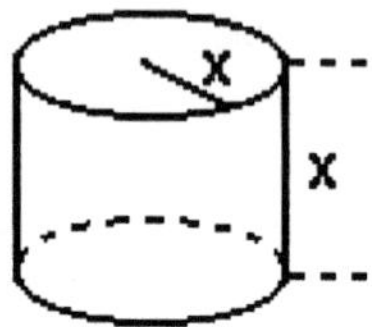

Surface Area $= 100\pi$

$$2\pi x^2 + 2\pi x(x) = 100\pi$$
$$2\pi x^2 + 2\pi x^2 = 100\pi$$
$$4\pi x^2 = 100\pi$$
$$x^2 = 25$$
$$x^2 - 25 = 0$$
$$(x-5)(x+5) = 0$$
$$x - 5 = 0 \quad \text{or} \quad x + 5 = 0$$
$$x = 5 \quad \text{or} \quad x = -5 \left(\begin{array}{c}\text{Discard this}\\ \text{solution.}\end{array}\right)$$

The length of a radius is 5 centimeters.

PROBLEM SET 6.3 Factoring Trinomials of the Form $x^2 + bx + c$

1. We need two integers whose product is 24 and whose sum is 10. They are 4 and 6.
$x^2 + 10x + 24 = (x+4)(x+6)$

3. We need two integers whose product is 40 and whose sum is 13. They are 5 and 8.
$x^2 + 13x + 40 = (x+5)(x+8)$

5. We need two integers whose product is 18 and whose sum is -11. They are -2 and -9.
$x^2 - 11x + 18 = (x-2)(x-9)$

7. We need two integers whose product is 28 and whose sum is -11. They are -4 and -7.
$n^2 - 11n + 28 = (n-4)(n-7)$

9. We need two integers whose product is -27 and whose sum is 6. They are 9 and -3.
$n^2 + 6n - 27 = (n-3)(n+9)$

11. We need two integers whose product is -40 and whose sum is -6. They are 4 and -10.
$n^2 - 6n - 40 = (n+4)(n-10)$

13. We need two integers whose product is 24 and whose sum is 12. No such integers exist, therefore $t^2 + 12t + 24$ is not factorable.

15. We need two integers whose product is 72 and whose sum is -18. They are -6 and -12.
$x^2 - 18x + 72 = (x-6)(x-12)$

17. We need two integers whose product is -66 and whose sum is 5. They are -6 and 11.
$x^2 + 5x - 66 = (x-6)(x+11)$

19. We need two integers whose product is -72 and whose sum is -1. They are 8 and -9.
$y^2 - y - 72 = (y+8)(y-9)$

21. We need two integers whose product is 80 and whose sum is 21. They are 5 and 16.
$x^2 + 21x + 80 = (x+5)(x+16)$

23. We need two integers whose product is -72 and whose sum is 6. They are -6 and 12.
$x^2 + 6x - 72 = (x-6)(x+12)$

25. We need two integers whose product is -48 and whose sum is -10. No such integers exist, therefore $x^2 - 10x - 48$ is not factorable.

27. We need two integers whose product is -10 and whose sum is 3. They are -2 and 5.
$x^2 + 3xy - 10y^2 = (x-2y)(x+5y)$

29. We need two integers whose product is -32 and whose sum is -4. They are 4 and -8.
$a^2 - 4ab - 32b^2 = (a+4b)(a-8b)$

31. $x^2 + 10x + 21 = 0$
$(x+3)(x+7) = 0$
$x+3=0$ or $x+7=0$
$x=-3$ or $x=-7$
The solution set is $\{-7, -3\}$.

33. $x^2 - 9x + 18 = 0$
$(x-3)(x-6) = 0$
$x-3=0$ or $x-6=0$
$x=3$ or $x=6$
The solution set is $\{3, 6\}$.

35. $x^2 - 3x - 10 = 0$
$(x-5)(x+2) = 0$
$x-5=0$ or $x+2=0$
$x=5$ or $x=-2$
The solution set is $\{-2, 5\}$.

37. $n^2 + 5n - 36 = 0$
$(n+9)(n-4) = 0$
$n+9=0$ or $n-4=0$
$n=-9$ or $n=4$
The solution set is $\{-9, 4\}$.

39. $n^2 - 6n - 40 = 0$
$(n-10)(n+4) = 0$
$n-10=0$ or $n+4=0$
$n=10$ or $n=-4$
The solution set is $\{-4, 10\}$.

41. $t^2 + t - 56 = 0$
$(t+8)(t-7) = 0$
$t+8=0$ or $t-7=0$
$t=-8$ or $t=7$
The solution set is $\{-8, 7\}$.

43. $x^2 - 16x + 28 = 0$
$(x - 14)(x - 2) = 0$
$x - 14 = 0$ or $x - 2 = 0$
$x = 14$ or $x = 2$
The solution set is $\{2, 14\}$.

45. $x^2 + 11x = 12$
$x^2 + 11x - 12 = 0$
$(x + 12)(x - 1) = 0$
$x + 12 = 0$ or $x - 1 = 0$
$x = -12$ or $x = 1$
The solution set is $\{-12, 1\}$.

47. $x(x - 10) = -16$
$x^2 - 10x = -16$
$x^2 - 10x + 16 = 0$
$(x - 8)(x - 2) = 0$
$x - 8 = 0$ or $x - 2 = 0$
$x = 8$ or $x = 2$
The solution set is $\{2, 8\}$.

49. $-x^2 - 2x + 24 = 0$
Multiplied both sides by -1.
$x^2 + 2x - 24 = 0$
$(x + 6)(x - 4) = 0$
$x + 6 = 0$ or $x - 4 = 0$
$x = -6$ or $x = 4$
The solution set is $\{-6, 4\}$.

51. Let n represent one integer, then $n + 1$ represents the next integer.
$n(n + 1) = 56$
$n^2 + n = 56$
$n^2 + n - 56 = 0$
$(n + 8)(n - 7) = 0$
$n + 8 = 0$ or $n - 7 = 0$
$n = -8$ or $n = 7$
If $n = -8$, then $n + 1 = -7$. If $n = 7$, then $n + 1 = 8$. Thus, the consecutive integers are either -8 and -7, or 7 and 8.

53. Let n represent one integer, then $n + 2$ represents the next consecutive even integer.
$n(n + 2) = 168$
$n^2 + 2n = 168$
$n^2 + 2n - 168 = 0$
$(n + 14)(n - 12) = 0$
$n + 14 = 0$ or $n - 12 = 0$
$n = -14$ or $n = 12$
$n + 2 = -12$ or $n + 2 = 14$
Since -14 and -12 are not whole numbers, the consecutive even whole numbers would be 12 and 14.

55. Let n represent the first integer, then $n + 1$, $n + 2$, and $n + 3$ represent the other three consecutive integers.
$(n + 2)(n + 3) = 2[n(n + 1)] - 22$
$n^2 + 5n + 6 = 2(n^2 + n) - 22$
$n^2 + 5n + 6 = 2n^2 + 2n - 22$
$-n^2 + 3n + 28 = 0$
$n^2 - 3n - 28 = 0$
$(n - 7)(n + 4) = 0$
$n - 7 = 0$ or $n + 4 = 0$
$n = 7$ or $n = -4$
If the first integer is -4, then the four consecutive integers are -4, -3, -2, and -1. If the first integer is 7, then the four consecutive integers are 7, 8, 9, and 10.

57. Let n represent the larger number, then $n - 3$ is the smaller number.
$n^2 = 10(n - 3) + 9$
$n^2 = 10n - 30 + 9$
$n^2 = 10n - 21$
$n^2 - 10n + 21 = 0$
$(n - 7)(n - 3) = 0$
$n - 7 = 0$ or $n - 3 = 0$
$n = 7$ or $n = 3$
If $n = 7$, then $n - 3 = 4$. If $n = 3$, then $n - 3 = 0$. Thus the numbers are 4 and 7, or 0 and 3.

59. Let l represent the length of the rectangle, then $l-3$ represents the width of the rectangle.

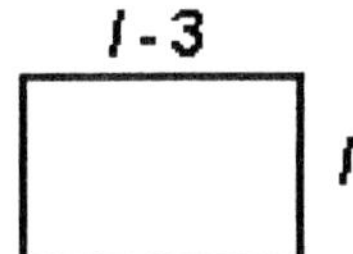

$$\begin{aligned} \text{Area} &= (2)\text{Perimeter} - 6 \\ l(l-3) &= 2[2l + 2(l-3)] - 6 \\ l^2 - 3l &= 2[2l + 2l - 6] - 6 \\ l^2 - 3l &= 2[4l - 6] - 6 \\ l^2 - 3l &= 8l - 12 - 6 \\ l^2 - 3l &= 8l - 18 \end{aligned}$$

$l^2 - 11l + 18 = 0$

$(l-2)(l-9) = 0$

$l - 2 = 0$ or $l - 9 = 0$

$l = 2$ or $l = 9$

If $l = 2,$ then $l - 3 = -1$ which is not possible. If $l = 9,$ then $l - 3 = 6.$ Therefore, the rectangle is 9 inches by 6 inches.

61. Let w represent the width of the rectangle, then one-half the perimeter less the width, $15 - w$, represents the length of the rectangle.

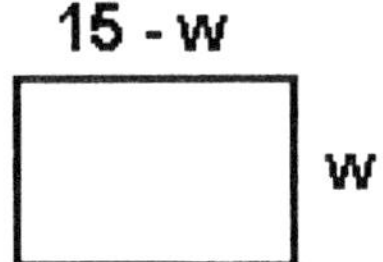

$$\begin{aligned} \text{Area} &= 54 \\ w(15-w) &= 54 \\ 15w - w^2 &= 54 \\ -w^2 + 15w - 54 &= 0 \\ w^2 - 15w + 54 &= 0 \\ (w-6)(w-9) &= 0 \end{aligned}$$

$w - 6 = 0$ or $w - 9 = 0$

$w = 6$ or $w = 9$

If the width is 6, then $15 - 6 = 9.$ If the width is 9, then $15 - 9 = 6.$ Thus, the length and width of the rectangle must be 6 centimeters by 9 centimeters.

63. Let r represent the number of rows in the orchard, then $r + 5$ represents the number of trees per row.

(trees per row)(rows) = number of trees

$$\begin{aligned} (r+5)(r) &= 84 \\ r^2 + 5r &= 84 \\ r^2 + 5r - 84 &= 0 \\ (r+12)(r-7) &= 0 \end{aligned}$$

$r + 12 = 0$ or $r - 7 = 0$

$r = -12$ or $r = 7$

The number of rows cannot be negative, so $r = -12$ is discarded. Thus there would be 7 rows of trees.

65. Let x represent the length of one leg of the triangle, then $x - 7$ represents the other leg and $x + 2$ represents the length of the hypotenuse.

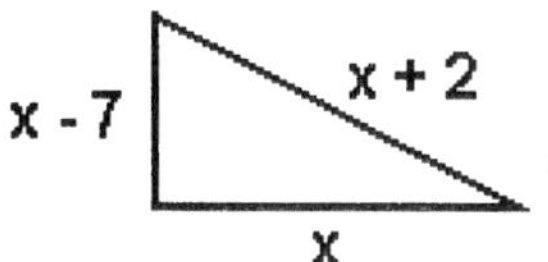

$$x^2 + (x-7)^2 = (x+2)^2$$

Pythagorean Theorem

$$\begin{aligned} x^2 + x^2 - 14x + 49 &= x^2 + 4x + 4 \\ 2x^2 - 14x + 49 &= x^2 + 4x + 4 \\ x^2 - 18x + 45 &= 0 \\ (x-3)(x-15) &= 0 \end{aligned}$$

$x - 3 = 0$ or $x - 15 = 0$

$x = 3$ or $x = 15$

If $x = 3$, then $x - 7 = -4$ which is not possible. If $x = 15$ then $x - 7 = 8$ and $x + 2 = 17.$ Thus, the sides of the right triangle are 8 feet, 15 feet and 17 feet.

67. Let x represent the length of one leg of the triangle, then $x-2$ represents the other leg.

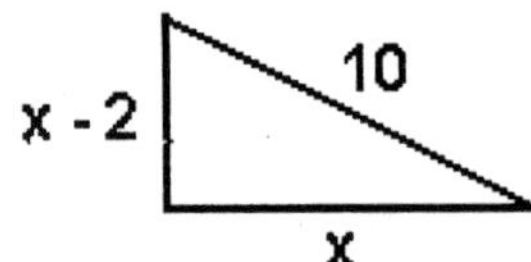

$$x^2+(x-2)^2=10^2$$
Pythagorean Theorem
$$x^2+x^2-4x+4=100$$
$$2x^2-4x+4=100$$
$$2x^2-4x-96=0$$
$$x^2-2x-48=0$$
$$(x-8)(x+6)=0$$
$$x-8=0 \quad \text{or} \quad x+6=0$$
$$x=8 \quad \text{or} \quad x=-6$$
The solution of $x=-6$ is not possible. If $x=8$, then $x-2=6$. Thus, the legs of the triangle are 6 inches and 8 inches.

PROBLEM SET 6.4 Factoring Trinomials of the Form ax^2+bx+c

1. We need two integers whose product is 6 and whose sum is 7. They are 6 and 1.
$3x^2+7x+2=$
$3x^2+1x+6x+2=$
$x(3x+1)+2(3x+1)=$
$(3x+1)(x+2)$

3. We need two integers whose product is 60 and whose sum is 19. They are 4 and 15.
$6x^2+19x+10=$
$6x^2+4x+15x+10=$
$2x(3x+2)+5(3x+2)=$
$(3x+2)(2x+5)$

5. We need two integers whose product is 24 and whose sum is -25. They are -1 and -24.
$4x^2-25x+6=$
$4x^2-1x-24x+6=$
$x(4x-1)-6(4x-1)=$
$(4x-1)(x-6)$

7. We need two integers whose product is 240 and whose sum is -31. They are -15 and -16.
$12x^2-31x+20=$
$12x^2-15x-16x+20=$
$3x(4x-5)-4(4x-5)=$
$(4x-5)(3x-4)$

9. We need two integers whose product is -70 and whose sum is -33. They are 2 and -35.
$5y^2-33y-14=$
$5y^2+2y-35y-14=$
$y(5y+2)-7(5y+2)=$
$(5y+2)(y-7)$

11. We need two integers whose product is -48 and whose sum is 13. They are -3 and 16.
$2n^2+13n-24=$
$2n^2-3n+16n-24=$
$n(2n-3)+8(2n-3)=$
$(2n-3)(n+8)$

13. We need two integers whose product is 14 and whose sum is 1. No such integers exist, therefore $2x^2+x+7$ is not factorable.

15. We need two integers whose product is 126 and whose sum is 45. They are 3 and 42.
$18x^2+45x+7=$
$18x^2+3x+42x+7=$
$3x(6x+1)+7(6x+1)=$
$(6x+1)(3x+7)$

17. We need two integers whose product is 56 and whose sum is -30. They are -2 and -28.
$7x^2 - 30x + 8 =$
$7x^2 - 2x - 28x + 8 =$
$x(7x - 2) - 4(7x - 2) =$
$(7x - 2)(x - 4)$

19. We need two integers whose product is -168 and whose sum is 2. They are -12 and 14.
$8x^2 + 2x - 21 =$
$8x^2 - 12x + 14x - 21 =$
$4x(2x - 3) + 7(2x - 3) =$
$(2x - 3)(4x + 7)$

21. We need two integers whose product is -126 and whose sum is -15. They are 6 and -21.
$9t^2 - 15t - 14 =$
$9t^2 + 6t - 21t - 14 =$
$3t(3t + 2) - 7(3t + 2) =$
$(3t + 2)(3t - 7)$

23. We need two integers whose product is -420 and whose sum is 79. They are -5 and 84.
$12y^2 + 79y - 35 =$
$12y^2 - 5y + 84y - 35 =$
$y(12y - 5) + 7(12y - 5) =$
$(12y - 5)(y + 7)$

25. We need two integers whose product is -30 and whose sum is 2. No such integers exist, therefore $6n^2 + 2n - 5$ is not factorable.

27. We need two integers whose product is 294 and whose sum is 55. They are 6 and 49.
$14x^2 + 55x + 21 =$
$14x^2 + 6x + 49x + 21 =$
$2x(7x + 3) + 7(7x + 3) =$
$(7x + 3)(2x + 7)$

29. We need two integers whose product is 240 and whose sum is -31. They are -15 and -16.
$20x^2 - 31x + 12 =$
$20x^2 - 15x - 16x + 12 =$
$5x(4x - 3) - 4(4x - 3) =$
$(4x - 3)(5x - 4)$

31. We need two integers whose product is -240 and whose sum is -8. They are 12 and -20.
$16n^2 - 8n - 15 =$
$16n^2 + 12n - 20n - 15 =$
$4n(4n + 3) - 5(4n + 3) =$
$(4n + 3)(4n - 5)$

33. We need two integers whose product is 600 and whose sum is -50. They are -20 and -30.
$24x^2 - 50x + 25 =$
$24x^2 - 20x - 30x + 25 =$
$4x(6x - 5) - 5(6x - 5) =$
$(6x - 5)(4x - 5)$

35. We need two integers whose product is 144 and whose sum is 25. They are 9 and 16.
$2x^2 + 25x + 72 =$
$2x^2 + 9x + 16x + 72 =$
$x(2x + 9) + 8(2x + 9) =$
$(2x + 9)(x + 8)$

37. We need two integers whose product is -42 and whose sum is 1. They are -6 and 7.
$21a^2 + a - 2 =$
$21a^2 - 6a + 7a - 2 =$
$3a(7a - 2) + 1(7a - 2) =$
$(7a - 2)(3a + 1)$

39. We need two integers whose product is -180 and whose sum is -31. They are 5 and -36.
$12a^2 - 31a - 15 =$
$12a^2 + 5a - 36a - 15 =$
$a(12a + 5) - 3(12a + 5) =$
$(12a + 5)(a - 3)$

Problem Set 6.4

41. We need two integers whose product is 36 and whose sum is 12. They are 6 and 6.
$4x^2 + 12x + 9 =$
$4x^2 + 6x + 6x + 9 =$
$2x(2x + 3) + 3(2x + 3) =$
$(2x + 3)(2x + 3)$

43. We need two integers whose product is 6 and whose sum is -5. They are -2 and -3.
$6x^2 - 5xy + y^2 =$
$6x^2 - 2xy - 3xy + y^2 =$
$2x(3x - y) - y(3x - y)$
$(3x - y)(2x - y)$

45. We need two integers whose product is -120 and whose sum is 7. They are -8 and 15.
$20x^2 + 7xy - 6y^2 =$
$20x^2 - 8xy + 15xy - 6y^2 =$
$4x(5x - 2y) + 3y(5x - 2y) =$
$(5x - 2y)(4x + 3y)$

47. We need two integers whose product is 60 and whose sum is -32. They are -2 and -30.
$5x^2 - 32x + 12 =$
$5x^2 - 2x - 30x + 12 =$
$x(5x - 2) - 6(5x - 2) =$
$(5x - 2)(x - 6)$

49. We need two integers whose product is -56 and whose sum is -55. They are 1 and -56.
$8x^2 - 55x - 7 =$
$8x^2 + 1x - 56x - 7 =$
$x(8x + 1) - 7(8x + 1) =$
$(8x + 1)(x - 7)$

51. $2x^2 + 13x + 6 = 0$
$(2x + 1)(x + 6) = 0$
$2x + 1 = 0$ or $x + 6 = 0$
$2x = -1$ or $x = -6$
$x = -\frac{1}{2}$ or $x = -6$
The solution set is $\left\{-6, -\frac{1}{2}\right\}$.

53. $12x^2 + 11x + 2 = 0$
$(4x + 1)(3x + 2) = 0$
$4x + 1 = 0$ or $3x + 2 = 0$
$4x = -1$ or $3x = -2$
$x = -\frac{1}{4}$ or $x = -\frac{2}{3}$
The solution set is $\left\{-\frac{2}{3}, -\frac{1}{4}\right\}$.

55. $3x^2 - 25x + 8 = 0$
$(3x - 1)(x - 8) = 0$
$3x - 1 = 0$ or $x - 8 = 0$
$3x = 1$ or $x = 8$
$x = \frac{1}{3}$ or $x = 8$
The solution set is $\left\{\frac{1}{3}, 8\right\}$.

57. $15n^2 - 41n + 14 = 0$
$(3n - 7)(5n - 2) = 0$
$3n - 7 = 0$ or $5n - 2 = 0$
$3n = 7$ or $5n = 2$
$n = \frac{7}{3}$ or $n = \frac{2}{5}$
The solution set is $\left\{\frac{2}{5}, \frac{7}{3}\right\}$.

59. $6t^2 + 37t - 35 = 0$
$(6t - 5)(t + 7) = 0$
$6t - 5 = 0$ or $t + 7 = 0$
$6t = 5$ or $t = -7$
$t = \frac{5}{6}$ or $t = -7$
The solution set is $\left\{-7, \frac{5}{6}\right\}$.

61. $16y^2 - 18y - 9 = 0$
$(8y + 3)(2y - 3) = 0$
$8y + 3 = 0$ or $2y - 3 = 0$
$8y = -3$ or $2y = 3$
$y = -\frac{3}{8}$ or $y = \frac{3}{2}$
The solution set is $\left\{-\frac{3}{8}, \frac{3}{2}\right\}$.

63. $9x^2 - 6x - 8 = 0$

$(3x - 4)(3x + 2) = 0$

$3x - 4 = 0$ or $3x + 2 = 0$

$3x = 4$ or $3x = -2$

$x = \frac{4}{3}$ or $x = -\frac{2}{3}$

The solution set is $\left\{-\frac{2}{3}, \frac{4}{3}\right\}$.

65. $10x^2 - 29x + 10 = 0$

$(2x - 5)(5x - 2) = 0$

$2x - 5 = 0$ or $5x - 2 = 0$

$2x = 5$ or $5x = 2$

$x = \frac{5}{2}$ or $x = \frac{2}{5}$

The solution set is $\left\{\frac{2}{5}, \frac{5}{2}\right\}$.

67. $6x^2 + 19x = -10$

$6x^2 + 19x + 10 = 0$

$(2x + 5)(3x + 2) = 0$

$2x + 5 = 0$ or $3x + 2 = 0$

$2x = -5$ or $3x = -2$

$x = -\frac{5}{2}$ or $x = -\frac{2}{3}$

The solution set is $\left\{-\frac{5}{2}, -\frac{2}{3}\right\}$.

69. $16x(x + 1) = 5$

$16x^2 + 16x = 5$

$16x^2 + 16x - 5 = 0$

$(4x - 1)(4x + 5) = 0$

$4x - 1 = 0$ or $4x + 5 = 0$

$4x = 1$ or $4x = -5$

$x = \frac{1}{4}$ or $x = -\frac{5}{4}$

The solution set is $\left\{-\frac{5}{4}, \frac{1}{4}\right\}$.

71. $35n^2 - 34n - 21 = 0$

$(7n + 3)(5n - 7) = 0$

$7n + 3 = 0$ or $5n - 7 = 0$

$7n = -3$ or $5n = 7$

$n = -\frac{3}{7}$ or $n = \frac{7}{5}$

The solution set is $\left\{-\frac{3}{7}, \frac{7}{5}\right\}$.

73. $4x^2 - 45x + 50 = 0$

$(4x - 5)(x - 10) = 0$

$4x - 5 = 0$ or $x - 10 = 0$

$4x = 5$ or $x = 10$

$x = \frac{5}{4}$ or $x = 10$

The solution set is $\left\{\frac{5}{4}, 10\right\}$.

75. $7x^2 + 46x - 21 = 0$

$(7x - 3)(x + 7) = 0$

$7x - 3 = 0$ or $x + 7 = 0$

$7x = 3$ or $x = -7$

$x = \frac{3}{7}$ or $x = -7$

The solution set is $\left\{-7, \frac{3}{7}\right\}$.

77. $12x^2 - 43x - 20 = 0$

$(12x + 5)(x - 4) = 0$

$12x + 5 = 0$ or $x - 4 = 0$

$12x = -5$ or $x = 4$

$x = -\frac{5}{12}$ or $x = 4$

The solution set is $\left\{-\frac{5}{12}, 4\right\}$.

79. $18x^2 + 55x - 28 = 0$

$(9x - 4)(2x + 7) = 0$

$9x - 4 = 0$ or $2x + 7 = 0$

$9x = 4$ or $2x = -7$

$x = \frac{4}{9}$ or $x = -\frac{7}{2}$

The solution set is $\left\{-\frac{7}{2}, \frac{4}{9}\right\}$.

Problem Set 6.5

PROBLEM SET 6.5 Factoring, Solving Equations, and Problem Solving

1. $x^2 + 4x + 4 =$
$(x)^2 + 2(x)(2) + (2)^2 =$
$(x + 2)^2$

3. $x^2 - 10x + 25 =$
$(x)^2 - 2(x)(5) + (5)^2 =$
$(x - 5)^2$

5. $9n^2 + 12n + 4 =$
$(3n)^2 + 2(3n)(2) + (2)^2 =$
$(3n + 2)^2$

7. $16a^2 - 8a + 1 =$
$(4a)^2 - 2(4a)(1) + (1)^2 =$
$(4a - 1)^2$

9. $4 + 36x + 81x^2 =$
$(2)^2 + 2(2)(9x) + (9x)^2 =$
$(2 + 9x)^2$

11. $16x^2 - 24xy + 9y^2 =$
$(4x)^2 - 2(4x)(3y) + (3y)^2 =$
$(4x - 3y)^2$

13. We need two integers whose product is 16 and whose sum is 17. They are 1 and 16.
$2x^2 + 17x + 8 =$
$2x^2 + 1x + 16x + 8 =$
$x(2x + 1) + 8(2x + 1) =$
$(2x + 1)(x + 8)$

15. $2x^3 - 72x =$
$2x(x^2 - 36) =$
$2x(x - 6)(x + 6)$

17. We need two integers whose product is -60 and whose sum is -7. They are 5 and -12.
$n^2 - 7y - 60 = (n + 5)(n - 12)$

19. We need two integers whose product is -12 and whose sum is -7. No such integers exist, therefore $3a^2 - 7a - 4$ is not factorable.

21. $8x^2 + 72 = 8(x^2 + 9)$

23. $9x^2 + 30x + 25 =$
$(3x)^2 + 2(3x)(5) + (5)^2 =$
$(3x + 5)^2$

25. We need two integers whose product is 42 and whose sum is 13. They are 6 and 7.
$15x^2 + 65x + 70 =$
$5(3x^2 + 13x + 14) =$
$5(3x^2 + 6x + 7x + 14) =$
$5[3x(x + 2) + 7(x + 2)] =$
$5(x + 2)(3x + 7)$

27. We need two integers whose product is -360 and whose sum is 2. They are -18 and 20.
$24x^2 + 2x - 15 =$
$24x^2 - 18x + 20x - 15 =$
$6x(4x - 3) + 5(4x - 3) =$
$(4x - 3)(6x + 5)$

29. $xy + 5y - 8x - 40 =$
$y(x + 5) - 8(x + 5) =$
$(x + 5)(y - 8)$

31. We need two integers whose product is -140 and whose sum is 31. They are -4 and 35.
$20x^2 + 31xy - 7y^2 =$
$20x^2 - 4xy + 35xy - 7y^2 =$
$4x(5x - y) + 7y(5x - y) =$
$(5x - y)(4x + 7y)$

33. $24x^2 + 18x - 81 =$
$3(8x^2 + 6x - 27) =$
$3(2x - 3)(4x + 9)$
We need two integers whose product is -216 and whose sum is 6. They are -12 and 18.

$8x^2 + 6x - 27 =$
$8x^2 - 12x + 18x - 27 =$
$4x(2x - 3) + 9(2x - 3) =$
$(2x - 3)(4x + 9)$

35. $12x^2 + 6x + 30 = 6(2x^2 + x + 5)$
[$2x^2 + x + 5$ is not factorable.]

37. $5x^4 - 80 =$
$5(x^4 - 16) =$
$5[(x^2 - 4)(x^2 + 4)] =$
$5(x - 2)(x + 2)(x^2 + 4)$

39. $x^2 + 12xy + 36y^2 =$
$(x)^2 + 2(x)(6y) + (6y)^2 =$
$(x + 6y)^2$

41. $4x^2 - 20x = 0$
$4(x^2 - 5x) = 0$
$4x(x - 5) = 0$
$4x = 0$ or $x - 5 = 0$
$x = 0$ or $x = 5$
The solution set is $\{0, 5\}$.

43. $x^2 - 9x - 36 = 0$
$(x - 12)(x + 3) = 0$
$x - 12 = 0$ or $x + 3 = 0$
$x = 12$ or $x = -3$
The solution set is $\{-3, 12\}$.

45. $-2x^3 + 8x = 0$
Divided by -2.
$x^3 - 4x = 0$
$x(x^2 - 4) = 0$
$x(x - 2)(x + 2) = 0$
$x = 0$ or $x - 2 = 0$ or $x + 2 = 0$
$x = 0$ or $x = 2$ or $x = -2$
The solution set is $\{-2, 0, 2\}$.

47. $6n^2 - 29n - 22 = 0$
$(2n - 11)(3n + 2) = 0$
$2n - 11 = 0$ or $3n + 2 = 0$
$2n = 11$ or $3n = -2$
$n = \frac{11}{2}$ or $n = -\frac{2}{3}$
The solution set is $\left\{-\frac{2}{3}, \frac{11}{2}\right\}$.

49. $(3n - 1)(4n - 3) = 0$
$3n - 1 = 0$ or $4n - 3 = 0$
$3n = 1$ or $4n = 3$
$n = \frac{1}{3}$ or $n = \frac{3}{4}$
The solution set is $\left\{\frac{1}{3}, \frac{3}{4}\right\}$.

51. $(n - 2)(n + 6) = -15$
$n^2 + 4n - 12 = -15$
$n^2 + 4n + 3 = 0$
$(n + 3)(n + 1) = 0$
$n + 3 = 0$ or $n + 1 = 0$
$n = -3$ or $n = -1$
The solution set is $\{-3, -1\}$.

53. $2x^2 = 12x$
Divided by 2.
$x^2 = 6x$
$x^2 - 6x = 0$
$x(x - 6) = 0$
$x = 0$ or $x - 6 = 0$
$x = 0$ or $x = 6$
The solution set is $\{0, 6\}$.

55. $t^3 - 2t^2 - 24t = 0$
$t(t^2 - 2t - 24) = 0$
$t(t - 6)(t + 4) = 0$
$t = 0$ or $t - 6 = 0$ or $t + 4 = 0$
$t = 0$ or $t = 6$ or $t = -4$
The solution set is $\{-4, 0, 6\}$.

57. $12 - 40x + 25x^2 = 0$
$(2 - 5x)(6 - 5x) = 0$
$2 - 5x = 0$ or $6 - 5x = 0$
$-5x = -2$ or $-5x = -6$
$x = \frac{2}{5}$ or $x = \frac{6}{5}$
The solution set is $\left\{\frac{2}{5}, \frac{6}{5}\right\}$.

59. $n^2 - 28n + 192 = 0$
$(n-12)(n-16) = 0$
$n - 12 = 0$ or $n - 16 = 0$
$n = 12$ or $n = 16$
The solution set is $\{12, 16\}$.

61. $(3n+1)(n+2) = 12$
$3n^2 + 7n + 2 = 12$
$3n^2 + 7n - 10 = 0$
$(3n+10)(n-1) = 0$
$3n + 10 = 0$ or $n - 1 = 0$
$3n = -10$ or $n = 1$
$n = -\frac{10}{3}$ or $n = 1$
The solution set is $\left\{-\frac{10}{3}, 1\right\}$.

63. $x^3 = 6x^2$
$x^3 - 6x^2 = 0$
$x^2(x-6) = 0$
$x(x)(x-6) = 0$
$x = 0$ or $x = 0$ or $x - 6 = 0$
$x = 0$ or $x = 0$ or $x = 6$
The solution set is $\{0, 6\}$.

65. $9x^2 - 24x + 16 = 0$
$(3x-4)(3x-4) = 0$
$3x - 4 = 0$ or $3x - 4 = 0$
$3x = 4$ or $3x = 4$
$x = \frac{4}{3}$ or $x = \frac{4}{3}$
The solution set is $\left\{\frac{4}{3}\right\}$.

67. $x^3 + 10x^2 + 25x = 0$
$x(x^2 + 10x + 25) = 0$
$x(x+5)(x+5) = 0$
$x = 0$ or $x + 5 = 0$ or $x + 5 = 0$
$x = 0$ or $x = -5$ or $x = -5$
The solution set is $\{-5, 0\}$.

69. $24x^2 + 17x - 20 = 0$
$(3x+4)(8x-5) = 0$
$3x + 4 = 0$ or $8x - 5 = 0$
$3x = -4$ or $8x = 5$
$x = -\frac{4}{3}$ or $x = \frac{5}{8}$
The solution set is $\left\{-\frac{4}{3}, \frac{5}{8}\right\}$.

71. Let n represent one of the numbers, then $4n + 7$ represents the other number.
$n(4n+7) = 15$
$4n^2 + 7n = 15$
$4n^2 + 7n - 15 = 0$
$(4n-5)(n+3) = 0$
$4n - 5 = 0$ or $n + 3 = 0$
$4n = 5$ or $n = -3$
$n = \frac{5}{4}$ or $n = -3$
If $n = \frac{5}{4}$, then $4n + 7 =$
$4\left(\frac{5}{4}\right) + 7 = 12.$
If $n = -3$, then $4n + 7 =$
$4(-3) + 7 = -5.$
Thus, then numbers are $\frac{5}{4}$ and 12, or -3 and -5.

73. Let n represent one number, then $2n + 3$ represents the other number.
$n(2n+3) = -1$
$2n^2 + 3n = -1$
$2n^2 + 3n + 1 = 0$
$(2n+1)(n+1) = 0$
$2n + 1 = 0$ or $n + 1 = 0$
$2n = -1$ or $n = -1$
$n = -\frac{1}{2}$ or $n = -1$
If $n = -\frac{1}{2}$, then $2n + 3 = 2\left(-\frac{1}{2}\right) + 3 = 2.$
If $n = -1$, then $2n + 3 = 2(-1) + 3 = 1.$
Thus, the numbers are $-\frac{1}{2}$ and 2, or -1 and 1.

75. Let n represent one number, then $2n+1$ represents the other number.

$$n^2+(2n+1)^2=97$$
$$n^2+4n^2+4n+1=97$$
$$5n^2+4n+1=97$$
$$5n^2+4n-96=0$$
$$(5n+24)(n-4)=0$$
$$5n+24=0 \quad \text{or} \quad n-4=0$$
$$5n=-24 \quad \text{or} \quad n=4$$
$$n=-\frac{24}{5} \quad \text{or} \quad n=4$$

If $n=-\frac{24}{5}$, then $2n+1=$
$$2\left(-\frac{24}{5}\right)+1=-\frac{48}{5}+\frac{5}{5}=-\frac{43}{5}.$$
If $n=4$, then $2n+1=2(4)+1=9$

Thus, the numbers are $-\frac{24}{5}$ and $-\frac{43}{5}$, or 4 and 9.

77. Let r represent the number of rows, then $2r-3$ represents the number of chairs per row.

$$(\text{rows})(\text{chairs per row})=54$$
$$r(2r-3)=54$$
$$2r^2-3r=54$$
$$2r^2-3r-54=0$$
$$(2r+9)(r-6)=0$$
$$2r+9=0 \quad \text{or} \quad r-6=0$$
$$2r=-9 \quad \text{or} \quad r=6$$
$$r=-\frac{9}{2} \quad \text{or} \quad r=6$$

The solution of $r=-\frac{9}{2}$ is not reasonable.

If $r=6$, then $2r-3=2(6)-3=9$.

Thus, there would be 6 rows with 9 chairs per row.

79. Let s represent the length of a side of the smaller square, then $3s$ represents the length of a side of the larger square.

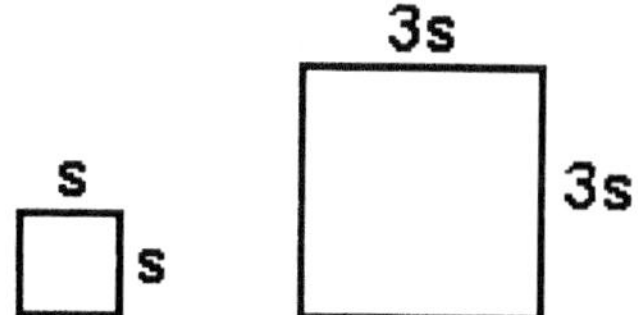

Smaller Area + Larger Area = 360

$$s^2+(3s)^2=360$$
$$s^2+9s^2=360$$
$$10s^2=360$$
$$s^2=36$$
$$s^2-36=0$$
$$(s-6)(s+6)=0$$
$$s-6=0 \quad \text{or} \quad s+6=0$$
$$s=6 \quad \text{or} \quad s=-6$$

The negative solution must be discarded. Therefore, the smaller square is 6 feet by 6 feet and the larger square is 18 feet by 18 feet.

81. Let w represent the width of the rectangle, then $2w+1$ represents the length.

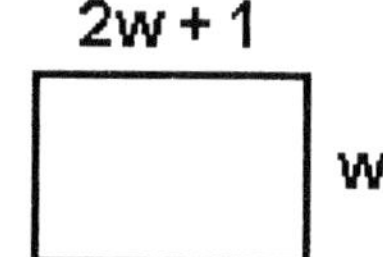

$$w(2w+1)=55$$
$$2w^2+w=55$$
$$2w^2+w-55=0$$
$$(2w+11)(w-5)=0$$
$$2w+11=0 \quad \text{or} \quad w-5=0$$
$$2w=-11 \quad \text{or} \quad w=5$$
$$w=-\frac{11}{2} \quad \text{or} \quad w=5$$

The negative solution must be discarded. Therefore, the rectangle is 5 centimeters by 11 centimeters.

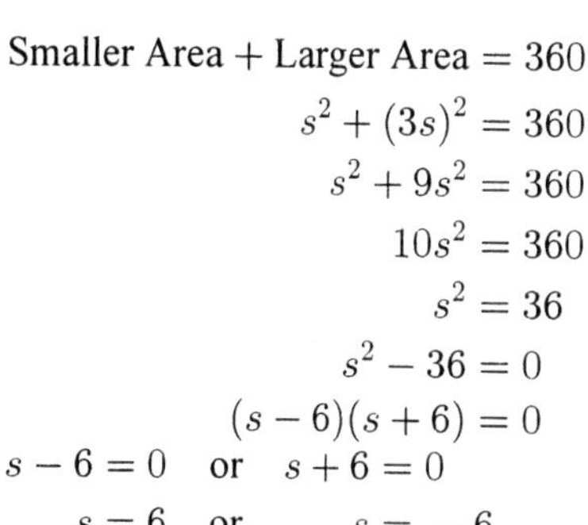

83. Let h represent the length of an altitude to a side of the triangle, then $3h-1$ represents the side.

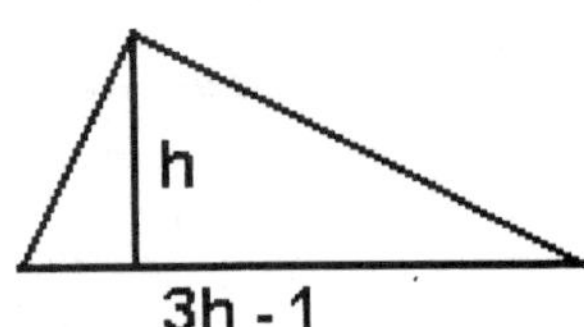

$$\frac{1}{2}h(3h-1)=51$$
$$h(3h-1)=102$$
$$3h^2-h=102$$
$$3h^2-h-102=0$$
$$(3h+17)(h-6)=0$$
$$3h+17=0 \quad \text{or} \quad h-6=0$$
$$3h=-17 \quad \text{or} \quad h=6$$
$$h=-\frac{17}{3} \quad \text{or} \quad h=6$$

The negative solution must be discarded. Therefore, the length of the side is 17 inches and the altitude to that side is 6 inches long.

85. Let x represent the width of the strip, then $8-2x$ represents the reduced width of the paper, and $11-2x$ represents the reduced length of the paper.

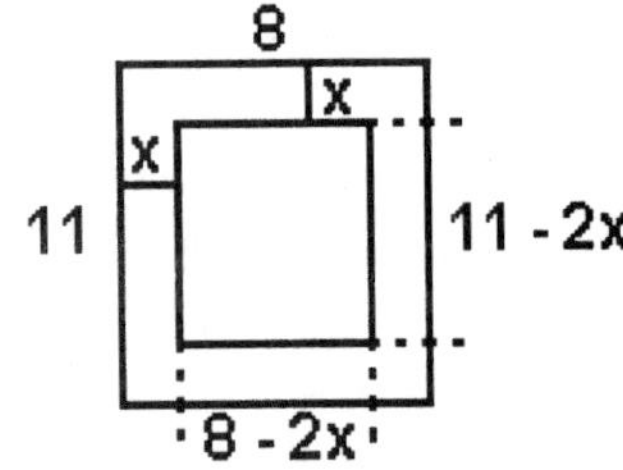

$$(8-2x)(11-2x)=40$$
$$88-38x+4x^2=40$$
$$48-38x+4x^2=0$$
$$24-19x+2x^2=0$$
$$(3-2x)(8-x)=0$$
$$3-2x=0 \quad \text{or} \quad 8-x=0$$
$$-2x=-3 \quad \text{or} \quad 8=x$$
$$x=\frac{3}{2} \quad \text{or} \quad x=8$$

The solution $x=8$ must be discarded since the width of the paper is 8 inches. Therefore, the width of the strip is $1\frac{1}{2}$ inches.

87. Let r represent a radius of the larger circle, then $r-6$ represents a radius of the smaller circle.

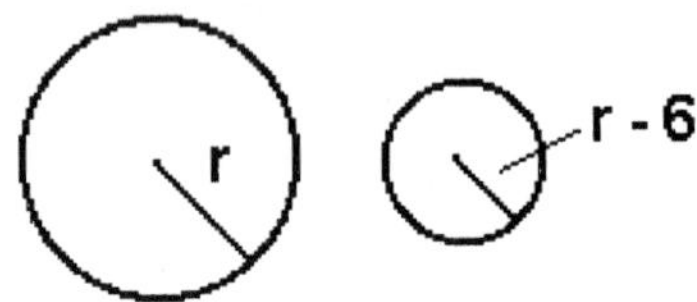

$$\text{Larger Area} + \text{Smaller Area} = 180\pi$$
$$\pi r^2+\pi(r-6)^2=180\pi$$
$$\pi r^2+\pi\left(r^2-12r+36\right)=180\pi$$
$$\pi r^2+\pi r^2-12\pi r+36\pi=180\pi$$
$$2\pi r^2-12\pi r+36\pi=180\pi$$
$$2\pi r^2-12\pi r-144\pi=0$$

Divided both sides by 2π.

$$r^2-6r-72=0$$
$$(r-12)(r+6)=0$$
$$r-12=0 \quad \text{or} \quad r+6=0$$
$$r=12 \quad \text{or} \quad r=-6$$

Discard the negative solution. The length of a radius of the larger circle is 12 inches and a radius of the smaller circle is 6 inches.

CHAPTER 6 Review Problem Set

1. We need two integers whose product is 14 and whose sum is -9. They are -2 and -7.
$x^2 - 9x + 14 = (x-2)(x-7)$

2. $3x^2 + 21x = 3x(x+7)$

3. $9x^2 - 4 =$
$(3x)^2 - (2)^2 =$
$(3x-2)(3x+2)$

4. We need two integers whose product is -20 and whose sum is 8. They are -2 and 10.
$4x^2 + 8x - 5 =$
$4x^2 - 2x + 10x - 5 =$
$2x(2x-1) + 5(2x-1) =$
$(2x-1)(2x+5)$

5. $25x^2 - 60x + 36 =$
$(5x)^2 - 2(5x)(6) + (6)^2 =$
$(5x-6)^2$

6. $n^3 + 13n^2 + 40n =$
$n(n^2 + 13n + 40) =$
$n(n+5)(n+8)$

We need two integers whose product is 40 and whose sum is 13. They are 5 and 8.
$n^2 + 13n + 40 =$
$(n+5)(n+8)$

7. We need two integers whose product is -12 and whose sum is 11. They are -1 and 12.
$y^2 + 11y - 12 = (y-1)(y+12)$

8. $3xy^2 + 6x^2y = 3xy(y+2x)$

9. $x^4 - 1 =$
$(x^2)^2 - (1)^2 =$
$(x^2-1)(x^2+1) =$
$(x-1)(x+1)(x^2+1)$

10. We need two integers whose product is -90 and whose sum is 9. They are -6 and 15.
$18n^2 + 9n - 5 =$
$18n^2 - 6n + 15n - 5 =$
$6n(3n-1) + 5(3n-1) =$
$(3n-1)(6n+5)$

11. We need two integers whose product is 24 and whose sum is 7. No such integers exist, therefore $x^2 + 7x + 24$ is not factorable.

12. We need two integers whose product is -28 and whose sum is -3. They are 4 and -7.
$4x^2 - 3x - 7 =$
$4x^2 + 4x - 7x - 7 =$
$4x(x+1) - 7(x+1) =$
$(x+1)(4x-7)$

13. $3n^2 + 3n - 90 =$
$3(n^2 + n - 30) =$
$3(n-5)(n+6)$
We need two integers whose product is -30 and whose sum is 1. They are -5 and 6.
$n^2 + n - 30 = (n-5)(n+6)$

14. $x^3 - xy^2 = x(x^2 - y^2) = x(x-y)(x+y)$

15. We need two integers whose product is -4 and whose sum is 3. They are -1 and 4.
$2x^2 + 3xy - 2y^2 =$
$2x^2 - xy + 4xy - 2y^2 =$
$x(2x-y) + 2y(2x-y) =$
$(2x-y)(x+2y)$

16. $4n^2 - 6n - 40 =$
$2(2n^2 - 3n - 20) =$
$2(2n + 5)(n - 4)$
We need two integers whose product is -40 and whose sum is -3. They are 5 and -8.
$2n^2 - 3n - 20 =$
$2n^2 + 5n - 8n - 20 =$
$n(2n + 5) - 4(2n + 5) =$
$(2n + 5)(n - 4)$

17. $5x + 5y + ax + ay =$
$5(x + y) + a(x + y) =$
$(x + y)(5 + a)$

18. We need two integers whose product is -84 and whose sum is -5. They are 7 and -12.
$21t^2 - 5t - 4 =$
$21t^2 + 7t - 12t - 4 =$
$7t(3t + 1) - 4(3t + 1) =$
$(3t + 1)(7t - 4)$

19. $2x^3 - 2x =$
$2x(x^2 - 1) =$
$2x(x - 1)(x + 1)$

20. $3x^3 - 108x =$
$3x(x^2 - 36) =$
$3x(x - 6)(x + 6)$

21. $16x^2 + 40x + 25 =$
$(4x)^2 + 2(4x)(5) + (5)^2 =$
$(4x + 5)^2$

22. $xy - 3x - 2y + 6 =$
$x(y - 3) - 2(y - 3) =$
$(y - 3)(x - 2)$

23. We need two integers whose product is -30 and whose sum is -7. They are 3 and -10.
$15x^2 - 7xy - 2y^2 =$
$15x^2 + 3xy - 10xy - 2y^2 =$
$3x(5x + y) - 2y(5x + y) =$
$(5x + y)(3x - 2y)$

24. $6n^4 - 5n^3 + n^2 =$
$n^2(6n^2 - 5n + 1) =$
$n^2(3n - 1)(2n - 1)$

We need two integers whose product is 6 and whose sum is -5. They are -2 and -3.
$6n^2 - 5n + 1 =$
$6n^2 - 2n - 3n + 1 =$
$2n(3n - 1) - 1(3n - 1) =$
$(3n - 1)(2n - 1)$

25. $x^2 + 4x - 12 = 0$
$(x + 6)(x - 2) = 0$
$x + 6 = 0$ or $x - 2 = 0$
$x = -6$ or $x = 2$
The solution set is $\{-6, 2\}$.

26. $x^2 = 11x$
$x^2 - 11x = 0$
$x(x - 11) = 0$
$x = 0$ or $x - 11 = 0$
$x = 0$ or $x = 11$
The solution set is $\{0, 11\}$.

27. $2x^2 + 3x - 20 = 0$
$(2x - 5)(x + 4) = 0$
$2x - 5 = 0$ or $x + 4 = 0$
$2x = 5$ or $x = -4$
$x = \frac{5}{2}$ or $x = -4$
The solution set is $\left\{-4, \frac{5}{2}\right\}$.

28. $9n^2 + 21n - 8 = 0$
$(3n - 1)(3n + 8) = 0$
$3n - 1 = 0$ or $3n + 8 = 0$
$3n = 1$ or $3n = -8$
$n = \frac{1}{3}$ or $n = -\frac{8}{3}$
The solution set is $\left\{-\frac{8}{3}, \frac{1}{3}\right\}$.

29.
$$6n^2 = 24$$
$$6n^2 - 24 = 0$$
$$n^2 - 4 = 0$$
$$(n-2)(n+2) = 0$$
$$n - 2 = 0 \quad \text{or} \quad n + 2 = 0$$
$$n = 2 \quad \text{or} \quad n = -2$$
The solution set is $\{-2, 2\}$.

30.
$$16y^2 + 40y + 25 = 0$$
$$(4y+5)(4y+5) = 0$$
$$4y + 5 = 0 \quad \text{or} \quad 4y + 5 = 0$$
$$4y = -5 \quad \text{or} \quad 4y = -5$$
$$y = -\frac{5}{4} \quad \text{or} \quad y = -\frac{5}{4}$$
The solution set is $\left\{-\frac{5}{4}\right\}$.

31.
$$t^3 - t = 0$$
$$t(t^2 - 1) = 0$$
$$t(t-1)(t+1) = 0$$
$$t = 0 \quad \text{or} \quad t - 1 = 0 \quad \text{or} \quad t + 1 = 0$$
$$t = 0 \quad \text{or} \quad t = 1 \quad \text{or} \quad t = -1$$
The solution set is $\{-1, 0, 1\}$.

32.
$$28x^2 + 71x + 18 = 0$$
$$(4x+9)(7x+2) = 0$$
$$4x + 9 = 0 \quad \text{or} \quad 7x + 2 = 0$$
$$4x = -9 \quad \text{or} \quad 7x = -2$$
$$x = -\frac{9}{4} \quad \text{or} \quad x = -\frac{2}{7}$$
The solution set is $\left\{-\frac{9}{4}, -\frac{2}{7}\right\}$.

33.
$$x^2 + 3x - 28 = 0$$
$$(x+7)(x-4) = 0$$
$$x + 7 = 0 \quad \text{or} \quad x - 4 = 0$$
$$x = -7 \quad \text{or} \quad x = 4$$
The solution set is $\{-7, 4\}$.

34.
$$(x-2)(x+2) = 21$$
$$x^2 - 4 = 21$$
$$x^2 - 25 = 0$$
$$(x+5)(x-5) = 0$$
$$x + 5 = 0 \quad \text{or} \quad x - 5 = 0$$
$$x = -5 \quad \text{or} \quad x = 5$$
The solution set is $\{-5, 5\}$.

35.
$$5n^2 + 27n = 18$$
$$5n^2 + 27n - 18 = 0$$
$$(5n-3)(n+6) = 0$$
$$5n - 3 = 0 \quad \text{or} \quad n + 6 = 0$$
$$5n = 3 \quad \text{or} \quad n = -6$$
$$n = \frac{3}{5} \quad \text{or} \quad n = -6$$
The solution set is $\left\{-6, \frac{3}{5}\right\}$.

36.
$$4n^2 + 10n = 14$$
$$4n^2 + 10n - 14 = 0$$
$$2n^2 + 5n - 7 = 0$$
$$(2n+7)(n-1) = 0$$
$$2n + 7 = 0 \quad \text{or} \quad n - 1 = 0$$
$$2n = -7 \quad \text{or} \quad n = 1$$
$$n = -\frac{7}{2} \quad \text{or} \quad n = 1$$
The solution set is $\left\{-\frac{7}{2}, 1\right\}$.

37.
$$2x^3 - 8x = 0$$
Divided both sides by 2.
$$x^3 - 4x = 0$$
$$x(x^2 - 4) = 0$$
$$x(x-2)(x+2) = 0$$
$$x = 0 \quad \text{or} \quad x - 2 = 0 \quad \text{or} \quad x + 2 = 0$$
$$x = 0 \quad \text{or} \quad x = 2 \quad \text{or} \quad x = -2$$
The solution set is $\{-2, 0, 2\}$.

38.
$$x^2 - 20x + 96 = 0$$
$$(x-8)(x-12) = 0$$
$$x - 8 = 0 \quad \text{or} \quad x - 12 = 0$$
$$x = 8 \quad \text{or} \quad x = 12$$
The solution set is $\{8, 12\}$.

39.
$$4t^2 + 17t - 15 = 0$$
$$(4t-3)(t+5) = 0$$
$$4t - 3 = 0 \quad \text{or} \quad t + 5 = 0$$
$$4t = 3 \quad \text{or} \quad t = -5$$
$$t = \frac{3}{4} \quad \text{or} \quad t = -5$$

The solution set is $\left\{-5, \frac{3}{4}\right\}$.

40. $3(x+2) - x(x+2) = 0$
$(x+2)(3-x) = 0$
$x+2=0$ or $3-x=0$
$x=-2$ or $3=x$
The solution set is $\{-2, 3\}$.

41. $(2x-5)(3x+7) = 0$
$2x-5=0$ or $3x+7=0$
$2x=5$ or $3x=-7$
$x=\frac{5}{2}$ or $t=-\frac{7}{3}$
The solution set is $\left\{-\frac{7}{3}, \frac{5}{2}\right\}$.

42. $(x+4)(x-1) = 50$
$x^2+3x-4=50$
$x^2+3x-54=0$
$(x+9)(x-6)=0$
$x+9=0$ or $x-6=0$
$x=-9$ or $x=6$
The solution set is $\{-9, 6\}$.

43. $-7n-2n^2=-15$
$-2n^2-7n=-15$
$-2n^2-7n+15=0$
$2n^2+7n-15=0$
$(2n-3)(n+5)=0$
$2n-3=0$ or $n+5=0$
$2n=3$ or $n=-5$
$n=\frac{3}{2}$ or $n=-5$
The solution set is $\left\{-5, \frac{3}{2}\right\}$.

44. $-23x+6x^2=-20$
$6x^2-23x=-20$
$6x^2-23x+20=0$
$(2x-5)(3x-4)=0$
$2x-5=0$ or $3x-4=0$
$2x=5$ or $3x=4$
$x=\frac{5}{2}$ or $x=\frac{4}{3}$

The solution set is $\left\{\frac{4}{3}, \frac{5}{2}\right\}$.

45. Let n represent the smaller number, then $2n-1$ is the larger number.
$(2n-1)^2-n^2=33$
$4n^2-4n+1-n^2=33$
$3n^2-4n+1=33$
$3n^2-4n-32=0$
$(3n+8)(n-4)=0$
$3n+8=0$ or $n-4=0$
$3n=-8$ or $n=4$
$n=-\frac{8}{3}$ or $n=4$
If $n=-\frac{8}{3}$, then $2n-1=2\left(-\frac{8}{3}\right)-1=$
$-\frac{16}{3}-\frac{3}{3}=-\frac{19}{3}$.
If $n=4$, then $2n-1=2(4)-1=7$
The numbers are $-\frac{8}{3}$ and $-\frac{19}{3}$, or 4 and 7.

46. Let w represent the width of the rectangle, then $5w-2$ represents the length of the rectangle.

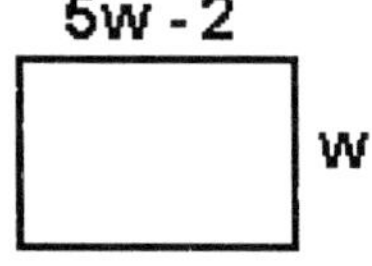

$w(5w-2)=16$
$5w^2-2w=16$
$5w^2-2w-16=0$
$(5w+8)(w-2)=0$
$5w+8=0$ or $w-2=0$
$5w=-8$ or $w=2$
$w=-\frac{8}{5}$ or $w=2$
Discard the negative solution. The width of the rectangle is 2 centimeters and the length is $5(2)-2=8$ centimeters.

47. Let s represent the length of a side of the smaller square, then $5s$ represents the length of a side of the larger square.

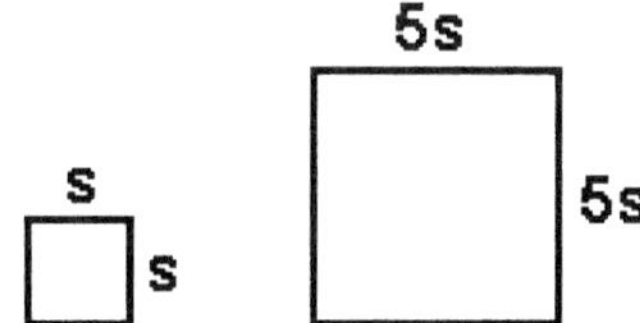

Smaller Area + Larger Area = 104

$$s^2 + (5s)^2 = 104$$
$$s^2 + 25s^2 = 104$$
$$26s^2 = 104$$
$$s^2 = 4$$
$$s^2 - 4 = 0$$
$$(s-2)(s+2) = 0$$
$$s - 2 = 0 \quad \text{or} \quad s + 2 = 0$$
$$s = 2 \quad \text{or} \quad s = -2$$

Discard the negative solution. The smaller square is 2 inches by 2 inches and the larger square is 10 inches by 10 inches.

48. Let x represent the length of the shorter leg, then $2x - 1$ represents the length of the longer leg, and $2x + 1$ represents the hypotenuse of the triangle.

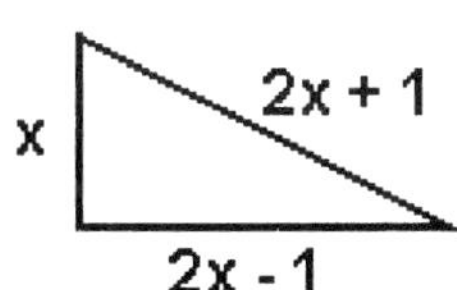

$$x^2 + (2x-1)^2 = (2x+1)^2$$

Pythagorean Theorem

$$x^2 + 4x^2 - 4x + 1 = 4x^2 + 4x + 1$$
$$5x^2 - 4x + 1 = 4x^2 + 4x + 1$$
$$x^2 - 8x = 0$$
$$x(x-8) = 0$$
$$x = 0 \quad \text{or} \quad x - 8 = 0$$
$$x = 0 \quad \text{or} \quad x = 8$$

Discard the zero solution. The sides of the triangle are 8 units, 15 units, and 17 units.

49. Let n represent one number, then $6n + 1$ represents the other number.

$$n(6n+1) = 26$$
$$6n^2 + n = 26$$
$$6n^2 + n - 26 = 0$$
$$(6n+13)(n-2) = 0$$
$$6n + 13 = 0 \quad \text{or} \quad n - 2 = 0$$
$$6n = -13 \quad \text{or} \quad n = 2$$
$$n = -\frac{13}{6} \quad \text{or} \quad n = 2$$

If $n = -\frac{13}{6}$, then $6n + 1 =$

$$6\left(-\frac{13}{6}\right) + 1 = -13 + 1 = -12.$$

If $n = 2$, then $6n + 1 = 6(2) + 1 = 13$

The numbers are $-\frac{13}{6}$ and -12, or 2 and 13.

50. Let n represent the first whole number, then $n + 2$ and $n + 4$ represent the next two consecutive odd whole numbers.

$$n^2 + (n+2)^2 = (n+4)^2 + 9$$
$$n^2 + n^2 + 4n + 4 = n^2 + 8n + 16 + 9$$
$$2n^2 + 4n + 4 = n^2 + 8n + 25$$
$$n^2 - 4n - 21 = 0$$
$$(n+3)(n-7) = 0$$
$$n + 3 = 0 \quad \text{or} \quad n - 7 = 0$$
$$n = -3 \quad \text{or} \quad n = 7$$

Discard the negative solution because only positive whole numbers are wanted. Thus, the positive odd whole numbers are 7, 9, and 11.

51. Let s represent the number of shelves, then $9s - 1$ represents the number of books per shelf.

$$\begin{pmatrix}\text{Number of}\\ \text{shelves}\end{pmatrix}\begin{pmatrix}\text{books per}\\ \text{shelf}\end{pmatrix} = 140$$
$$s(9s-1) = 140$$
$$9s^2 - s = 140$$
$$9s^2 - s - 140 = 0$$
$$(9s+35)(s-4) = 0$$
$$9s + 35 = 0 \quad \text{or} \quad s - 4 = 0$$
$$9s = -35 \quad \text{or} \quad s = 4$$
$$s = -\frac{35}{9} \quad \text{or} \quad s = 4$$

Discard the negative solution. There would be 4 shelves in the bookcase.

52. Let w represent the width of the rectangle, then $8w$ represents the length of the rectangle and w represents a side of the square.

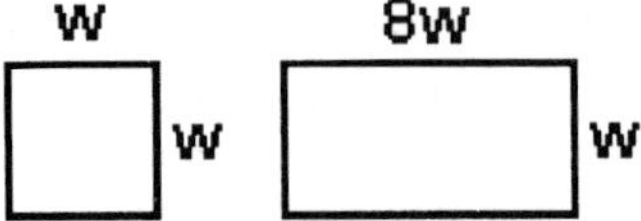

$$\text{Square Area} + \text{Rectangle Area} = 225$$
$$w^2 + w(8w) = 225$$
$$w^2 + 8w^2 = 225$$
$$9w^2 = 225$$
$$w^2 = 25$$
$$w^2 - 25 = 0$$
$$(w+5)(w-5) = 0$$
$$w + 5 = 0 \quad \text{or} \quad w - 5 = 0$$
$$w = -5 \quad \text{or} \quad w = 5$$

Discard the negative solution. The square would be 5 yards by 5 yards and the rectangle would be 5 yards by 40 yards.

53. Let n represent the first integer, then $n+1$ represents the next integer.

$$n^2 + (n+1)^2 = 613$$
$$n^2 + n^2 + 2n + 1 = 613$$
$$2n^2 + 2n - 612 = 0$$
$$n^2 + n - 306 = 0$$
$$(n+18)(n-17) = 0$$
$$n + 18 = 0 \quad \text{or} \quad n - 17 = 0$$
$$n = -18 \quad \text{or} \quad n = 17$$

The numbers would be -18 and -17, or 17 and 18.

54. Let x represent the length of a side of the cube.

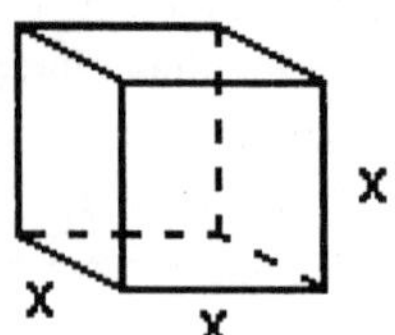

$$\text{Volume of Cube} = \text{Surface Area}$$
$$x^3 = 2x(x) + 2x(x) + 2x(x)$$
$$x^3 = 2x^2 + 2x^2 + 2x^2$$
$$x^3 = 6x^2$$
$$x^3 - 6x^2 = 0$$
$$x^2(x-6) = 0$$
$$x = 0 \quad \text{or} \quad x = 0 \quad \text{or} \quad x - 6 = 0$$
$$x = 0 \quad \text{or} \quad x = 0 \quad \text{or} \quad x = 6$$

Discard the zero solution. The length of a side of the cube is 6 units.

55. Let r represent a radius of the smaller circle, then $3r+1$ represents a radius of the larger circle.

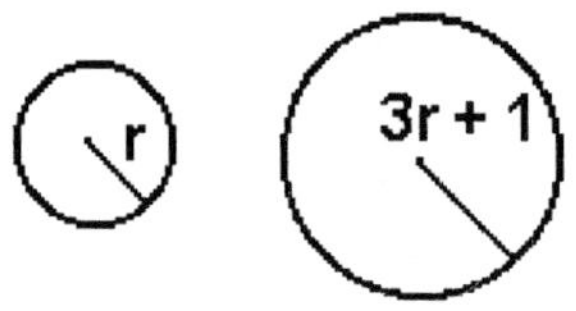

$$\text{Smaller Area} + \text{Larger Area} = 53\pi$$
$$\pi r^2 + \pi(3r+1)^2 = 53\pi$$
$$\pi r^2 + \pi(9r^2 + 6r + 1) = 53\pi$$
$$\pi r^2 + 9\pi r^2 + 6\pi r + \pi = 53\pi$$
$$10\pi r^2 + 6\pi r + \pi = 53\pi$$
$$10\pi r^2 + 6\pi r - 52\pi = 0$$

Divided both sides by 2π.

$$5r^2 + 3r - 26 = 0$$
$$(5r+13)(r-2) = 0$$
$$5r + 13 = 0 \quad \text{or} \quad r - 2 = 0$$
$$5r = -13 \quad \text{or} \quad r = 2$$
$$r = -\frac{13}{5} \quad \text{or} \quad r = 2$$

Discard the negative solution. The length of a radius of the smaller circle is 2 meters and a radius of the larger circle is 7 meters.

56. Let n represent the first odd whole number, then $n+2$ represents the next odd whole number. The sum of the two numbers is $n+(n+2)=2n+2$.

$$n(n+2)=5(2n+2)-1$$
$$n^2+2n=10n+10-1$$
$$n^2+2n=10n+9$$
$$n^2-8n-9=0$$
$$(n+1)(n-9)=0$$
$$n+1=0 \quad \text{or} \quad n-9=0$$
$$n=-1 \quad \text{or} \quad n=9$$

Discard the negative solution because whole numbers are positive. The numbers would be 9 and 11.

57. Let x represent the amount that the width and length are both reduced. The original area of the photograph is $8(14)=112$ square centimeters. The new area is $112-40=72$ square centimeters.

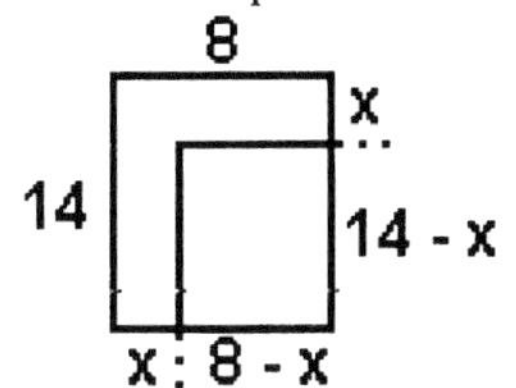

$$(8-x)(14-x)=72$$
$$112-22x+x^2=72$$
$$40-22x+x^2=0$$
$$(20-x)(2-x)=0$$
$$20-x=0 \quad \text{or} \quad 2-x=0$$
$$20=x \quad \text{or} \quad 2=x$$

The solution $x=20$ is not reasonable. Thus, the length and the width must be reduced by 2 centimeters.

58. Let x represent the width of the plowed strip. Then $120-2x$ represents the length of the unplowed garden, and $90-2x$ represents the width. The original area of the garden is $90(120)=10,800$ square feet. Thus, the unplowed garden is 5400 square feet.

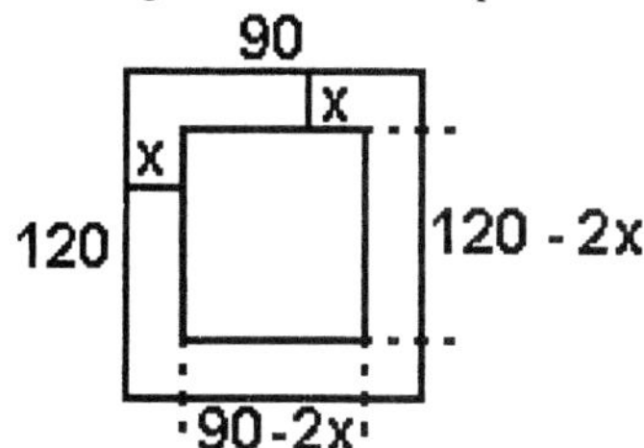

$$(90-2x)(120-2x)=5400$$
$$10,800-420x+4x^2=5400$$
$$5400-420x+4x^2=0$$
$$1350-105x+x^2=0$$
$$(90-x)(15-x)=0$$
$$90-x=0 \quad \text{or} \quad 15-x=0$$
$$90=x \quad \text{or} \quad 15=x$$

The solution $x=90$ is not reasonable. Thus, the width of the strip must be 15 feet.

CHAPTER 6 Test

1. We need two integers whose product is -10 and whose sum is 3. They are -2 and 5.
$x^2+3x-10=(x-2)(x+5)$

2. We need two integers whose product is -24 and whose sum is -5. They are 3 and -8.
$x^2-5x-24=(x+3)(x-8)$

3. $2x^3-2x=$
$2x(x^2-1)=$
$2x(x-1)(x+1)$

4. We need two integers whose product is 108 and whose sum is 21. They are 9 and 12.
$x^2+21x+108=(x+9)(x+12)$

5. $18n^2+21n+6=$
$3(6n^2+7n+2)=$
$3(2n+1)(3n+2)$
We need two integers whose product is 12 and whose sum is 7. They are 3 and 4.

$6n^2 + 7n + 2 =$
$6n^2 + 3n + 4n + 2 =$
$3n(2n + 1) + 2(2n + 1) =$
$(2n + 1)(3n + 2)$

6. $ax + ay + 2bx + 2by =$
$a(x + y) + 2b(x + y) =$
$(x + y)(a + 2b)$

7. We need two integers whose product is -60 and whose sum is 17. They are -3 and 20.
$4x^2 + 17x - 15 =$
$4x^2 - 3x + 20x - 15 =$
$x(4x - 3) + 5(4x - 3) =$
$(4x - 3)(x + 5)$

8. $6x^2 + 24 = 6(x^2 + 4)$

9. $30x^3 - 76x^2 + 48x =$
$2x(15x^2 - 38x + 24) =$
$2x(5x - 6)(3x - 4)$
We need two integers whose product is 360 and whose sum is -38. They are -18 and -20.
$15x^2 - 38x + 24 =$
$15x^2 - 18x - 20x + 24 =$
$3x(5x - 6) - 4(5x - 6) =$
$(5x - 6)(3x - 4)$

10. We need two integers whose product is -168 and whose sum is 13. They are -8 and 21.
$28 + 13x - 6x^2 =$
$28 - 8x + 21x - 6x^2 =$
$4(7 - 2x) + 3x(7 - 2x) =$
$(7 - 2x)(4 + 3x)$

11. $7x^2 = 63$
Divided both sides by 7.
$x^2 = 9$
$x^2 - 9 = 0$
$(x - 3)(x + 3) = 0$
$x - 3 = 0$ or $x + 3 = 0$
$x = 3$ or $x = -3$
The solution set is $\{-3, 3\}$.

12. $x^2 + 5x - 6 = 0$
$(x - 1)(x + 6) = 0$
$x - 1 = 0$ or $x + 6 = 0$
$x = 1$ or $x = -6$
The solution set is $\{-6, 1\}$.

13. $4n^2 = 32n$
$4n^2 - 32n = 0$
$n^2 - 8n = 0$
$n(n - 8) = 0$
$n = 0$ or $n - 8 = 0$
$n = 0$ or $n = 8$
The solution set is $\{0, 8\}$.

14. $(3x - 2)(2x + 5) = 0$
$3x - 2 = 0$ or $2x + 5 = 0$
$3x = 2$ or $2x = -5$
$x = \frac{2}{3}$ or $x = -\frac{5}{2}$
The solution set is $\left\{-\frac{5}{2}, \frac{2}{3}\right\}$.

15. $(x - 3)(x + 7) = -9$
$x^2 + 4x - 21 = -9$
$x^2 + 4x - 12 = 0$
$(x + 6)(x - 2) = 0$
$x + 6 = 0$ or $x - 2 = 0$
$n = -6$ or $x = 2$
The solution set is $\{-6, 2\}$.

16. $x^3 + 16x^2 + 48x = 0$
$x(x^2 + 16x + 48) = 0$
$x(x + 4)(x + 12) = 0$
$x = 0$ or $x + 4 = 0$ or $x + 12 = 0$
$x = 0$ or $x = -4$ or $x = -12$
The solution set is $\{-12, -4, 0\}$.

17. $9(x - 5) - x(x - 5) = 0$
$(x - 5)(9 - x) = 0$
$x - 5 = 0$ or $9 - x = 0$
$x = 5$ or $9 = x$
The solution set is $\{5, 9\}$.

18.
$$3t^2 + 35t = 12$$
$$3t^2 + 35t - 12 = 0$$
$$(3t - 1)(t + 12) = 0$$
$$3t - 1 = 0 \quad \text{or} \quad t + 12 = 0$$
$$3t = 1 \quad \text{or} \quad t = -12$$
$$t = \frac{1}{3} \quad \text{or} \quad t = -12$$
The solution set is $\left\{-12, \frac{1}{3}\right\}$.

19.
$$8 - 10x - 3x^2 = 0$$
$$(4 + x)(2 - 3x) = 0$$
$$4 + x = 0 \quad \text{or} \quad 2 - 3x = 0$$
$$x = -4 \quad \text{or} \quad -3x = -2$$
$$x = -4 \quad \text{or} \quad x = \frac{2}{3}$$
The solution set is $\left\{-4, \frac{2}{3}\right\}$.

20.
$$3x^3 = 75x$$
$$x^3 = 25x$$
$$x^3 - 25x = 0$$
$$x(x^2 - 25) = 0$$
$$x(x + 5)(x - 5) = 0$$
$$x = 0 \quad \text{or} \quad x + 5 = 0 \quad \text{or} \quad x - 5 = 0$$
$$x = 0 \quad \text{or} \quad x = -5 \quad \text{or} \quad x = 5$$
The solution set is $\{-5, 0, 5\}$.

21.
$$25n^2 - 70n + 49 = 0$$
$$(5n - 7)(5n - 7) = 0$$
$$5n - 7 = 0 \quad \text{or} \quad 5n - 7 = 0$$
$$5n = 7 \quad \text{or} \quad 5n = 7$$
$$x = \frac{7}{5} \quad \text{or} \quad x = \frac{7}{5}$$
The solution set is $\left\{\frac{7}{5}\right\}$.

22. Let w represent the width of the rectangle, then $2w - 2$ represents the length of the rectangle.

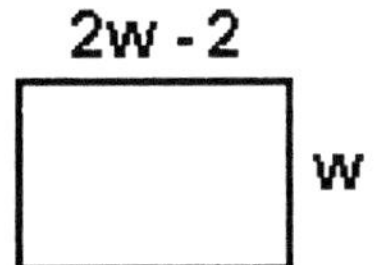

$$w(2w - 2) = 112$$
$$2w^2 - 2w = 112$$
$$2w^2 - 2w - 112 = 0$$
$$w^2 - w - 56 = 0$$
$$(w - 8)(w + 7) = 0$$
$$w - 8 = 0 \quad \text{or} \quad w + 7 = 0$$
$$w = 8 \quad \text{or} \quad w = -7$$
Discard the negative solution. If $w = 8$, then the length of the rectangle is $2(8) - 2 = 14$ inches.

23. Let x represent the length of the shorter leg, then $x + 4$ represents the length of the longer leg and $x + 8$ represents the length of the hypotenuse.

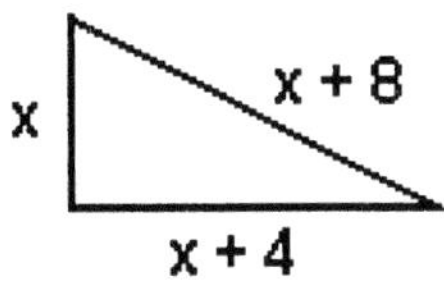

$$x^2 + (x + 4)^2 = (x + 8)^2$$
Pythagorean Theorem
$$x^2 + x^2 + 8x + 16 = x^2 + 16x + 64$$
$$2x^2 + 8x + 16 = x^2 + 16x + 64$$
$$x^2 - 8x - 48 = 0$$
$$(x - 12)(x + 4) = 0$$
$$x - 12 = 0 \quad \text{or} \quad x + 4 = 0$$
$$x = 12 \quad \text{or} \quad x = -4$$
Discard the negative solution. The length of the shorter leg is 12 centimeters.

24. Let r represent the number of rows, then $3r - 5$ represents the number of chairs per row.

$$\begin{aligned}(\text{rows})(\text{chairs per rows}) &= 112\\ r(3r-5) &= 112\\ 3r^2 - 5r &= 112\\ 3r^2 - 5r - 112 &= 0\\ (3r+16)(r-7) &= 0\end{aligned}$$

$3r + 16 = 0$ or $r - 7 = 0$

$3r = -16$ or $r = 7$

$r = -\dfrac{16}{3}$ or $r = 7$

Discard the negative solution. There are 7 rows with 16 chairs per row.

25. Let x represent the length of a side of the cube.

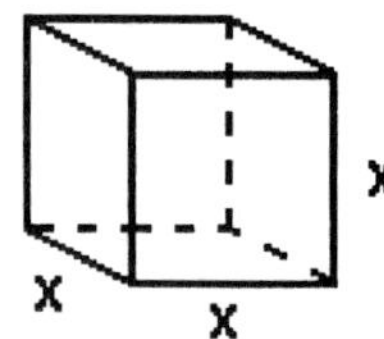

$$\begin{aligned}\text{Volume of Cube} &= 2[\text{Surface Area}]\\ x^3 &= 2[2x(x) + 2x(x) + 2x(x)]\\ x^3 &= 2\left[2x^2 + 2x^2 + 2x^2\right]\\ x^3 &= 2\left[6x^2\right]\\ x^3 &= 12x^2\\ x^3 - 12x^2 &= 0\\ x^2(x-12) &= 0\end{aligned}$$

$x = 0$ or $x = 0$ or $x - 12 = 0$

$x = 0$ or $x = 0$ or $x = 12$

Discard the zero solution. The length of a side of the cube is 12 units.

Chapter 7 Algebraic Fractions

PROBLEM SET 7.1 Simplifying Algebraic Fractions

1. $\dfrac{6x}{14y} = \dfrac{\cancel{2}\cdot 3\cdot x}{\cancel{2}\cdot 7\cdot y} = \dfrac{3x}{7y}$

3. $\dfrac{9xy}{24x} = \dfrac{3\cdot\cancel{3}\cdot\cancel{x}\cdot y}{\cancel{3}\cdot 8\cdot\cancel{x}} = \dfrac{3y}{8}$

5. $\dfrac{-15x^2y}{25x} = -\dfrac{3\cdot\cancel{5}\cdot\overset{x}{\cancel{x^2}}\cdot y}{5\cdot\cancel{5}\cdot\cancel{x}} = -\dfrac{3xy}{5}$

7. $\dfrac{-36x^4y^3}{-48x^6y^2} = +\dfrac{\overset{3}{\cancel{36}}\cdot\cancel{x^4}\cdot\overset{y}{\cancel{y^3}}}{\underset{4}{\cancel{48}}\cdot\underset{x^2}{\cancel{x^6}}\cdot\cancel{y^2}} = \dfrac{3y}{4x^2}$

9. $\dfrac{12a^2b^5}{-54a^2b^3} = -\dfrac{\cancel{6}\cdot 2\cdot\cancel{a^2}\cdot\overset{b^2}{\cancel{b^5}}}{\cancel{6}\cdot 9\cdot\cancel{a^2}\cdot\cancel{b^3}} = -\dfrac{2b^2}{9}$

11. $\dfrac{32xy^2z^3}{72yz^4} = \dfrac{\overset{4}{\cancel{32}}\cdot x\cdot\overset{y}{\cancel{y^2}}\cdot\cancel{z^3}}{\underset{9}{\cancel{72}}\cdot\cancel{y}\cdot\underset{z}{\cancel{z^4}}} = \dfrac{4xy}{9z}$

13. $\dfrac{xy}{x^2-2x} = \dfrac{\cancel{x}y}{\cancel{x}(x-2)} = \dfrac{y}{x-2}$

15. $\dfrac{8x+12y}{12} = \dfrac{\cancel{4}(2x+3y)}{\underset{3}{\cancel{12}}} = \dfrac{2x+3y}{3}$

17. $\dfrac{x^2+2x}{x^2-7x} = \dfrac{\cancel{x}(x+2)}{\cancel{x}(x-7)} = \dfrac{x+2}{x-7}$

19. $\dfrac{7-x}{x-7} = -1$ because $7-x$ and $x-7$ are opposites.

21. $\dfrac{15-3n}{n-5} = \dfrac{3(5-n)}{n-5} = 3(-1) = -3$

Remember $\dfrac{5-n}{n-5} = -1$

23. $\dfrac{4x^3-4x}{1-x^2} = \dfrac{4x(x^2-1)}{1-x^2} = -4x$

Remember $\dfrac{x^2-1}{1-x^2} = -1$

25. $\dfrac{x^2-1}{3x^2-3x} = \dfrac{\cancel{(x-1)}(x+1)}{3x\cancel{(x-1)}} = \dfrac{x+1}{3x}$

27. $\dfrac{x^2+xy}{x^2} = \dfrac{\cancel{x}(x+y)}{\underset{x}{\cancel{x^2}}} = \dfrac{x+y}{x}$

29. $\dfrac{6x^3-15x^2y}{6x^2+24xy} = \dfrac{\cancel{3}\overset{x}{\cancel{x^2}}(2x-5y)}{\underset{2}{\cancel{6}}\cancel{x}(x+4y)} =$

$\dfrac{x(2x-5y)}{2(x+4y)}$

31. $\dfrac{n^2+2n}{n^2+3n+2} = \dfrac{n\cancel{(n+2)}}{(n+1)\cancel{(n+2)}} =$

$\dfrac{n}{n+1}$

33. $\dfrac{2n^2+5n-3}{n^2-9} = \dfrac{(2n-1)\cancel{(n+3)}}{(n-3)\cancel{(n+3)}} =$

$\dfrac{2n-1}{n-3}$

35. $\dfrac{2x^2+17x+35}{3x^2+19x+20} = \dfrac{(2x+7)\cancel{(x+5)}}{(3x+4)\cancel{(x+5)}} =$

$\dfrac{2x+7}{3x+4}$

37. $\dfrac{9(x-1)^2}{12(x-1)^3} =$

$\dfrac{\overset{3}{\cancel{9}}\cancel{(x-1)}\cancel{(x-1)}}{\underset{4}{\cancel{12}}\cancel{(x-1)}\cancel{(x-1)}(x-1)} =$

$\dfrac{3}{4(x-1)}$

39. $\dfrac{7x^2+61x-18}{7x^2+19x-6} =$

$\dfrac{\cancel{(7x-2)}(x+9)}{\cancel{(7x-2)}(x+3)} =$

$\dfrac{x+9}{x+3}$

41. $\dfrac{10a^2+a-3}{15a^2+4a-3} =$

$\dfrac{\cancel{(5a+3)}(2a-1)}{\cancel{(5a+3)}(3a-1)} =$

$\dfrac{2a-1}{3a-1}$

43. $\dfrac{x^2+2xy-3y^2}{2x^2-xy-y^2} =$

$\dfrac{\cancel{(x-y)}(x+3y)}{\cancel{(x-y)}(2x+y)} =$

$\dfrac{x+3y}{2x+y}$

45. $\dfrac{x^2-9}{-x^2-3x} =$

$\dfrac{(x-3)\cancel{(x+3)}}{-x\cancel{(x+3)}} =$

$-\dfrac{x-3}{x}$

47. $\dfrac{n^2+14n+49}{8n+56} =$

$\dfrac{(n+7)\cancel{(n+7)}}{8\cancel{(n+7)}} =$

$\dfrac{n+7}{8}$

49. $\dfrac{4n^2-12n+9}{2n^2-n-3} =$

$\dfrac{\cancel{(2n-3)}(2n-3)}{\cancel{(2n-3)}(n+1)} =$

$\dfrac{2n-3}{n+1}$

51. $\dfrac{y^2-6y-72}{y^2-8y-84} =$

$\dfrac{(y-12)\cancel{(y+6)}}{(y-14)\cancel{(y+6)}} =$

$\dfrac{y-12}{y-14}$

53. $\dfrac{1-x^2}{x-x^2} =$

$\dfrac{\cancel{(1-x)}(1+x)}{x\cancel{(1-x)}} =$

$\dfrac{1+x}{x}$

55. $\dfrac{6-x-2x^2}{12+7x-10x^2} =$

$\dfrac{(2+x)\cancel{(3-2x)}}{(4+5x)\cancel{(3-2x)}} =$

$\dfrac{2+x}{4+5x}$

57. $\dfrac{x^2+7x-18}{12-4x-x^2} = \dfrac{(x+9)(x-2)}{(6+x)(2-x)}$

$= -\dfrac{x+9}{x+6}$

Remember $\dfrac{x-2}{2-x} = -1$

59. $\dfrac{5x-40}{80-10x} = \dfrac{\cancel{5}(x-8)}{\underset{2}{\cancel{10}}(8-x)} = -\dfrac{1}{2}$

Remember $\dfrac{x-8}{8-x} = -1$

PROBLEM SET 7.2 Multiplying and Dividing Algebraic Fractions

1. $\frac{5}{9}\cdot\frac{3}{10} = \frac{\cancel{5}\cdot\cancel{3}}{\underset{3}{\cancel{9}}\cdot\underset{2}{\cancel{10}}} = \frac{1}{6}$

3. $\left(-\frac{3}{4}\right)\left(\frac{6}{7}\right) = -\frac{3\cdot\overset{3}{\cancel{6}}}{\underset{2}{\cancel{4}}\cdot 7} = -\frac{9}{14}$

5. $\left(\frac{17}{9}\right)\div\left(-\frac{19}{9}\right) =$

$\left(\frac{17}{9}\right)\cdot\left(-\frac{9}{19}\right) =$

$-\frac{17\cdot\cancel{9}}{\cancel{9}\cdot 19} = -\frac{17}{19}$

7. $\frac{8xy}{12y}\cdot\frac{6x}{14y} = \frac{\overset{2}{\cancel{8}}\cdot\overset{2}{\cancel{6}}\cdot x\cdot x\cdot\cancel{y}}{\underset{\cancel{3}}{\cancel{12}}\cdot\underset{7}{\cancel{14}}\cdot y\cdot\cancel{y}} = \frac{2x^2}{7y}$

9. $\left(-\frac{5n^2}{18n}\right)\left(\frac{27n}{25}\right) =$

$-\frac{\cancel{5}\cdot\overset{3}{\cancel{27}}\cdot n^2\cdot\cancel{n}}{\underset{2}{\cancel{18}}\cdot\underset{5}{\cancel{25}}\cdot\cancel{n}} =$

$-\frac{3n^2}{10}$

11. $\frac{3a^2}{7}\div\frac{6a}{28} = \frac{3a^2}{7}\cdot\frac{28}{6a} =$

$\frac{\cancel{3}\cdot\overset{\overset{2}{\cancel{4}}}{\cancel{28}}\cdot\overset{a}{\cancel{a^2}}}{\cancel{7}\cdot\underset{\cancel{2}}{\cancel{6}}\cdot\cancel{a}} = 2a$

13. $\frac{18a^2b^2}{-27a}\div\frac{-9a}{5b} = \frac{18a^2b^2}{-27a}\cdot\frac{5b}{-9a} =$

$\frac{\overset{2}{\cancel{18}}\cdot 5\cdot\cancel{a^2}\cdot b^2\cdot b}{27\cdot\cancel{9}\cdot\cancel{a}\cdot\cancel{a}} = \frac{10b^3}{27}$

15. $24x^3\div\frac{16x}{y} = \frac{24x^3}{1}\cdot\frac{y}{16x} =$

$\frac{\overset{3}{\cancel{24}}\cdot\overset{x^2}{\cancel{x^3}}\cdot y}{\underset{2}{\cancel{16}}\cdot\cancel{x}} = \frac{3x^2y}{2}$

17. $\frac{1}{15ab^3}\div\frac{-1}{12a} = \frac{1}{15ab^3}\cdot\frac{12a}{-1} =$

$-\frac{\overset{4}{\cancel{12}}\cdot\cancel{a}}{\underset{5}{\cancel{15}}\cdot\cancel{a}\cdot b^3} = -\frac{4}{5b^3}$

19. $\frac{18rs}{34}\div\frac{9r}{1} = \frac{18rs}{34}\cdot\frac{1}{9r} =$

$\frac{\overset{\cancel{2}}{\cancel{18}}\cdot\cancel{r}\cdot s}{\underset{17}{\cancel{34}}\cdot\cancel{9}\cdot\cancel{r}} = \frac{s}{17}$

21. $\frac{y}{x+y}\cdot\frac{x^2-y^2}{xy} =$

$\frac{\cancel{y}(x-y)\cancel{(x+y)}}{\cancel{(x+y)}(x)(\cancel{y})} = \frac{x-y}{x}$

Problem Set 7.2

23. $\frac{2x^2+xy}{xy} \cdot \frac{y}{10x+5y} =$

$\frac{\cancel{x}(\cancel{2x+y})(\cancel{y})}{(\cancel{x})(\cancel{y})(5)(\cancel{2x+y})} = \frac{1}{5}$

25. $\frac{6ab}{4ab+4b^2} \div \frac{7a-7b}{a^2-b^2} =$

$\frac{6ab}{4ab+4b^2} \cdot \frac{a^2-b^2}{7a-7b} =$

$\frac{\overset{3}{\cancel{6}}a\cancel{b}(\cancel{a-b})(\cancel{a+b})}{\underset{2}{\cancel{4}}\cancel{b}(\cancel{a+b})(7)(\cancel{a-b})} = \frac{3a}{14}$

27. $\frac{x^2+11x+30}{x^2+4} \cdot \frac{5x^2+20}{x^2+14x+45} =$

$\frac{(\cancel{x+5})(x+6)(5)(\cancel{x^2+4})}{(\cancel{x^2+4})(\cancel{x+5})(x+9)} =$

$\frac{5(x+6)}{x+9}$

29. $\frac{2x^2-3xy+y^2}{4x^2y} \div \frac{x^2-y^2}{6x^2y^2} =$

$\frac{2x^2-3xy+y^2}{4x^2y} \cdot \frac{6x^2y^2}{x^2-y^2} =$

$\frac{(2x-y)(\cancel{x-y})(\overset{3}{\cancel{6}}\cancel{x^2}\overset{y}{\cancel{y^2}})}{\underset{2}{\cancel{4}}\cancel{x^2}\cancel{y}(\cancel{x-y})(x+y)} =$

$\frac{3y(2x-y)}{2(x+y)}$

31. $\frac{a+a^2}{15a^2+11a+2} \cdot \frac{1-a}{1-a^2} =$

$\frac{a(\cancel{1+a})(\cancel{1-a})}{(5a+2)(3a+1)(\cancel{1-a})(\cancel{1+a})} =$

$\frac{a}{(5a+2)(3a+1)}$

33. $\frac{2x^2-2xy}{x^2+4x-32} \cdot \frac{x^2-16}{5xy-5y^2} =$

$\frac{2x(\cancel{x-y})(\cancel{x-4})(x+4)}{(x+8)(\cancel{x-4})(5y)(\cancel{x-y})} =$

$\frac{2x(x+4)}{5y(x+8)}$

35. $\frac{2x^2-xy-3y^2}{(x+y)^2} \div \frac{4x^2-12xy+9y^2}{10x-15y} =$

$\frac{2x^2-xy-3y^2}{(x+y)^2} \cdot \frac{10x-15y}{4x^2-12xy+9y^2} =$

$\frac{(\cancel{2x-3y})(\cancel{x+y})(5)(\cancel{2x-3y})}{(\cancel{x+y})(x+y)(\cancel{2x-3y})(\cancel{2x-3y})} =$

$\frac{5}{x+y}$

37. $\frac{(3t-1)^2}{45t-15} \div \frac{12t^2+5t-3}{20t+5} =$

$\frac{(3t-1)^2}{45t-15} \cdot \frac{20t+5}{12t^2+5t-3} =$

$\frac{(\cancel{3t-1})(\cancel{3t-1})(\cancel{5})(4t+1)}{\underset{3}{\cancel{15}}(\cancel{3t-1})(\cancel{3t-1})(4t+3)} =$

$\frac{4t+1}{3(4t+3)}$

39. $\frac{n^3-n}{n^2+7n+6} \cdot \frac{4n+24}{n^2-n} =$

$\frac{\cancel{n}(\cancel{n-1})(\cancel{n+1})(4)(\cancel{n+6})}{(\cancel{n+1})(\cancel{n+6})(\cancel{n})(\cancel{n-1})} =$

4

41. $\frac{6}{9y} \div \frac{30x}{12y^2} \cdot \frac{5xy}{4} =$

$\left(\frac{6}{9y} \cdot \frac{12y^2}{30x}\right) \cdot \frac{5xy}{4} =$

$$\left(\frac{\cancel{6}\cdot\cancel{12}^{4}\cdot\cancel{y^2}^{y}}{\cancel{9}_{3}\cdot\cancel{30}_{5}\cdot x\cdot\cancel{y}}\right)\cdot\frac{5xy}{4}=$$

$$\frac{\cancel{4}y}{\cancel{15}_{3}\,\cancel{x}}\cdot\frac{\cancel{5}\cancel{x}y}{\cancel{4}}=\frac{y^2}{3}$$

$$\frac{\cancel{x}(x-1)\cancel{(x+y)}}{y(1-y)(x-y)\cancel{(x+y)}\,(\cancel{x}y)}=$$

$$\frac{x-1}{y^2(1-y)(x-y)}$$

43. $\dfrac{8x^2}{xy-xy^2}\cdot\dfrac{x-1}{8x^2-8y^2}\div\dfrac{xy}{x+y}=$

$$\left[\frac{\cancel{8x^2}^{\cancel{x}}\,(x-1)}{\cancel{x}y(1-y)(\cancel{8})(x^2-y^2)}\right]\cdot\frac{x+y}{xy}=$$

45. $\dfrac{x^2+9x+18}{x^2+3x}\cdot\dfrac{x^2+5x}{x^2-25}\div\dfrac{x^2+8x}{x^2+3x-40}=$

$$\frac{\cancel{(x+3)}\,(x+6)(\cancel{x})\cancel{(x+5)}}{(\cancel{x})\cancel{(x+3)}\,(x-5)\cancel{(x+5)}}\div\frac{x^2+8}{x^2+3x-40}=$$

$$\frac{(x+6)}{\cancel{(x-5)}}\cdot\frac{\cancel{(x+8)}\,\cancel{(x-5)}}{x\cancel{(x+8)}}=\frac{x+6}{x}$$

PROBLEM SET 7.3 Adding and Subtracting Algebraic Fractions

1. $\dfrac{5}{x}+\dfrac{12}{x}=\dfrac{5+12}{x}=\dfrac{17}{x}$

3. $\dfrac{7}{3x}-\dfrac{5}{3x}=\dfrac{7-5}{3x}=\dfrac{2}{3x}$

5. $\dfrac{7}{2n}+\dfrac{1}{2n}=\dfrac{7+1}{2n}=\dfrac{8}{2n}=\dfrac{4}{n}$

7. $\dfrac{9}{4x^2}-\dfrac{13}{4x^2}=\dfrac{9-13}{4x^2}=\dfrac{-4}{4x^2}=-\dfrac{1}{x^2}$

9. $\dfrac{x+1}{x}+\dfrac{3}{x}=\dfrac{(x+1)+(3)}{x}=\dfrac{x+4}{x}$

11. $\dfrac{3}{x-1}-\dfrac{6}{x-1}=\dfrac{3-6}{x-1}=$

$$\frac{-3}{x-1}=-\frac{3}{x-1}$$

13. $\dfrac{x+1}{x}-\dfrac{1}{x}=\dfrac{(x+1)-(1)}{x}=\dfrac{x}{x}=1$

15. $\dfrac{3t-1}{4}+\dfrac{2t+3}{4}=$

$$\frac{(3t-1)+(2t+3)}{4}=$$

$$\frac{5t+2}{4}$$

17. $\dfrac{7a+2}{3}-\dfrac{4a-6}{3}=$

$$\frac{(7a+2)-(4a-6)}{3}=$$

$$\frac{7a+2-4a+6}{3}=$$

$$\frac{3a+8}{3}$$

19. $\dfrac{4n+3}{8}+\dfrac{6n+5}{8}=$

$$\frac{(4n+3)+(6n+5)}{8}=$$

$$\frac{10n+8}{8}=\frac{2(5n+4)}{8}=$$

$$\frac{5n+4}{4}$$

21. $\dfrac{3n-7}{6}-\dfrac{9n-1}{6}=$

$$\frac{(3n-7)-(9n-1)}{6}=$$

Problem Set 7.3

$$\frac{3n-7-9n+1}{6}=$$
$$\frac{-6n-6}{6}=\frac{6(-n-1)}{6}=$$
$$-n-1$$

23. $$\frac{5x-2}{7x}-\frac{8x+3}{7x}=$$
$$\frac{(5x-2)-(8x+3)}{7x}=$$
$$\frac{5x-2-8x-3}{7x}=$$
$$\frac{-3x-5}{7x}$$

25. $$\frac{3(x+2)}{4x}+\frac{6(x-1)}{4x}=$$
$$\frac{3(x+2)+6(x-1)}{4x}=$$
$$\frac{3x+6+6x-6}{4x}=$$
$$\frac{9x}{4x}=\frac{9}{4}$$

27. $$\frac{6(n-1)}{3n}+\frac{3(n+2)}{3n}=$$
$$\frac{6(n-1)+3(n+2)}{3n}=$$
$$\frac{6n-6+3n+6}{3n}=$$
$$\frac{9n}{3n}=3$$

29. $$\frac{2(3x-4)}{7x^2}-\frac{7x-8}{7x^2}=$$
$$\frac{2(3x-4)-(7x-8)}{7x^2}=$$
$$\frac{6x-8-7x+8}{7x^2}=$$
$$\frac{-x}{7x^2}=-\frac{1}{7x}$$

31. $$\frac{a^2}{a+2}-\frac{4}{a+2}=\frac{a^2-4}{a+2}$$
$$\frac{(a-2)(a+2)}{a+2}=a-2$$

33. $$\frac{3x}{(x-6)^2}-\frac{18}{(x-6)^2}=$$
$$\frac{3x-18}{(x-6)^2}=$$
$$\frac{3(x-6)}{(x-6)(x-6)}=\frac{3}{x-6}$$

35. $$\frac{3x}{8}+\frac{5x}{4}=$$
$$\frac{3x}{8}+\left(\frac{5x}{4}\right)\left(\frac{2}{2}\right)=$$
$$\frac{3x}{8}+\frac{2(5x)}{8}=$$
$$\frac{3x+10x}{8}=\frac{13x}{8}$$

37. $$\frac{7n}{12}-\frac{4n}{3}=$$
$$\frac{7n}{12}-\left(\frac{4n}{3}\right)\left(\frac{4}{4}\right)=$$
$$\frac{7n}{12}-\frac{16n}{12}=\frac{7n-16n}{12}=$$
$$\frac{-9n}{12}=-\frac{3n}{4}$$

39. $$\frac{y}{6}+\frac{3y}{4}=$$
$$\left(\frac{y}{6}\right)\left(\frac{2}{2}\right)+\left(\frac{3y}{4}\right)\left(\frac{3}{3}\right)=$$
$$\frac{2y}{12}+\frac{9y}{12}=$$
$$\frac{2y+9y}{12}=\frac{11y}{12}$$

41. $$\frac{8x}{3}-\frac{3x}{7}=$$
$$\left(\frac{8x}{3}\right)\left(\frac{7}{7}\right)-\left(\frac{3x}{7}\right)\left(\frac{3}{3}\right)=$$

$$\frac{56x}{21} - \frac{9x}{21} = \frac{56x - 9x}{21} = \frac{47x}{21}$$

43. $\frac{2x}{6} + \frac{3x}{5} =$

$$\left(\frac{2x}{6}\right)\left(\frac{5}{5}\right) + \left(\frac{3x}{5}\right)\left(\frac{6}{6}\right) =$$

$$\frac{10x}{30} + \frac{18x}{30} = \frac{10x + 18x}{30} =$$

$$\frac{28x}{30} = \frac{14x}{15}$$

45. $\frac{7n}{8} - \frac{3n}{9} =$

$$\left(\frac{7n}{8}\right)\left(\frac{9}{9}\right) - \left(\frac{3n}{9}\right)\left(\frac{8}{8}\right) =$$

$$\frac{63n}{72} - \frac{24n}{72} = \frac{63n - 24n}{72} =$$

$$\frac{39n}{72} = \frac{13n}{24}$$

47. $\frac{x+3}{5} + \frac{x-4}{2} =$

$$\left(\frac{x+3}{5}\right)\left(\frac{2}{2}\right) + \left(\frac{x-4}{2}\right)\left(\frac{5}{5}\right) =$$

$$\frac{2(x+3)}{10} + \frac{5(x-4)}{10} =$$

$$\frac{2(x+3) + 5(x-4)}{10} =$$

$$\frac{2x + 6 + 5x - 20}{10} = \frac{7x - 14}{10}$$

49. $\frac{x-6}{9} + \frac{x+2}{3} =$

$$\frac{x-6}{9} + \left(\frac{x+2}{3}\right)\left(\frac{3}{3}\right) =$$

$$\frac{x-6}{9} + \frac{3(x+2)}{9} =$$

$$\frac{x - 6 + 3(x+2)}{9} =$$

$$\frac{x - 6 + 3x + 6}{9} = \frac{4x}{9}$$

51. $\frac{3n-1}{3} + \frac{2n+5}{4} =$

$$\left(\frac{3n-1}{3}\right)\left(\frac{4}{4}\right) + \left(\frac{2n+5}{4}\right)\left(\frac{3}{3}\right) =$$

$$\frac{4(3n-1)}{12} + \frac{3(2n+5)}{12} =$$

$$\frac{4(3n-1) + 3(2n+5)}{12} =$$

$$\frac{12n - 4 + 6n + 15}{12} = \frac{18n + 11}{12}$$

53. $\frac{4n-3}{6} - \frac{3n+5}{18} =$

$$\left(\frac{4n-3}{6}\right)\left(\frac{3}{3}\right) - \frac{3n+5}{18} =$$

$$\frac{3(4n-3)}{18} - \frac{3n+5}{18} =$$

$$\frac{3(4n-3) - (3n+5)}{18} =$$

$$\frac{12n - 9 - 3n - 5}{18} = \frac{9n - 14}{18}$$

55. $\frac{3x}{4} + \frac{x}{6} - \frac{5x}{8} =$

$$\left(\frac{3x}{4}\right)\left(\frac{6}{6}\right) + \left(\frac{x}{6}\right)\left(\frac{4}{4}\right) - \left(\frac{5x}{8}\right)\left(\frac{3}{3}\right) =$$

$$\frac{18x}{24} + \frac{4x}{24} - \frac{15x}{24} =$$

$$\frac{18x + 4x - 15x}{24} = \frac{7x}{24}$$

57. $\frac{x}{5} - \frac{3}{10} - \frac{7x}{12} =$

$$\left(\frac{x}{5}\right)\left(\frac{12}{12}\right) - \left(\frac{3}{10}\right)\left(\frac{6}{6}\right) - \left(\frac{7x}{12}\right)\left(\frac{5}{5}\right) =$$

$$\frac{12x}{60} - \frac{18}{60} - \frac{35x}{60} =$$

$$\frac{12x - 18 - 35x}{60} = \frac{-23x - 18}{60}$$

59. $\dfrac{5}{8x}+\dfrac{1}{6x}=$

$\left(\dfrac{5}{8x}\right)\left(\dfrac{3}{3}\right)+\left(\dfrac{1}{6x}\right)\left(\dfrac{4}{4}\right)=$

$\dfrac{15}{24x}+\dfrac{4}{24x}=\dfrac{15+4}{24x}=\dfrac{19}{24x}$

61. $\dfrac{5}{6y}-\dfrac{7}{9y}=$

$\left(\dfrac{5}{6y}\right)\left(\dfrac{3}{3}\right)-\left(\dfrac{7}{9y}\right)\left(\dfrac{2}{2}\right)=$

$\dfrac{15}{18y}-\dfrac{14}{18y}=\dfrac{15-14}{18y}=\dfrac{1}{18y}$

63. $\dfrac{5}{12x}-\dfrac{11}{16x^2}=$

$\left(\dfrac{5}{12x}\right)\left(\dfrac{4x}{4x}\right)-\left(\dfrac{11}{16x^2}\right)\left(\dfrac{3}{3}\right)=$

$\dfrac{20x}{48x^2}-\dfrac{33}{48x^2}=\dfrac{20x-33}{48x^2}$

65. $\dfrac{3}{2x}-\dfrac{2}{3x}+\dfrac{5}{4x}=$

$\left(\dfrac{3}{2x}\right)\left(\dfrac{6}{6}\right)-\left(\dfrac{2}{3x}\right)\left(\dfrac{4}{4}\right)+\left(\dfrac{5}{4x}\right)\left(\dfrac{3}{3}\right)=$

$\dfrac{18-8+15}{12x}=\dfrac{25}{12x}$

67. $\dfrac{3}{x-5}+\dfrac{7}{x}=$

$\left(\dfrac{3}{x-5}\right)\left(\dfrac{x}{x}\right)+\left(\dfrac{7}{x}\right)\left(\dfrac{x-5}{x-5}\right)=$

$\dfrac{3x}{x(x-5)}+\dfrac{7(x-5)}{x(x-5)}=$

$\dfrac{3x+7(x-5)}{x(x-5)}=$

$\dfrac{3x+7x-35}{x(x-5)}=\dfrac{10x-35}{x(x-5)}$

69. $\dfrac{2}{n-1}-\dfrac{3}{n}=$

$\left(\dfrac{2}{n-1}\right)\left(\dfrac{n}{n}\right)-\left(\dfrac{3}{n}\right)\left(\dfrac{n-1}{n-1}\right)=$

$\dfrac{2n}{n(n-1)}-\dfrac{3(n-1)}{n(n-1)}=$

$\dfrac{2n-3(n-1)}{n(n-1)}=$

$\dfrac{2n-3n+3}{n(n-1)}=\dfrac{-n+3}{n(n-1)}$

71. $\dfrac{4}{n}-\dfrac{6}{n+4}=$

$\left(\dfrac{4}{n}\right)\left(\dfrac{n+4}{n+4}\right)-\left(\dfrac{6}{n+4}\right)\left(\dfrac{n}{n}\right)=$

$\dfrac{4(n+4)}{n(n+4)}-\dfrac{6n}{n(n+4)}=\dfrac{4(n+4)-6n}{n(n+4)}=$

$\dfrac{4n+16-6n}{n(n+4)}=\dfrac{-2n+16}{n(n+4)}$

73. $\dfrac{6}{x}-\dfrac{12}{2x+1}=$

$\left(\dfrac{6}{x}\right)\left(\dfrac{2x+1}{2x+1}\right)-\left(\dfrac{12}{2x+1}\right)\left(\dfrac{x}{x}\right)=$

$\dfrac{6(2x+1)}{x(2x+1)}-\dfrac{12x}{x(2x+1)}=$

$\dfrac{6(2x+1)-12x}{x(2x+1)}=\dfrac{12x+6-12x}{x(2x+1)}=$

$\dfrac{6}{x(2x+1)}$

75. $\dfrac{4}{x+4}+\dfrac{6}{x-3}=$

$\left(\dfrac{4}{x+4}\right)\left(\dfrac{x-3}{x-3}\right)+\left(\dfrac{6}{x-3}\right)\left(\dfrac{x+4}{x+4}\right)=$

$\dfrac{4(x-3)}{(x+4)(x-3)}+\dfrac{6(x+4)}{(x-3)(x+4)}=$

$\dfrac{4(x-3)+6(x+4)}{(x+4)(x-3)}=$

$\dfrac{4x-12+6x+24}{(x+4)(x-3)}=\dfrac{10x+12}{(x+4)(x-3)}$

77. $\frac{3}{x-2} - \frac{9}{x+1} =$

$\left(\frac{3}{x-2}\right)\left(\frac{x+1}{x+1}\right) - \left(\frac{9}{x+1}\right)\left(\frac{x-2}{x-2}\right) =$

$\frac{3(x+1)}{(x-2)(x+1)} - \frac{9(x-2)}{(x+1)(x-2)} =$

$\frac{3(x+1) - 9(x-2)}{(x-2)(x+1)} =$

$\frac{3x+3-9x+18}{(x-2)(x+1)} = \frac{-6x+21}{(x-2)(x+1)}$

79. $\frac{3}{2x-1} - \frac{4}{3x+1} =$

$\left(\frac{3}{2x-1}\right)\left(\frac{3x+1}{3x+1}\right) - \left(\frac{4}{3x+1}\right)\left(\frac{2x-1}{2x-1}\right) =$

$\frac{3(3x+1)}{(2x-1)(3x+1)} - \frac{4(2x-1)}{(3x+1)(2x-1)} =$

$\frac{3(3x+1) - 4(2x-1)}{(2x-1)(3x+1)} =$

$\frac{9x+3-8x+4}{(2x-1)(3x+1)} = \frac{x+7}{(2x-1)(3x+1)}$

PROBLEM SET 7.4 More on Addition and Subtraction of Algebraic Fractions

1. $x^2 - 4x = x(x-4)$

$x = x$

LCD is $x(x-4)$

$\frac{4}{x^2-4x} + \frac{3}{x} = \frac{4}{x(x-4)} + \frac{3}{x} =$

$\frac{4}{x(x-4)} + \left(\frac{3}{x}\right)\left(\frac{x-4}{x-4}\right) =$

The following step will be eliminated in future problems.

$\frac{4}{x(x-4)} + \frac{3(x-4)}{x(x-4)} =$

$\frac{4+3(x-4)}{x(x-4)} = \frac{4+3x-12}{x(x-4)} = \frac{3x-8}{x(x-4)}$

3. $x^2 + 2x = x(x+2)$

$x = x$

LCD is $x(x+2)$

$\frac{7}{x^2+2x} - \frac{5}{x} = \frac{7}{x(x+2)} - \frac{5}{x} =$

$\frac{7}{x(x+2)} - \left(\frac{5}{x}\right)\left(\frac{x+2}{x+2}\right) =$

$\frac{7-5(x+2)}{x(x+2)} = \frac{7-5x-10}{x(x+2)} = \frac{-5x-3}{x(x+2)}$

5. $n = n$

$n^2 - 6n = n(n-6)$

LCD is $n(n-6)$

$\frac{8}{n} - \frac{2}{n^2-6n} = \frac{8}{n} - \frac{2}{n(n-6)} =$

$\left(\frac{8}{n}\right)\left(\frac{n-6}{n-6}\right) - \frac{2}{n(n-6)} =$

$\frac{8(n-6)-2}{n(n-6)} = \frac{8n-48-2}{n(n-6)} = \frac{8n-50}{n(n-6)}$

7. $n^2 + n = n(n+1)$

$n = n$

LCD is $n(n+1)$

$\frac{4}{n^2+n} - \frac{4}{n} = \frac{4}{n(n+1)} - \frac{4}{n} =$

$\frac{4}{n(n+1)} - \left(\frac{4}{n}\right)\left(\frac{n+1}{n+1}\right) =$

$\frac{4-4(n+1)}{n(n+1)} = \frac{4-4n-4}{n(n+1)} =$

$\frac{-4n}{n(n+1)} = -\frac{4}{n+1}$

Problem Set 7.4

9. $2x = 2x$

$x^2 - x = x(x-1)$

LCD is $2x(x-1)$

$$\frac{7}{2x} - \frac{x}{x^2 - x} = \frac{7}{2x} - \frac{x}{x(x-1)} =$$

$$\left(\frac{7}{2x}\right)\left(\frac{x-1}{x-1}\right) - \left[\frac{x}{x(x-1)}\right]\left(\frac{2}{2}\right) =$$

$$\frac{7(x-1) - 2x}{2x(x-1)} = \frac{7x - 7 - 2x}{2x(x-1)} =$$

$$\frac{5x-7}{2x(x-1)}$$

11. $x^2 - 16 = (x-4)(x+4)$

$x + 4 = x + 4$

LCD is $(x-4)(x+4)$

$$\frac{3}{x^2-16} + \frac{5}{x+4} =$$

$$\frac{3}{(x-4)(x+4)} + \frac{5}{x+4} =$$

$$\frac{3}{(x-4)(x+4)} + \left(\frac{5}{x+4}\right)\left(\frac{x-4}{x-4}\right) =$$

$$\frac{3 + 5(x-4)}{(x-4)(x+4)} = \frac{3 + 5x - 20}{(x-4)(x+4)} =$$

$$\frac{5x-17}{(x-4)(x+4)}$$

13. $x^2 - 1 = (x-1)(x+1)$

$x - 1 = x - 1$

LCD is $(x-1)(x+1)$

$$\frac{8x}{x^2-1} - \frac{4}{x-1} =$$

$$\frac{8x}{(x-1)(x+1)} - \frac{4}{x-1} =$$

$$\frac{8x}{(x-1)(x+1)} - \left(\frac{4}{x-1}\right)\left(\frac{x+1}{x+1}\right) =$$

$$\frac{8x - 4(x+1)}{(x-1)(x+1)} = \frac{8x - 4x - 4}{(x-1)(x+1)} =$$

$$\frac{4x-4}{(x-1)(x+1)} = \frac{4(x-1)}{(x-1)(x+1)} = \frac{4}{x+1}$$

15. $a^2 - 2a = a(a-2)$

$a^2 + 2a = a(a+2)$

LCD is $a(a-2)(a+2)$

$$\frac{4}{a^2-2a} + \frac{7}{a^2+2a} =$$

$$\frac{4}{a(a-2)} + \frac{7}{a(a+2)} =$$

$$\left[\frac{4}{a(a-2)}\right]\left(\frac{a+2}{a+2}\right) + \left[\frac{7}{a(a+2)}\right]\left(\frac{a-2}{a-2}\right) =$$

$$\frac{4(a+2) + 7(a-2)}{a(a-2)(a+2)} =$$

$$\frac{4a + 8 + 7a - 14}{a(a-2)(a+2)} =$$

$$\frac{11a-6}{a(a-2)(a+2)}$$

17. $x^2 - 6x = x(x-6)$

$x^2 + 6x = x(x+6)$

LCD is $x(x-6)(x+6)$

$$\frac{1}{x^2-6x} - \frac{1}{x^2+6x} =$$

$$\frac{1}{x(x-6)} - \frac{1}{x(x+6)} =$$

$$\left[\frac{1}{x(x-6)}\right]\left(\frac{x+6}{x+6}\right) - \left[\frac{1}{x(x+6)}\right]\left(\frac{x-6}{x-6}\right) =$$

$$\frac{1(x+6) - 1(x-6)}{x(x-6)(x+6)} = \frac{x + 6 - x + 6}{x(x-6)(x+6)}$$

$$\frac{12}{x(x-6)(x+6)} =$$

19. $n^2 - 16 = (n-4)(n+4)$

$3n + 12 = 3(n+4)$

LCD is $3(n-4)(n+4)$

$$\frac{n}{n^2-16} - \frac{2}{3n+12} =$$

$$\frac{n}{(n-4)(n+4)} - \frac{2}{3(n+4)} =$$

$$\left[\frac{n}{(n-4)(n+4)}\right]\left(\frac{3}{3}\right) - \left[\frac{2}{3(n+4)}\right]\left(\frac{n-4}{n-4}\right) =$$

$$\frac{3n-2(n-4)}{3(n-4)(n+4)} = \frac{3n-2n+8}{3(n-4)(n+4)} =$$

$$\frac{n+8}{3(n-4)(n+4)}$$

21. $6x+4 = 2(3x+2)$
$9x+6 = 3(3x+2)$
LCD is $6(3x+2)$

$$\frac{5x}{6x+4} + \frac{2x}{9x+6} =$$

$$\frac{5x}{2(3x+2)} + \frac{2x}{3(3x+2)} =$$

$$\left[\frac{5x}{2(3x+2)}\right]\left(\frac{3}{3}\right) + \left[\frac{2x}{3(3x+2)}\right]\left(\frac{2}{2}\right) =$$

$$\frac{15x+4x}{6(3x+2)} = \frac{19x}{6(3x+2)}$$

23. $5x+5 = 5(x+1)$
$3x+3 = 3(x+1)$
LCD is $15(x+1)$

$$\frac{x-1}{5x+5} - \frac{x-4}{3x+3} =$$

$$\frac{x-1}{5(x+1)} - \frac{x-4}{3(x+1)} =$$

$$\left[\frac{x-1}{5(x+1)}\right]\left(\frac{3}{3}\right) - \left[\frac{x-4}{3(x+1)}\right]\left(\frac{5}{5}\right) =$$

$$\frac{3(x-1)-5(x-4)}{15(x+1)} =$$

$$\frac{3x-3-5x+20}{15(x+1)} = \frac{-2x+17}{15(x+1)}$$

25. $x^2+7x+12 = (x+3)(x+4)$
$x^2-9 = (x-3)(x+3)$
LCD is $(x+3)(x+4)(x-3)$

$$\frac{2}{x^2+7x+12} + \frac{3}{x^2-9} =$$

$$\frac{2}{(x+3)(x+4)} + \frac{3}{(x-3)(x+3)} =$$

$$\left[\frac{2}{(x+3)(x+4)}\right]\left(\frac{x-3}{x-3}\right) + \left[\frac{3}{(x-3)(x+3)}\right]\left(\frac{x+4}{x+4}\right) =$$

$$\frac{2(x-3)+3(x+4)}{(x+3)(x+4)(x-3)} =$$

$$\frac{2x-6+3x+12}{(x+3)(x+4)(x-3)} =$$

$$\frac{5x+6}{(x+3)(x+4)(x-3)}$$

27. $x^2+6x+8 = (x+2)(x+4)$
$x^2-3x-10 = (x-5)(x+2)$
LCD is $(x+2)(x+4)(x-5)$

$$\frac{x}{x^2+6x+8} - \frac{5}{x^2-3x-10} =$$

$$\frac{x}{(x+2)(x+4)} - \frac{5}{(x-5)(x+2)} =$$

$$\left[\frac{x}{(x+2)(x+4)}\right]\left(\frac{x-5}{x-5}\right) - \left[\frac{5}{(x-5)(x+2)}\right]\left(\frac{x+4}{x+4}\right) =$$

$$\frac{x(x-5)-5(x+4)}{(x+2)(x+4)(x-5)} =$$

$$\frac{x^2-5x-5x-20}{(x+2)(x+4)(x-5)} =$$

$$\frac{x^2-10x-20}{(x+2)(x+4)(x-5)}$$

29. $ab+b^2 = b(a+b)$
$a^2+ab = a(a+b)$
LCD is $ab(a+b)$

$$\frac{a}{ab+b^2} - \frac{b}{a^2+ab} =$$

$$\frac{a}{b(a+b)} - \frac{b}{a(a+b)} =$$

$$\left[\frac{a}{b(a+b)}\right]\left(\frac{a}{a}\right)-\left[\frac{b}{a(a+b)}\right]\left(\frac{b}{b}\right)=$$
$$\frac{a^2-b^2}{ab(a+b)}=\frac{(a-b)(a+b)}{ab(a+b)}=$$
$$\frac{a-b}{ab}$$

31. $x-5=x-5$
$x^2-25=(x-5)(x+5)$
$x+5=x+5$
LCD is $(x-5)(x+5)$

$$\frac{3}{x-5}-\frac{4}{x^2-25}+\frac{5}{x+5}=$$
$$\frac{3}{x-5}-\frac{4}{(x-5)(x+5)}+\frac{5}{x+5}=$$
$$\left(\frac{3}{x-5}\right)\left(\frac{x+5}{x+5}\right)-\frac{4}{(x-5)(x+5)}+\left(\frac{5}{x+5}\right)\left(\frac{x-5}{x-5}\right)=$$
$$\frac{3(x+5)-4+5(x-5)}{(x-5)(x+5)}=$$
$$\frac{3x+15-4+5x-25}{(x-5)(x+5)}=\frac{8x-14}{(x-5)(x+5)}$$

33. $x^2-2x=x(x-2)$
$x^2+2x=x(x+2)$
$x^2-4=(x-2)(x+2)$
LCD is $x(x-2)(x+2)$

$$\frac{10}{x^2-2x}+\frac{8}{x^2+2x}-\frac{3}{x^2-4}=$$
$$\frac{10}{x(x-2)}+\frac{8}{x(x+2)}-\frac{3}{(x-2)(x+2)}=$$
$$\left[\frac{10}{x(x-2)}\right]\left(\frac{x+2}{x+2}\right)+\left[\frac{8}{x(x+2)}\right]\left(\frac{x-2}{x-2}\right)-\left[\frac{3}{(x-2)(x+2)}\right]\left(\frac{x}{x}\right)=$$
$$\frac{10(x+2)+8(x-2)-3(x)}{x(x-2)(x+2)}=$$
$$\frac{10x+20+8x-16-3x}{x(x-2)(x+2)}=$$
$$\frac{15x+4}{x(x-2)(x+2)}$$

35. $x^2+7x+10=(x+2)(x+5)$
$x+2=x+2$
$x+5=x+5$
LCD is $(x+2)(x+5)$

$$\frac{3x}{x^2+7x+10}-\frac{2}{x+2}+\frac{3}{x+5}=$$
$$\frac{3x}{(x+2)(x+5)}-\frac{2}{x+2}+\frac{3}{x+5}=$$
$$\frac{3x}{(x+2)(x+5)}-\left(\frac{2}{x+2}\right)\left(\frac{x+5}{x+5}\right)+\left(\frac{3}{x+5}\right)\left(\frac{x+2}{x+2}\right)=$$
$$\frac{3x-2(x+5)+3(x+2)}{(x+2)(x+5)}=$$
$$\frac{3x-2x-10+3x+6}{(x+2)(x+5)}=$$
$$\frac{4x-4}{(x+2)(x+5)}$$

37. $3x^2+7x-20=(3x-5)(x+4)$
$3x-5=3x-5$
$x+4=x+4$
LCD is $(3x-5)(x+4)$

$$\frac{5x}{3x^2+7x-20}-\frac{1}{3x-5}-\frac{2}{x+4}=$$
$$\frac{5x}{(3x-5)(x+4)}-\frac{1}{3x-5}-\frac{2}{x+4}=$$
$$\frac{5x}{(3x-5)(x+4)}-\left(\frac{1}{3x-5}\right)\left(\frac{x+4}{x+4}\right)$$

$$-\left(\frac{2}{x+4}\right)\left(\frac{3x-5}{3x-5}\right)=$$

$$\frac{5x-1(x+4)-2(3x-5)}{(3x-5)(x+4)}=$$

$$\frac{5x-x-4-6x+10}{(3x-5)(x+4)}=$$

$$\frac{-2x+6}{(3x-5)(x+4)}$$

39. $x+4=x+4$

$x-3=x-3$

$x^2+x-12=(x+4)(x-3)$

LCD is $(x+4)(x-3)$

$$\frac{2}{x+4}-\frac{1}{x-3}+\frac{2x+1}{x^2+x-12}=$$

$$\frac{2}{x+4}-\frac{1}{x-3}+\frac{2x+1}{(x+4)(x-3)}=$$

$$\left(\frac{2}{x+4}\right)\left(\frac{x-3}{x-3}\right)$$

$$-\left(\frac{1}{x-3}\right)\left(\frac{x+4}{x+4}\right)$$

$$+\frac{2x+1}{(x+4)(x-3)}=$$

$$\frac{2(x-3)-1(x+4)+2x+1}{(x+4)(x-3)}=$$

$$\frac{2x-6-x-4+2x+1}{(x+4)(x-3)}=$$

$$\frac{3x-9}{(x+4)(x-3)}=\frac{3(x-3)}{(x+4)(x-3)}=\frac{3}{x+4}$$

41. LCM of 2, 3, 4 and 6 is 12.

$$\frac{\frac{1}{2}-\frac{3}{4}}{\frac{1}{6}+\frac{1}{3}}=\left(\frac{12}{12}\right)\left[\frac{\frac{1}{2}-\frac{3}{4}}{\frac{1}{6}+\frac{1}{3}}\right]=$$

$$\frac{12\left(\frac{1}{2}-\frac{3}{4}\right)}{12\left(\frac{1}{6}+\frac{1}{3}\right)}=\frac{6-9}{2+4}=\frac{-3}{6}=-\frac{1}{2}$$

43. LCM of 3, 6 and 9 is 18.

$$\frac{\frac{2}{9}+\frac{1}{3}}{\frac{5}{6}-\frac{2}{3}}=\left(\frac{18}{18}\right)\left[\frac{\frac{2}{9}+\frac{1}{3}}{\frac{5}{6}-\frac{2}{3}}\right]=$$

$$\frac{18\left(\frac{2}{9}+\frac{1}{3}\right)}{18\left(\frac{5}{6}-\frac{2}{3}\right)}=\frac{4+6}{15-12}=\frac{10}{3}$$

45. LCM of 1, 3 and 4 is 12.

$$\frac{3-\frac{2}{3}}{2+\frac{1}{4}}=\left(\frac{12}{12}\right)\left[\frac{3-\frac{2}{3}}{2+\frac{1}{4}}\right]=$$

$$\frac{12\left(3-\frac{2}{3}\right)}{12\left(2+\frac{1}{4}\right)}=\frac{36-8}{24+3}=\frac{28}{27}$$

47. LCM of x and y is xy.

$$\frac{\frac{3}{x}}{\frac{9}{y}}=\left(\frac{xy}{xy}\right)\left[\frac{\frac{3}{x}}{\frac{9}{y}}\right]=$$

$$\frac{xy\left(\frac{3}{x}\right)}{xy\left(\frac{9}{y}\right)}=\frac{3y}{9x}=\frac{y}{3x}$$

49. LCM of x and y is xy.

$$\frac{\frac{2}{x}+\frac{3}{y}}{\frac{5}{x}-\frac{1}{y}}=\left(\frac{xy}{xy}\right)\left[\frac{\frac{2}{x}+\frac{3}{y}}{\frac{5}{x}-\frac{1}{y}}\right]=$$

$$\frac{xy\left(\frac{2}{x}+\frac{3}{y}\right)}{xy\left(\frac{5}{x}-\frac{1}{y}\right)}=\frac{2y+3x}{5y-x}$$

51. LCM of x^2 and y is x^2y.

$$\frac{\frac{1}{y}-\frac{4}{x^2}}{\frac{7}{x}-\frac{3}{y}}=\left(\frac{x^2y}{x^2y}\right)\left[\frac{\frac{1}{y}-\frac{4}{x^2}}{\frac{7}{x}-\frac{3}{y}}\right]=$$

$$\frac{x^2y\left(\frac{1}{y}-\frac{4}{x^2}\right)}{x^2y\left(\frac{7}{x}-\frac{3}{y}\right)}=\frac{x^2-4y}{7xy-3x^2}$$

53. LCM of x.

$$\frac{\frac{6}{x}+2}{\frac{3}{x}+4}=\left(\frac{x}{x}\right)\left[\frac{\frac{6}{x}+2}{\frac{3}{x}+4}\right]=$$

$$\frac{x\left(\frac{6}{x}+2\right)}{x\left(\frac{3}{x}+4\right)}=\frac{6+2x}{3+4x}$$

55. LCM of 2, 3, and x^2 is $6x^2$.

$$\frac{\frac{3}{2x^2}-\frac{4}{x}}{\frac{5}{3x}+\frac{7}{x^2}}=\left(\frac{6x^2}{6x^2}\right)\left[\frac{\frac{3}{2x^2}-\frac{4}{x}}{\frac{5}{3x}+\frac{7}{x^2}}\right]=$$

$$\frac{6x^2\left(\frac{3}{2x^2}-\frac{4}{x}\right)}{6x^2\left(\frac{5}{3x}+\frac{7}{x^2}\right)}=\frac{9-24x}{10x+42}$$

57. LCM of 2, 4 and x is $4x$.

$$\frac{\frac{x+2}{4}}{\frac{1}{x}+\frac{3}{2}}=\left(\frac{4x}{4x}\right)\left[\frac{\frac{x+2}{4}}{\frac{1}{x}+\frac{3}{2}}\right]=$$

$$\frac{4x\left(\frac{x+2}{4}\right)}{4x\left(\frac{1}{x}+\frac{3}{2}\right)}=\frac{x(x+2)}{4+6x}=$$

$$\frac{x^2+2x}{6x+4}$$

59. LCM is $x-1$.

$$\frac{\frac{1}{x-1}-2}{\frac{3}{x-1}+4}=\left(\frac{x-1}{x-1}\right)\left[\frac{\frac{1}{x-1}-2}{\frac{3}{x-1}+4}\right]=$$

$$\frac{(x-1)\left(\frac{1}{x-1}-2\right)}{(x-1)\left(\frac{3}{x-1}+4\right)}=\frac{1-2(x-1)}{3+4(x-1)}=$$

$$\frac{1-2x+2}{3+4x-4}=\frac{-2x+3}{4x-1}$$

61. She would have completed

$$\frac{m\text{ minutes}}{40\text{ minutes}}=\frac{m}{40}\text{ of the course.}$$

63. Let t represent the time in hours. Use $rt=d$ as a guideline and solve for t.

$$\left(r\,\frac{\text{kilometers}}{\text{per hour}}\right)(t)=k\text{ kilometers}$$

$$t=\frac{k\text{ kilometers}}{r\,\frac{\text{kilometers}}{\text{per hour}}}$$

Her time would be $\frac{k}{r}$ hours.

65. The price per liter would be $\frac{d}{l}$ dollars per liter.

67. Let n represent the number, then $\frac{34}{n}$ represents the other number.

69. Let w represent the width of the rectangle. Use $A = lw$ as a guideline and solve for w.

$$lw = 47$$
$$w = \frac{47}{l}$$

The width of the rectangle would be $\frac{47}{l}$ inches.

71. Let h represent the altitude to the given side.

$$48 = \frac{1}{2}bh$$
$$96 = bh$$
$$\frac{96}{b} = h$$

The altitude to the given side would be $\frac{96}{b}$ feet.

PROBLEM SET 7.5 Fractional Equations and Problem Solving

1. $\frac{x}{2} + \frac{x}{3} = 10$

Multiply both sides by 6.

$$6\left(\frac{x}{2} + \frac{x}{3}\right) = 6(10)$$
$$3x + 2x = 60$$
$$5x = 60$$
$$x = 12$$

The solution set is $\{12\}$.

3. $\frac{x}{6} - \frac{4x}{3} = \frac{1}{9}$

Multiply both sides by 18.

$$18\left(\frac{x}{6} - \frac{4x}{3}\right) = 18\left(\frac{1}{9}\right)$$
$$3x - 24x = 2$$
$$-21x = 2$$
$$x = -\frac{2}{21}$$

The solution set is $\left\{-\frac{2}{21}\right\}$.

5. $\frac{n}{2} + \frac{n-1}{6} = \frac{5}{2}$

Multiply both sides by 6.

$$6\left(\frac{n}{2} + \frac{n-1}{6}\right) = 6\left(\frac{5}{2}\right)$$
$$3n + (n-1) = 15$$
$$4n - 1 = 15$$
$$4n = 16$$
$$n = 4$$

The solution set is $\{4\}$.

7. $\frac{t-3}{4} + \frac{t+1}{9} = -1$

Multiply both sides by 36.

$$36\left(\frac{t-3}{4} + \frac{t+1}{9}\right) = 36(-1)$$
$$9(t-3) + 4(t+1) = -36$$
$$9t - 27 + 4t + 4 = -36$$
$$13t - 23 = -36$$
$$13t = -13$$
$$t = -1$$

The solution set is $\{-1\}$.

Problem Set 7.5

9. $$\frac{2x+3}{3}+\frac{3x-4}{4}=\frac{17}{4}$$

Multiply both sides by 12.

$$12\left(\frac{2x+3}{3}+\frac{3x-4}{4}\right)=12\left(\frac{17}{4}\right)$$
$$4(2x+3)+3(3x-4)=3(17)$$
$$8x+12+9x-12=51$$
$$17x=51$$
$$x=3$$

The solution set is $\{3\}$.

11. $$\frac{x-4}{8}-\frac{x+5}{4}=3$$

Multiply both sides by 8.

$$8\left(\frac{x-4}{8}-\frac{x+5}{4}\right)=8(3)$$
$$x-4-2(x+5)=24$$
$$x-4-2x-10=24$$
$$-x-14=24$$
$$-x=38$$
$$x=-38$$

The solution set is $\{-38\}$.

13. $$\frac{3x+2}{5}-\frac{2x-1}{6}=\frac{2}{15}$$

Multiply both sides by 30.

$$30\left(\frac{3x+2}{5}-\frac{2x-1}{6}\right)=30\left(\frac{2}{15}\right)$$
$$6(3x+2)-5(2x-1)=2(2)$$
$$18x+12-10x+5=4$$
$$8x+17=4$$
$$8x=-13$$
$$x=-\frac{13}{8}$$

The solution set is $\left\{-\frac{13}{8}\right\}$.

15. $$\frac{1}{x}+\frac{2}{3}=\frac{7}{6},\ x\neq 0$$

Multiply both sides by $6x$.

$$6x\left(\frac{1}{x}+\frac{2}{3}\right)=6x\left(\frac{7}{6}\right)$$
$$6(1)+2x(2)=x(7)$$
$$6+4x=7x$$
$$6=3x$$
$$2=x$$

The solution set is $\{2\}$.

17. $$\frac{5}{3n}-\frac{1}{9}=\frac{1}{n},\ n\neq 0$$

Multiply both sides by $9n$.

$$9n\left(\frac{5}{3n}-\frac{1}{9}\right)=9n\left(\frac{1}{n}\right)$$
$$3(5)-n(1)=9(1)$$
$$15-n=9$$
$$-n=-6$$
$$n=6$$

The solution set is $\{6\}$.

19. $$\frac{1}{2x}+3=\frac{4}{3x},\ x\neq 0$$

Multiply both sides by $6x$.

$$6x\left(\frac{1}{2x}+3\right)=6x\left(\frac{4}{3x}\right)$$
$$3(1)+6x(3)=2(4)$$
$$3+18x=8$$
$$18x=5$$
$$x=\frac{5}{18}$$

The solution set is $\left\{\frac{5}{18}\right\}$.

21. $\frac{4}{5t} - 1 = \frac{3}{2t},\ t \neq 0$

Multiply both sides by $10t$.

$$10t\left(\frac{4}{5t} - 1\right) = 10t\left(\frac{3}{2t}\right)$$
$$2(4) - 10t(1) = 5(3)$$
$$8 - 10t = 15$$
$$-10t = 7$$
$$t = -\frac{7}{10}$$

The solution set is $\left\{-\frac{7}{10}\right\}$.

23. $\frac{-5}{4h} + \frac{7}{6h} = \frac{1}{4},\ h \neq 0$

Multiply both sides by $12h$.

$$12h\left(\frac{-5}{4h} + \frac{7}{6h}\right) = 12h\left(\frac{1}{4}\right)$$
$$3(-5) + 2(7) = 3h(1)$$
$$-15 + 14 = 3h$$
$$-1 = 3h$$
$$-\frac{1}{3} = h$$

The solution set is $\left\{-\frac{1}{3}\right\}$.

25. $\frac{90 - n}{n} = 10 + \frac{2}{n},\ n \neq 0$

Multiply both sides by n.

$$n\left(\frac{90 - n}{n}\right) = n\left(10 + \frac{2}{n}\right)$$
$$90 - n = 10n + 2$$
$$90 - 11n = 2$$
$$-11n = -88$$
$$n = 8$$

The solution set is $\{8\}$.

27. $\frac{n}{49 - n} = 3 + \frac{1}{49 - n},\ n \neq 49$

Multiply both sides by $(49 - n)$.

$$(49 - n)\left(\frac{n}{49 - n}\right) = (49 - n)\left(3 + \frac{1}{49 - n}\right)$$
$$n = 3(49 - n) + 1$$
$$n = 147 - 3n + 1$$
$$n = 148 - 3n$$
$$4n = 148$$
$$n = 37$$

The solution set is $\{37\}$.

29. $\frac{x}{x + 3} - 2 = \frac{-3}{x + 3},\ x \neq -3$

Multiply both sides by $(x + 3)$.

$$(x + 3)\left(\frac{x}{x + 3} - 2\right) = (x + 3)\left(\frac{-3}{x + 3}\right)$$
$$x - 2(x + 3) = -3$$
$$x - 2x - 6 = -3$$
$$-x - 6 = -3$$
$$-x = 3$$
$$x = -3$$

Since the initial restriction was $x \neq -3$, the solution is $\emptyset$.

31. $\frac{7}{x + 3} = \frac{5}{x - 9},\ x \neq -3$ and $x \neq 9$

Multiply both sides by $(x + 3)(x - 9)$.

$$(x + 3)(x - 9)\left(\frac{7}{x + 3}\right) = (x + 3)(x - 9)\left(\frac{5}{x - 9}\right)$$
$$7(x - 9) = 5(x + 3)$$
$$7x - 63 = 5x + 15$$
$$2x - 63 = 15$$
$$2x = 78$$
$$x = 39$$

The solution set is $\{39\}$.

Problem Set 7.5

33. $$\frac{x}{x+2}+3=\frac{1}{x+2},\ x\neq -2$$
Multiply both sides by $(x+2)$.
$$(x+2)\left(\frac{x}{x+2}+3\right)=(x+2)\left(\frac{1}{x+2}\right)$$
$$x+3(x+2)=1$$
$$x+3x+6=1$$
$$4x+6=1$$
$$4x=-5$$
$$x=-\frac{5}{4}$$
The solution set is $\left\{-\frac{5}{4}\right\}$.

35. $$-1-\frac{5}{x-2}=\frac{3}{x-2},\ x\neq 2$$
Multiply both sides by $(x-2)$.
$$(x-2)\left(-1-\frac{5}{x-2}\right)=(x-2)\left(\frac{3}{x-2}\right)$$
$$-1(x-2)-5=3$$
$$-x+2-5=3$$
$$-x-3=3$$
$$-x=6$$
$$x=-6$$
The solution set is $\{-6\}$.

37. $$1+\frac{n+1}{2n}=\frac{3}{4},\ n\neq 0$$
Multiply both sides by $4n$.
$$4n\left(1+\frac{n+1}{2n}\right)=4n\left(\frac{3}{4}\right)$$
$$4n+2(n+1)=n(3)$$
$$4n+2n+2=3n$$
$$6n+2=3n$$
$$2=-3n$$
$$-\frac{2}{3}=n$$
The solution set is $\left\{-\frac{2}{3}\right\}$.

39. $$\frac{h}{2}-\frac{h}{4}+\frac{h}{3}=1$$
Multiply both sides by 12.
$$12\left(\frac{h}{2}-\frac{h}{4}+\frac{h}{3}\right)=12(1)$$
$$6h-3h+4h=12$$
$$7h=12$$
$$h=\frac{12}{7}$$
The solution set is $\left\{\frac{12}{7}\right\}$.

41. Let x represent the denominator of the fraction, then $x-8$ represents the numerator of the fraction.
$$\frac{x-8}{x}=\frac{5}{6}$$
Cross products are equal.
$$6(x-8)=5x$$
$$6x-48=5x$$
$$x-48=0$$
$$x=48$$
The fraction is $\frac{48-8}{48}=\frac{40}{48}$.

43. Let n represent the number to be added.
$$\frac{2+n}{5+n}=\frac{4}{5}$$
$$5(2+n)=4(5+n)$$
$$10+5n=20+4n$$
$$10+n=20$$
$$n=10$$
The number to be added would be 10.

45. Let n represent the smaller number, then $65-n$ represents the larger number.
$$\frac{65-n}{n}=8+\frac{2}{n}$$
$$n\left(\frac{65-n}{n}\right)=n\left(8+\frac{2}{n}\right)$$
$$65-n=8n+2$$
$$65-9n=2$$
$$-9n=-63$$
$$n=7$$
The smaller number is 7 and the larger number is $65-7=58$.

47. Let x represent the numerator of the fraction, then $x-4$ represents the denominator.

$$\frac{x+6}{2(x-4)} = 1$$
$$2(x-4)\left[\frac{x+6}{2(x-4)}\right] = 2(x-4)(1)$$
$$x+6 = 2x-8$$
$$x+14 = 2x$$
$$14 = x$$

The original fraction was $\frac{14}{14-4} = \frac{14}{10}$.

49. Let x represent the rate at which they both rode. Heidi's time was $3\frac{1}{3} = \frac{10}{3}$ hours more than Abby's time.

	Rate	Time	Distance
Heidi	x	$\frac{125}{x}$	125
Abby	x	$\frac{75}{x}$	75

Heidi's time $=$ Abby's time plus $3\frac{1}{3}$

$$\frac{125}{x} = \frac{75}{x} + \frac{10}{3}$$
$$3x\left(\frac{125}{x}\right) = 3x\left(\frac{75}{x} + \frac{10}{3}\right)$$
$$3(125) = 3(75) + x(10)$$
$$375 = 225 + 10x$$
$$150 = 10x$$
$$15 = x$$

They both rode at a rate of 15 miles per hour.

51. Let r represent Dave's rate, then $r+4$ represents Kent's rate. Their times are equal.

$$\text{time} = \frac{\text{distance Dave rides}}{\text{Dave's time}} = \frac{\text{distance Kent rides}}{\text{Kent's time}}$$
$$\frac{250}{r} = \frac{270}{r+4}$$

Cross products are equal.

$$250(r+4) = 270r$$
$$250r + 1000 = 270r$$
$$1000 = 20r$$
$$50 = r$$

Dave drove at a rate of 50 miles per hour and Kent's rate is 54 miles per hour.

PROBLEM SET 7.6 More Fractional Equations anf Problem Solving

1. $$\frac{4}{x} + \frac{7}{6} = \frac{1}{x} + \frac{2}{3x},\ x \neq 0$$

Multiply both sides by $6x$.

$$6x\left(\frac{4}{x} + \frac{7}{6}\right) = 6x\left(\frac{1}{x} + \frac{2}{3x}\right)$$
$$6(4) + x(7) = 6(1) + 2(2)$$
$$24 + 7x = 6 + 4$$
$$24 + 7x = 10$$
$$7x = -14$$
$$x = -2$$

The solution set is $\{-2\}$.

3. $$\frac{3}{2(x+1)} + \frac{4}{x+1} = \frac{11}{12},\ x \neq -1$$

Factor the first denominator.

Multiply both sides by $12(x+1)$.

$$12(x+1)\left[\frac{3}{2(x+1)} + \frac{4}{x+1}\right] = 12(x+1)\left(\frac{11}{12}\right)$$
$$6(3) + 12(4) = 11(x+1)$$
$$18 + 48 = 11x + 11$$
$$66 = 11x + 11$$
$$55 = 11x$$
$$5 = x$$

The solution set is $\{5\}$.

Problem Set 7.6

5. $$\frac{5}{2(n-5)} - \frac{3}{n-5} = 1,\ n \neq 5$$

Factor the first denominator.

Multiply both sides by $2(n-5)$.

$$2(n-5)\left[\frac{5}{2(n-5)} - \frac{3}{n-5}\right] = 2(n-5)(1)$$
$$5 - 2(3) = 2n - 10$$
$$5 - 6 = 2n - 10$$
$$-1 = 2n - 10$$
$$9 = 2n$$
$$\frac{9}{2} = n$$

The solution set is $\left\{\frac{9}{2}\right\}$.

7. $$\frac{3}{2t} - \frac{5}{t} = \frac{7}{5t} + 1,\ t \neq 0$$

Multiply by $10t$.

$$10t\left(\frac{3}{2t} - \frac{5}{t}\right) = 10t\left(\frac{7}{5t} + 1\right)$$
$$5(3) - 10(5) = 2(7) + 10t$$
$$15 - 50 = 14 + 10t$$
$$-35 = 14 + 10t$$
$$-49 = 10t$$
$$-\frac{49}{10} = t$$

The solution set is $\left\{-\frac{49}{10}\right\}$.

9. $$\frac{x}{x-2} + \frac{4}{x+2} = 1,\ x \neq -2 \text{ and } x \neq 2$$

Multiply by the LCD.

$$(x-2)(x+2)\left(\frac{x}{x-2} + \frac{4}{x+2}\right) = (x-2)(x+2)(1)$$
$$(x+2)(x) + (x-2)(4) = x^2 - 4$$
$$x^2 + 2x + 4x - 8 = x^2 - 4$$
$$6x - 8 = -4$$
$$6x = 4$$
$$x = \frac{4}{6} = \frac{2}{3}$$

The solution set is $\left\{\frac{2}{3}\right\}$.

11. $$\frac{x}{x-4} - \frac{2x}{x+4} = -1,\ x \neq -4 \text{ and } x \neq 4$$

Multiply by the LCD.

$$(x-4)(x+4)\left(\frac{x}{x-4} - \frac{2x}{x+4}\right) = (x-4)(x+4)(-1)$$
$$x(x+4) - 2x(x-4) = -1(x^2 - 16)$$
$$x^2 + 4x - 2x^2 + 8x = -x^2 + 16$$
$$-x^2 + 12x = -x^2 + 16$$
$$12x = 16$$
$$x = \frac{16}{12} = \frac{4}{3}$$

The solution set is $\left\{\frac{4}{3}\right\}$.

13. $$\frac{3n}{n+3} - \frac{n}{n-3} = 2,\ n \neq -3 \text{ and } n \neq 3$$

Multiply by the LCD.

$$(n+3)(n-3)\left(\frac{3n}{n+3} - \frac{n}{n-3}\right) = (n+3)(n-3)(2)$$

$$3n(n-3) - n(n+3) = 2\left(n^2 - 9\right)$$

$$3n^2 - 9n - n^2 - 3n = 2n^2 - 18$$

$$2n^2 - 12n = 2n^2 - 18$$

$$-12n = -18$$

$$n = \frac{18}{12} = \frac{3}{2}$$

The solution set is $\left\{\frac{3}{2}\right\}$.

15. $$\frac{3}{t^2-4} + \frac{5}{t+2} = \frac{2}{t-2},\ t \neq -2 \text{ and } t \neq 2$$

Factor the denominator.

$$\frac{3}{(t-2)(t+2)} + \frac{5}{t+2} = \frac{2}{t-2}$$

$$(t-2)(t+2)\left[\frac{3}{(t-2)(t+2)} + \frac{5}{t+2}\right] = (t-2)(t+2)\left(\frac{2}{t-2}\right)$$

$$3 + 5(t-2) = 2(t+2)$$

$$3 + 5t - 10 = 2t + 4$$

$$5t - 7 = 2t + 4$$

$$3t - 7 = 4$$

$$3t = 11$$

$$t = \frac{11}{3}$$

The solution set is $\left\{\frac{11}{3}\right\}$.

17. $$\frac{4}{x-1}-\frac{2x-3}{x^2-1}=\frac{6}{x+1},\ x\neq -1 \text{ and } x\neq 1$$

Factor the denominator.

$$\frac{4}{x-1}-\frac{2x-3}{(x-1)(x+1)}=\frac{6}{x+1}$$
$$(x-1)(x+1)\left[\frac{4}{x-1}-\frac{2x-3}{(x-1)(x+1)}\right]=(x-1)(x+1)\left(\frac{6}{x+1}\right)$$
$$4(x+1)-(2x-3)=6(x-1)$$
$$4x+4-2x+3=6x-6$$
$$2x+7=6x-6$$
$$-4x+7=-6$$
$$-4x=-13$$
$$x=\frac{13}{4}$$

The solution set is $\left\{\frac{13}{4}\right\}$.

19. $$8+\frac{5}{y^2+2y}=\frac{3}{y+2},\ y\neq -2 \text{ and } y\neq 0$$

Factor the denominator.

$$8+\frac{5}{y(y+2)}=\frac{3}{(y+2)}$$
$$y(y+2)\left[8+\frac{5}{y(y+2)}\right]=y(y+2)\left(\frac{3}{y+2}\right)$$
$$8y(y+2)+5=3y$$
$$8y^2+16y+5=3y$$
$$8y^2+13y+5=0$$
$$(8y+5)(y+1)=0$$
$$8y+5=0 \quad \text{or} \quad y+1=0$$
$$8y=-5 \quad \text{or} \quad y=-1$$
$$y=-\frac{5}{8} \quad \text{or} \quad y=-1$$

The solution set is $\left\{-1,\ -\frac{5}{8}\right\}$.

21. $$n+\frac{1}{n}=\frac{17}{4},\ n\neq 0$$
$$4n\left(n+\frac{1}{n}\right)=4n\left(\frac{17}{4}\right)$$
$$4n^2+4=17n$$

Subtracted $17n$ from both sides.

$$4n^2-17n+4=0$$
$$(4n-1)(n-4)=0$$

$$\begin{aligned} 4n - 1 = 0 \quad &\text{or} \quad n - 4 = 0 \\ 4n = 1 \quad &\text{or} \quad n = 4 \\ n = \frac{1}{4} \quad &\text{or} \quad n = 4 \end{aligned}$$

The solution set is $\left\{\frac{1}{4}, 4\right\}$.

23.

$$\frac{15}{4n} + \frac{15}{4(n+4)} = 1,\ n \neq -4 \text{ and } n \neq 0$$

$$4n(n+4)\left[\frac{15}{4n} + \frac{15}{4(n+4)}\right] = 4n(n+4)(1)$$

$$15(n+4) + 15n = 4n^2 + 16n$$

$$15n + 60 + 15n = 4n^2 + 16n$$

$$30n + 60 = 4n^2 + 16n$$

$$-4n^2 + 14n + 60 = 0$$

Divided by -2.

$$2n^2 - 7n - 30 = 0$$

$$(2n+5)(n-6) = 0$$

$$\begin{aligned} 2n + 5 = 0 \quad &\text{or} \quad n - 6 = 0 \\ 2n = -5 \quad &\text{or} \quad n = 6 \\ n = -\frac{5}{2} \quad &\text{or} \quad n = 6 \end{aligned}$$

The solution set is $\left\{-\frac{5}{2}, 6\right\}$.

25.

$$x - \frac{5x}{x-2} = \frac{-10}{x-2},\ x \neq 2$$

$$(x-2)\left(x - \frac{5x}{x-2}\right) = (x-2)\left(\frac{-10}{x-2}\right)$$

$$x(x-2) - 5x = -10$$

$$x^2 - 2x - 5x = -10$$

$$x^2 - 7x = -10$$

$$x^2 - 7x + 10 = 0$$

$$(x-5)(x-2) = 0$$

$$\begin{aligned} x - 5 = 0 \quad &\text{or} \quad x - 2 = 0 \\ x = 5 \quad &\text{or} \quad x = 2 \end{aligned}$$

Since the original restriction was $x \neq 2$, the solution set is $\{5\}$.

27.

$$
\begin{aligned}
\frac{t}{4t-4}+\frac{5}{t^2-1}&=\frac{1}{4},\ t\neq -1 \text{ and } t\neq 1\\
\frac{t}{4(t-1)}+\frac{5}{(t-1)(t+1)}&=\frac{1}{4}\\
4(t+1)(t-1)\left[\frac{t}{4(t-1)}+\frac{5}{(t-1)(t+1)}\right]&=4(t-1)(t+1)\left(\frac{1}{4}\right)\\
t(t+1)+4(5)&=(t-1)(t+1)\\
t^2+t+20&=t^2-1\\
t+20&=-1\\
t&=-21
\end{aligned}
$$

The solution set is $\{-21\}$.

29.

$$
\begin{aligned}
\frac{3}{n-5}+\frac{4}{n+7}&=\frac{2n+11}{n^2+2n-35},\ n\neq -7 \text{ and } n\neq 5\\
\frac{3}{n-5}+\frac{4}{n+7}&=\frac{2n+11}{(n-5)(n+7)}\\
(n-5)(n+7)\left(\frac{3}{n-5}+\frac{4}{n+7}\right)&=(n-5)(n+7)\left[\frac{2n+11}{(n-5)(n+7)}\right]\\
3(n+7)+4(n-5)&=2n+11\\
3n+21+4n-20&=2n+11\\
7n+1&=2n+11\\
5n+1&=11\\
5n&=10\\
n&=2
\end{aligned}
$$

The solution set is $\{2\}$.

31.

$$
\begin{aligned}
\frac{a}{a+2}+\frac{3}{a+4}&=\frac{14}{a^2+6a+8},\ a\neq -4 \text{ and } a\neq -2\\
\frac{a}{a+2}+\frac{3}{a+4}&=\frac{14}{(a+2)(a+4)}\\
(a+2)(a+4)\left(\frac{a}{a+2}+\frac{3}{a+4}\right)&=(a+2)(a+4)\left[\frac{14}{(a+2)(a+4)}\right]\\
a(a+4)+3(a+2)&=14\\
a^2+4a+3a+6&=14\\
a^2+7a+6&=14\\
a^2+7a-8&=0\\
(a+8)(a-1)&=0
\end{aligned}
$$

$$a+8=0 \quad \text{or} \quad a-1=0$$

$$a=-8 \quad \text{or} \quad a=1$$

The solution set is $\{-8, 1\}$.

33. Let n represent the number, then $\frac{1}{n}$ represents the reciprocal.

$$n+2\left(\frac{1}{n}\right)=\frac{9}{2}$$
$$n+\frac{2}{n}=\frac{9}{2},\ n\neq 0$$
$$2n\left(n+\frac{2}{n}\right)=2n\left(\frac{9}{2}\right)$$
$$2n^2+4=9n$$
$$2n^2-9n+4=0$$
$$(2n-1)(n-4)=0$$
$$2n-1=0 \quad \text{or} \quad n-4=0$$
$$2n=1 \quad \text{or} \quad n=4$$
$$n=\frac{1}{2} \quad \text{or} \quad n=4$$

The number is $\frac{1}{2}$ or 4.

35. Let n represent the number, then $\frac{1}{n}$ represents the reciprocal.

$$n=\frac{1}{n}+\frac{21}{10},\ n\neq 0$$
$$10n(n)=10n\left(\frac{1}{n}+\frac{21}{10}\right)$$
$$10n^2=10+21n$$
$$10n^2-21n-10=0$$
$$(5n+2)(2n-5)=0$$
$$5n+2=0 \quad \text{or} \quad 2n-5=0$$
$$5n=-2 \quad \text{or} \quad 2n=5$$
$$n=-\frac{2}{5} \quad \text{or} \quad n=\frac{5}{2}$$

The number is $-\frac{2}{5}$ or $\frac{5}{2}$.

37. Let r represent Tom's rate, then $r+3$ represents Celia's rate. The information is recorded in the following table:.

	Distance	Rate	Time $\left(t=\frac{d}{r}\right)$
Tom	85	r	$\frac{85}{r}$
Celia	60	$r+3$	$\frac{60}{r+3}$

$$\text{Time of Celia} = \text{Time of Tom} - 2$$

$$\frac{60}{r+3} = \frac{85}{r} - 2,\ r \neq -3 \text{ and } r \neq 0$$

$$r(r+3)\left(\frac{60}{r+3}\right) = r(r+3)\left(\frac{85}{r} - 2\right)$$

$$60r = 85(r+3) - 2r(r+3)$$

$$60r = 85r + 255 - 2r^2 - 6r$$

$$60r = -2r^2 + 79r + 255$$

$$0 = -2r^2 + 19r + 255$$

$$0 = 2r^2 - 19r - 255$$

$$0 = (2r+15)(r-17)$$

$$2r + 15 = 0 \quad \text{or} \quad r - 17 = 0$$

$$2r = -15 \quad \text{or} \quad r = 17$$

$$r = -\frac{15}{2} \quad \text{or} \quad r = 17$$

The negative solution needs to be discarded. Tom's rate is 17 miles per hour and Celia's rate is 20 miles per hour..

39. Let r represent Jeff's rate back from the country, then $r+4$ represents his rate out to the country. The information is recorded in the following table:

	Distance	Rate	Time $\left(t = \frac{d}{r}\right)$
Trip Out	40	$r+4$	$\frac{40}{r+4}$
Trip back	42	r	$\frac{42}{r}$

$$\text{Time of Trip Back} = \text{Time of Trip Out} + 1$$

$$\frac{42}{r} = \frac{40}{r+4} + 1,\ r \neq -4 \text{ and } r \neq 0$$

$$r(r+4)\left(\frac{42}{r}\right) = r(r+4)\left(\frac{40}{r+4} + 1\right)$$

$$42(r+4) = 40r + r(r+4)$$

$$42r + 168 = 40r + r^2 + 4r$$

$$42r + 168 = r^2 + 44r$$

$$0 = r^2 + 2r - 168$$

$$0 = (r+14)(r-12)$$

$$r + 14 = 0 \quad \text{or} \quad r - 12 = 0$$

$$r = -14 \quad \text{or} \quad r = 12$$

The negative solution needs to be discarded. The rate back is 12 miles per hour and the rate out is 16 miles per hour.

41. Let t represent the time for the tank to overflow. The information is recorded in the following table:

	Rate	Time	Quantity
Fill	$\frac{1}{5}$	t	$\frac{t}{5}$
Drain	$\frac{1}{6}$	t	$\frac{t}{6}$

Fill − Drain = 1 tank Filled

$$\frac{t}{5} - \frac{t}{6} = 1$$
$$30\left(\frac{t}{5} - \frac{t}{6}\right) = 30(1)$$
$$6t - 5t = 30$$
$$t = 30$$

It would take 30 minutes for the tank to overflow.

43. Let t represent Mike's time, then $2t$ represents Barry's time. The sum of the individual rates equal the rate working together.

Mike's rate + Barry's rate = Rate together

$$\frac{1}{t} + \frac{1}{2t} = \frac{1}{40}$$
$$40t\left(\frac{1}{t} + \frac{1}{2t}\right) = 40t\left(\frac{1}{40}\right)$$
$$40 + 20 = t$$
$$60 = t$$

It would take Mike 60 minutes and Barry 120 minutes to deliver the papers alone.

45. Let t represent Mike's time to do the job by himself. The information is recorded in the following table:

	Rate	Time	Quantity
Pat	$\frac{1}{12}$	$3 + 5 = 8$	$\frac{8}{12} = \frac{2}{3}$
Mike	$\frac{1}{t}$	5	$\frac{5}{t}$

Pat's Portion + Mike's Portion = 1 Task

$$\frac{2}{3} + \frac{5}{t} = 1,\ t \neq 0$$
$$3t\left(\frac{2}{3} + \frac{5}{t}\right) = 3t(1)$$
$$2t + 15 = 3t$$
$$15 = t$$

It would take Mike 15 hours to complete the task.

47. Let t represent the time that Card reader B is used. The information is recorded in the following table:

	Rate	Time	Quantity
Reader A	600	$6+t$	$600(6+t)$
Reader B	850	t	$850t$

$$\begin{aligned} \text{Reader A} + \text{Reader B} &= 9400 \\ 600(6+t) + 850t &= 9400 \\ 3600 + 600t + 850t &= 9400 \\ 3600 + 1450t &= 9400 \\ 1450t &= 5800 \\ t &= 4 \end{aligned}$$

Card reader B was used for 4 minutes.

49. Let t represent Paul's rate, then $r + 20$ represents Amelia's rate. The information is recorded in the following table:

	Quantity	Rate	Time $(t = \frac{Q}{r})$
Paul	600	r	$\frac{600}{r}$
Amelia	600	$r+20$	$\frac{600}{r+20}$

$$\begin{aligned} \text{Amelia's time} &= \text{Paul's time} - 5 \\ \frac{600}{r+20} &= \frac{600}{r} - 5,\ r \neq -20 \text{ and } r \neq 0 \\ r(r+20)\left(\frac{600}{r+20}\right) &= r(r+20)\left(\frac{600}{r} - 5\right) \\ 600r &= 600(r+20) - 5r(r+20) \\ 600r &= 600r + 12{,}000 - 5r^2 - 100r \\ 600r &= -5r^2 + 500r + 12{,}000 \\ 0 &= -5r^2 - 100r + 12{,}000 \\ 0 &= r^2 + 20r - 2400 \\ 0 &= (r+60)(r-40) \end{aligned}$$

$$r + 60 = 0 \quad \text{or} \quad r - 40 = 0$$

$$r = -60 \quad \text{or} \quad r = 40$$

Discard the negative solution. Paul types at a rate of 40 words per minute and Amelia types at a rate of 60 words per minute.

CHAPTER 7 Review Problem Set

1. $$\frac{56x^3y}{72xy^3} = \frac{\overset{7}{\cancel{56}} \cdot \overset{x^2}{\cancel{x^3}} \cdot \cancel{y}}{\underset{9}{\cancel{72}} \cdot \cancel{x} \cdot \underset{y^2}{\cancel{y^3}}} = \frac{7x^2}{9y^2}$$

2. $$\frac{x^2 - 9x}{x^2 - 6x - 27} =$$
$$\frac{x\cancel{(x-9)}}{\cancel{(x-9)}(x+3)} =$$
$$\frac{x}{x+3}$$

3. $$\frac{3n^2 - n - 10}{n^2 - 4} =$$
$$\frac{(3n+5)\cancel{(n-2)}}{\cancel{(n-2)}(n+2)} =$$
$$\frac{3n+5}{n+2}$$

4. $$\frac{16a^2 + 24a + 9}{20a^2 + 7a - 6} =$$
$$\frac{(4a+3)\cancel{(4a+3)}}{\cancel{(4a+3)}(5a-2)} =$$
$$\frac{4a+3}{5a-2}$$

5. $$\frac{7x^2y^2}{12y^3} \cdot \frac{18y}{28x} = \frac{7 \cdot \overset{3}{\cancel{18}} \cdot \overset{x}{\cancel{x^2}} \cdot \cancel{y^2} \cdot \cancel{y}}{\underset{2}{\cancel{12}} \cdot \underset{4}{\cancel{28}} \cdot \underset{y}{\cancel{y^3}} \cdot \cancel{x}} = \frac{3x}{8}$$

6. $$\frac{x^2y}{x^2 + 2x} \cdot \frac{x^2 - x - 6}{y} =$$
$$\frac{\overset{x}{\cancel{x^2}} \cdot \cancel{y}(x-3)\cancel{(x+2)}}{\cancel{x}\cancel{(x+2)}(\cancel{y})} =$$
$$x(x-3)$$

7. $$\frac{n^2 - 2n - 24}{n^2 + 11n + 28} \div \frac{n^3 - 6n^2}{n^2 - 49} =$$
$$\frac{n^2 - 2n - 24}{n^2 + 11n + 28} \cdot \frac{n^2 - 49}{n^3 - 6n^2} =$$
$$\frac{\cancel{(n-6)}\cancel{(n+4)}}{\cancel{(n+4)}\cancel{(n+7)}} \cdot \frac{(n-7)\cancel{(n+7)}}{n^2\cancel{(n-6)}} =$$
$$\frac{n-7}{n^2}$$

8. $$\frac{4a^2 + 4a + 1}{(a+6)^2} \div \frac{6a^2 - 5a - 4}{3a^2 + 14a - 24} =$$
$$\frac{4a^2 + 4a + 1}{(a+6)^2} \cdot \frac{3a^2 + 14a - 24}{6a^2 - 5a - 4} =$$
$$\frac{\cancel{(2a+1)}(2a+1)}{\cancel{(a+6)}(a+6)} \cdot \frac{\cancel{(3a-4)}\cancel{(a+6)}}{\cancel{(2a+1)}\cancel{(3a-4)}} =$$
$$\frac{2a+1}{a+6}$$

9. $$\frac{3x+4}{5} + \frac{2x-7}{4} =$$
$$\left(\frac{3x+4}{4}\right)\left(\frac{4}{4}\right) + \left(\frac{2x-7}{4}\right)\left(\frac{5}{5}\right) =$$
$$\frac{4(3x+4)}{20} + \frac{5(2x-7)}{20} =$$
$$\frac{4(3x+4) + 5(2x-7)}{20} =$$
$$\frac{12x + 16 + 10x - 35}{20} = \frac{22x - 19}{20}$$

10. $$\frac{7}{3x} + \frac{5}{4x} - \frac{2}{8x^2} =$$
$$\frac{7}{3x} \cdot \frac{8x}{8x} + \frac{5}{4x} \cdot \frac{6x}{6x} - \frac{2}{8x^2} \cdot \frac{3}{3} =$$

$$\frac{7(8x)+5(6x)-2(3)}{24x^2}=$$
$$\frac{56x+30x-6}{24x^2}=\frac{86x-6}{24x^2}=$$
$$\frac{2(43x-3)}{24x^2}=\frac{43x-3}{12x^2}$$

11. $$\frac{7}{n}+\frac{3}{n-1}=$$
$$\left(\frac{7}{n}\right)\left(\frac{n-1}{n-1}\right)+\left(\frac{3}{n-1}\right)\left(\frac{n}{n}\right)=$$
$$\frac{7(n-1)}{n(n-1)}+\frac{3n}{n(n-1)}=$$
$$\frac{7(n-1)+3n}{n(n-1)}=\frac{7n-7+3n}{n(n-1)}=$$
$$\frac{10n-7}{n(n-1)}$$

12. $$\frac{2}{a-4}-\frac{3}{a-2}=$$
$$\frac{2}{a-4}\cdot\frac{a-2}{a-2}-\frac{3}{a-2}\cdot\frac{a-4}{a-4}=$$
$$\frac{2(a-2)-3(a-4)}{(a-2)(a-4)}=$$
$$\frac{2a-4-3a+12}{(a-2)(a-4)}=\frac{-a+8}{(a-2)(a-4)}$$

13. $$\frac{2x}{x^2-3x}-\frac{3}{4x}=\frac{2x}{x(x-3)}-\frac{3}{4x}=$$
$$\left[\frac{2x}{x(x-3)}\right]\left(\frac{4}{4}\right)-\left(\frac{3}{4x}\right)\left(\frac{x-3}{x-3}\right)=$$
$$\frac{2x(4)-3(x-3)}{4x(x-3)}=\frac{8x-3x+9}{4x(x-3)}=$$
$$\frac{5x+9}{4x(x-3)}$$

14. $$\frac{2}{x^2+7x+10}+\frac{3}{x^2-25}=$$
$$\frac{2}{(x+5)(x+2)}+\frac{3}{(x-5)(x+5)}=$$
$$\frac{2}{(x+5)(x+2)}\cdot\frac{x-5}{x-5}$$
$$+\frac{3}{(x-5)(x+5)}\cdot\frac{x+2}{x+2}=$$
$$\frac{2(x-5)+3(x+2)}{(x+5)(x+2)(x-5)}=$$
$$\frac{2x-10+3x+6}{(x+5)(x+2)(x-5)}=$$
$$\frac{5x-4}{(x+5)(x+2)(x-5)}$$

15. $$\frac{5x}{x^2-4x-21}-\frac{3}{x-7}+\frac{4}{x+3}=$$
$$\frac{5x}{(x-7)(x+3)}-\frac{3}{x-7}+\frac{4}{x+3}=$$
$$\frac{5x}{(x-7)(x+3)}-\left(\frac{3}{x-7}\right)\left(\frac{x+3}{x+3}\right)+$$
$$\left(\frac{4}{x+3}\right)\left(\frac{x-7}{x-7}\right)=$$
$$\frac{5x-3(x+3)+4(x-7)}{(x-7)(x+3)}=$$
$$\frac{5x-3x-9+4x-28}{(x-7)(x+3)}=$$
$$\frac{6x-37}{(x-7)(x+3)}$$

16. LCM of x, y and y^2 is xy^2.
$$\frac{\frac{3}{x}-\frac{4}{y^2}}{\frac{4}{y}+\frac{5}{x}}=\left(\frac{xy^2}{xy^2}\right)\left(\frac{\frac{3}{x}-\frac{4}{y^2}}{\frac{4}{y}+\frac{5}{x}}\right)=$$
$$\frac{xy^2\left(\frac{3}{x}-\frac{4}{y^2}\right)}{xy^2\left(\frac{4}{y}+\frac{5}{x}\right)}=\frac{3y^2-4x}{4xy+5y^2}$$

17. LCM of x and y is xy.

$$\frac{\frac{2}{x}-1}{3+\frac{5}{y}} = \left(\frac{xy}{xy}\right)\left(\frac{\frac{2}{x}-1}{3+\frac{5}{y}}\right) =$$

$$\frac{xy\left(\frac{2}{x}-1\right)}{xy\left(3+\frac{5}{y}\right)} = \frac{2y-xy}{3xy+5x}$$

18.

$$\frac{2x-1}{3}+\frac{3x-2}{4}=\frac{5}{6}$$
$$12\left(\frac{2x-1}{3}+\frac{3x-2}{4}\right)=12\left(\frac{5}{6}\right)$$
$$4(2x-1)+3(3x-2)=2(5)$$
$$8x-4+9x-6=10$$
$$17x-10=10$$
$$17x=20$$
$$x=\frac{20}{17}$$

The solution set is $\left\{\frac{20}{17}\right\}$.

19.

$$\frac{5}{3x}-2=\frac{7}{2x}+\frac{1}{5x},\ x\neq 0$$
$$30x\left(\frac{5}{3x}-2\right)=30x\left(\frac{7}{2x}+\frac{1}{5x}\right)$$
$$10(5)-60x=15(7)+6$$
$$50-60x=105+6$$
$$50-60x=111$$
$$-60x=61$$
$$x=-\frac{61}{60}$$

The solution set is $\left\{-\frac{61}{60}\right\}$.

20.

$$\frac{67-x}{x}=6+\frac{4}{x},\ x\neq 0$$
$$x\left(\frac{67-x}{x}\right)=x\left(6+\frac{4}{x}\right)$$
$$67-x=6x+4$$
$$67-7x=4$$
$$-7x=-63$$
$$x=9$$

The solution set is $\{9\}$.

21.

$$\frac{5}{2n+3}=\frac{6}{3n-2},\ n\neq -\frac{3}{2} \text{ and } n\neq \frac{2}{3}$$

Cross Products are Equal.

$$5(3n-2)=6(2n+3)$$
$$15n-10=12n+18$$
$$3n-10=18$$
$$3n=28$$
$$n=\frac{28}{3}$$

The solution set is $\left\{\frac{28}{3}\right\}$.

22.

$$\frac{x}{x-3} + \frac{5}{x+3} = 1,\ n \neq -3 \text{ and } n \neq 3$$
$$(x-3)(x+3)\left(\frac{x}{x-3} + \frac{5}{x+3}\right) = (x-3)(x+3)(1)$$
$$x(x+3) + 5(x-3) = x^2 - 9$$
$$x^2 + 3x + 5x - 15 = x^2 - 9$$
$$x^2 + 8x - 15 = x^2 - 9$$
$$8x - 15 = -9$$
$$8x = 6$$
$$x = \frac{6}{8} = \frac{3}{4}$$

The solution set is $\left\{\frac{3}{4}\right\}$.

23.

$$n + \frac{1}{n} = 2,\ n \neq 0$$
$$n\left(n + \frac{1}{n}\right) = n(2)$$
$$n^2 + 1 = 2n$$
$$n^2 - 2n + 1 = 0$$
$$(n-1)(n-1) = 0$$
$$n - 1 = 0 \quad \text{or} \quad n - 1 = 0$$
$$n = 1 \quad \text{or} \quad n = 1$$

The solution set is $\{1\}$.

24.

$$\frac{n-1}{n^2 + 8n - 9} - \frac{n}{n+9} = 4,\ n \neq -9 \text{ and } n \neq 1$$
$$\frac{(n-1)}{(n+9)(n-1)} - \frac{n}{n+9} = 4$$
$$\frac{1}{n+9} - \frac{n}{n+9} = 4$$
$$(n+9)\left(\frac{1-n}{n+9}\right) = (n+9)(4)$$
$$1 - n = 4n + 36$$
$$1 = 5n + 36$$
$$-35 = 5n$$
$$-7 = n$$

The solution set is $\{-7\}$.

25.
$$\frac{6}{7x} - \frac{1}{6} = \frac{5}{6x},\ x \neq 0$$
$$42x\left(\frac{6}{7x} - \frac{1}{6}\right) = 42x\left(\frac{5}{6x}\right)$$
$$6(6) - 7x = 7(5)$$
$$36 - 7x = 35$$
$$-7x = -1$$
$$x = \frac{1}{7}$$
The solution set is $\left\{\frac{1}{7}\right\}$.

26.
$$n + \frac{1}{n} = \frac{5}{2},\ n \neq 0$$
$$2n\left(n + \frac{1}{n}\right) = 2n\left(\frac{5}{2}\right)$$
$$2n^2 + 2 = 5n$$
$$2n^2 - 5n + 2 = 0$$
$$(2n - 1)(n - 2) = 0$$
$$2n - 1 = 0 \quad \text{or} \quad n - 2 = 0$$
$$2n = 1 \quad \text{or} \quad n = 2$$
$$n = \frac{1}{2} \quad \text{or} \quad n = 2$$
The solution set is $\left\{\frac{1}{2}, 2\right\}$.

27.
$$\frac{n}{5} = \frac{10}{n-5},\ n \neq 5$$
Cross products are equal.
$$n(n - 5) = 5(10)$$
$$n^2 - 5n = 50$$
$$n^2 - 5n - 50 = 0$$
$$(n - 10)(n + 5) = 0$$
$$n - 10 = 0 \quad \text{or} \quad n + 5 = 0$$
$$n = 10 \quad \text{or} \quad n = -5$$
The solution set is $\{-5, 10\}$.

28.

$$\frac{-1}{2x-5}+\frac{2x-4}{4x^2-25}=\frac{5}{6x+15},\ x\neq-\frac{5}{2}\text{ and }x\neq\frac{5}{2}$$
$$\frac{-1}{2x-5}+\frac{2x-4}{(2x-5)(2x+5)}=\frac{5}{3(2x+5)}$$
$$3(2x-5)(2x+5)\left[\frac{-1}{2x-5}+\frac{2x-4}{(2x-5)(2x+5)}\right]=3(2x-5)(2x+5)\left[\frac{5}{3(2x+5)}\right]$$
$$-3(2x+5)+3(2x-4)=5(2x-5)$$
$$-6x-15+6x-12=10x-25$$
$$-27=10x-25$$
$$-10x=2$$
$$x=-\frac{2}{10}=-\frac{1}{5}$$

The solution set is $\left\{-\frac{1}{5}\right\}$.

29.

$$1+\frac{1}{n-1}=\frac{1}{n^2-n},\ n\neq0\text{ and }n\neq1$$
$$1+\frac{1}{n-1}=\frac{1}{n(n-1)}$$
$$n(n-1)\left(1+\frac{1}{n-1}\right)=n(n-1)\left[\frac{1}{n(n-1)}\right]$$
$$n(n-1)+n=1$$
$$n^2-n+n=1$$
$$n^2=1$$
$$n^2-1=0$$
$$(n-1)(n+1)=0$$
$$n-1=0\quad\text{or}\quad n+1=0$$
$$n=1\quad\text{or}\quad n=-1$$

Since $n\neq1$ because of the original restriction, the solution set is $\{-1\}$.

30. Let n represent the smaller number, then $75-n$ represents the larger number.

$$\frac{75-n}{n}=9+\frac{5}{n},\ n\neq0$$
$$n\left(\frac{75-n}{n}\right)=n\left(9+\frac{5}{n}\right)$$
$$75-n=9n+5$$
$$75=10n+5$$
$$70=10n$$
$$7=n$$

The numbers are 7 and 68.

31. Let t represent Becky's time, then $3t$ represents Nancy's time. The sum of the individual rates equal the rate working together.

$$\frac{1}{t}+\frac{1}{3t}=\frac{1}{2}$$
$$6t\left(\frac{1}{t}+\frac{1}{3t}\right)=6t\left(\frac{1}{2}\right)$$
$$6+2=3t$$
$$8=3t$$
$$\frac{8}{3}=t$$

It would take Becky $\frac{8}{3}=2\frac{2}{3}$ hours to complete the task and it would take Nancy $3\left(\frac{8}{3}\right)=8$ hours to complete the task.

32. Let n represent the number, then $\frac{1}{n}$ represents the reciprocal.

$$n+(2)\frac{1}{n}=3,\ n\neq 0$$
$$n\left(n+\frac{2}{n}\right)=n(3)$$
$$n^2+2=3n$$
$$n^2-3n+2=0$$
$$(n-2)(n-1)=0$$
$$n-2=0 \quad \text{or} \quad n-1=0$$
$$n=2 \quad \text{or} \quad n=1$$

The number could be either 1 or 2.

33. Let n represent the numerator of the fraction, then $2n$ represents the denominator.

$$\frac{n+4}{2n+18}=\frac{4}{9},\ n\neq -9$$

Cross products are equal.

$$9(n+4)=4(2n+18)$$
$$9n+36=8n+72$$
$$n+36=72$$
$$n=36$$

The original fraction is $\frac{36}{72}$.

34. Let r represent Todd's rate, then $r+7$ represents Lanette's rate.

	Distance	Rate	Time $\left(t=\frac{d}{r}\right)$
Todd	30	r	$\frac{30}{r}$
Lanette	44	$r+7$	$\frac{44}{r+7}$

Todd's time = Lanette's time

$$\frac{30}{r} = \frac{44}{r+7}, \ r \neq -7 \text{ and } r \neq 0$$

$$r(r+7)\left(\frac{30}{r}\right) = r(r+7)\left(\frac{44}{r+7}\right)$$

$$30(r+7) = 44r$$

$$30r + 210 = 44r$$

$$210 = 14r$$

$$15 = r$$

Todd's rate is 15 miles per hour and Lanette's rate is 22 miles per hour.

35. Let r represent Jim's rate for the first 20 miles, then $r-2$ represents his rate on the last 16 miles. The information is recorded in the following table:

	Distance	Rate	Time $\left(t = \frac{d}{r}\right)$
First part	20	r	$\frac{20}{r}$
Last part	16	$r-2$	$\frac{16}{r-2}$

Time of the first part + Time of the last part = 4

$$\frac{20}{r} + \frac{16}{r-2} = 4, \ r \neq 2 \text{ and } r \neq 0$$

$$r(r-2)\left(\frac{20}{r} + \frac{16}{r-2}\right) = r(r-2)(4)$$

$$20(r-2) + 16r = 4r(r-2)$$

$$20r - 40 + 16r = 4r^2 - 8r$$

$$36r - 40 = 4r^2 - 8r$$

$$0 = 4r^2 - 44r + 40$$

$$0 = r^2 - 11r + 10$$

$$0 = (r-1)(r-10)$$

$$r - 1 = 0 \quad \text{or} \quad r - 10 = 0$$

$$r = 1 \quad \text{or} \quad r = 10$$

If the rate for the first part is 1 mile per hour, the rate for the second part would be -1 mile per hour which is not reasonable. Thus, the rate for the first part would be 10 miles per hour and the rate for the second part would be 8 miles per hour.

36. Let r represent the time for the tank to overflow.

	Rate	Time	Quantity
Fill	$\frac{1}{10}$	t	$\frac{t}{10}$
Drain	$\frac{1}{12}$	t	$\frac{t}{12}$

$$\text{Fill} - \text{Drain} = 1 \text{ Tank Filled}$$
$$\frac{t}{10} - \frac{t}{12} = 1$$
$$120\left(\frac{t}{10} - \frac{t}{12}\right) = 120(1)$$
$$12t - 10t = 120$$
$$2t = 120$$
$$t = 60$$
It would take 60 minutes for the tank to overflow.

37. Let r represent Corinne's rate, then $r - 6$ represents Sue's rate. The information is recorded in the following table:

	Quantity	Rate	Time $\left(t = \frac{Q}{r}\right)$
Corinne	840	r	$\frac{840}{r}$
Sue	1000	$r - 6$	$\frac{1000}{r-6}$

$$\text{Sue's time} = \text{Corinne's time} + 5$$
$$\frac{1000}{r-6} = \frac{840}{r} + 5, \; r \neq 0 \text{ and } r \neq 6$$
$$r(r-6)\left(\frac{1000}{r-6}\right) = r(r-6)\left(\frac{840}{r} + 5\right)$$
$$1000r = 840(r-6) + 5r(r-6)$$
$$1000r = 840r + 5040 + 5r^2 - 30r$$
$$1000r = 5r^2 + 810r + 5040$$
$$0 = 5r^2 - 190r + 5040$$
$$0 = r^2 - 38r + 1008$$
$$0 = (r+18)(r-56)$$
$$r + 18 = 0 \quad \text{or} \quad r - 56 = 0$$
$$r = -18 \quad \text{or} \quad r = 56$$
Discard the negative solution. Corinne types at a rate of 56 words per minute and Sue types at a rate of 50 words per minute.

CHAPTER 7 Test

1. $$\frac{72x^4y^5}{81x^2y^4} = \frac{\overset{8}{\cancel{72}} \cdot \overset{x^2}{\cancel{x^4}} \cdot \overset{y}{\cancel{y^5}}}{\underset{9}{\cancel{81}} \cdot \cancel{x^2} \cdot \cancel{y^4}} = \frac{8x^2y}{9}$$

2. $$\frac{x^2 + 6x}{x^2 - 36} =$$
$$\frac{x\cancel{(x+6)}}{(x-6)\cancel{(x+6)}} =$$
$$\frac{x}{x-6}$$

Chapter 7 Test

3. $\dfrac{2n^2-7n-4}{3n^2-8n-16}=$

$\dfrac{(2n+1)\cancel{(n-4)}}{(3n+4)\cancel{(n-4)}}=$

$\dfrac{2n+1}{3n+4}$

4. $\dfrac{2x^3+7x^2-15x}{x^3-25x}=$

$\dfrac{\cancel{x}(2x-3)\cancel{(x+5)}}{\cancel{x}(x-5)\cancel{(x+5)}}=$

$\dfrac{2x-3}{x-5}$

5. $\dfrac{8x^2y}{7x}\cdot\dfrac{21xy^3}{12y^2}=$

$\dfrac{\overset{2}{\cancel{8}}\cdot\overset{3}{\cancel{21}}\cdot x^2\cdot\cancel{x}\cdot y\cdot\overset{y}{\cancel{y^3}}}{\cancel{7}\cdot\underset{3}{\cancel{12}}\cdot\cancel{x}\cdot\cancel{y^2}}=$

$2x^2y^2$

6. $\dfrac{x^2-49}{x^2+7x}\div\dfrac{x^2-4x-21}{x^2-2x}=$

$\dfrac{x^2-49}{x^2+7x}\cdot\dfrac{x^2-2x}{x^2-4x-21}=$

$\dfrac{\cancel{(x-7)}\,\cancel{(x+7)}}{\cancel{x}\cancel{(x+7)}}\cdot\dfrac{\cancel{x}(x-2)}{\cancel{(x-7)}\,(x+3)}=$

$\dfrac{x-2}{x+3}$

7. $\dfrac{x^2-5x-36}{x^2-15x+54}\cdot\dfrac{x^2-2x-24}{x^2+7x}=$

$\dfrac{\cancel{(x-9)}\,(x+4)}{\cancel{(x-6)}\,\cancel{(x-9)}}\cdot\dfrac{\cancel{(x-6)}\,(x+4)}{x(x+7)}=$

$\dfrac{(x+4)(x+4)}{x(x+7)}=\dfrac{(x+4)^2}{x(x+7)}$

8. $\dfrac{3x-1}{6}-\dfrac{2x-3}{8}=$

$\left(\dfrac{3x-1}{6}\right)\left(\dfrac{4}{4}\right)-\left(\dfrac{2x-3}{8}\right)\left(\dfrac{3}{3}\right)=$

$\dfrac{4(3x-1)-3(2x-3)}{24}=$

$\dfrac{12x-4-6x+9}{24}=\dfrac{6x+5}{24}$

9. $\dfrac{n+2}{3}-\dfrac{n-1}{5}+\dfrac{n-6}{6}=$

$\dfrac{n+2}{3}\left(\dfrac{10}{10}\right)-\dfrac{n-1}{5}\left(\dfrac{6}{6}\right)+\dfrac{n-6}{6}\left(\dfrac{5}{5}\right)=$

$\dfrac{10(n+2)-6(n-1)+5(n-6)}{30}=$

$\dfrac{10n+20-6n+6+5n-30}{30}=\dfrac{9n-4}{30}$

10. $\dfrac{3}{2x}-\dfrac{5}{6}+\dfrac{7}{9x}=$

$\left(\dfrac{3}{2x}\right)\left(\dfrac{9}{9}\right)-\left(\dfrac{5}{6}\right)\left(\dfrac{3x}{3x}\right)+\left(\dfrac{7}{9x}\right)\left(\dfrac{2}{2}\right)=$

$\dfrac{27-15x+14}{18x}=\dfrac{-15x+41}{18x}$

11. $\dfrac{6}{n}-\dfrac{4}{n-1}=$

$\dfrac{6}{n}\left(\dfrac{n-1}{n-1}\right)-\dfrac{4}{n-1}\left(\dfrac{n}{n}\right)=$

$\dfrac{6(n-1)-4n}{n(n-1)}=\dfrac{6n-6-4n}{n(n-1)}=$

$\dfrac{2n-6}{n(n-1)}$

12. $\dfrac{2x}{x^2+6x} - \dfrac{3}{4x} = \dfrac{2x}{x(x+6)} - \dfrac{3}{4x} =$

$\left[\dfrac{2x}{x(x+6)}\right]\left(\dfrac{4}{4}\right) - \left(\dfrac{3}{4x}\right)\left(\dfrac{x+6}{x+6}\right) =$

$\dfrac{4(2x)-3(x+6)}{4x(x+6)} = \dfrac{8x-3x-18}{4x(x+6)} =$

$\dfrac{5x-18}{4x(x+6)}$

13. $\dfrac{9}{x^2+4x-32} + \dfrac{5}{x+8} =$

$\dfrac{9}{(x+8)(x-4)} + \dfrac{5}{x+8} =$

$\dfrac{9}{(x+8)(x-4)} + \dfrac{5}{x+8}\left(\dfrac{x-4}{x-4}\right) =$

$\dfrac{9+5(x-4)}{(x+8)(x-4)} = \dfrac{9+5x-20}{(x+8)(x-4)} =$

$\dfrac{5x-11}{(x+8)(x-4)}$

14. $\dfrac{-3}{6x^2-7x-20} - \dfrac{5}{3x^2-14x-24} =$

$\dfrac{-3}{(3x+4)(2x-5)} - \dfrac{5}{(3x+4)(x-6)} =$

$\left[\dfrac{-3}{(3x+4)(2x-5)}\right]\left(\dfrac{x-6}{x-6}\right)$

$- \left[\dfrac{5}{(3x+4)(x-6)}\right]\left(\dfrac{2x-5}{2x-5}\right) =$

$\dfrac{-3(x-6)-5(2x-5)}{(3x+4)(2x-5)(x-6)} =$

$\dfrac{-3x+18-10x+25}{(3x+4)(2x-5)(x-6)} =$

$\dfrac{-13x+43}{(3x+4)(2x-5)(x-6)}$

15. $\dfrac{x+3}{5} - \dfrac{x-2}{6} = \dfrac{23}{30}$

$30\left(\dfrac{x+3}{5} - \dfrac{x-2}{6}\right) = 30\left(\dfrac{23}{30}\right)$

$6(x+3) - 5(x-2) = 23$

$6x+18-5x+10 = 23$

$x+28 = 23$

$x = -5$

The solution set is $\{-5\}$.

16. $\dfrac{5}{8x} - 2 = \dfrac{3}{x},\ x \neq 0$

$8x\left(\dfrac{5}{8x} - 2\right) = 8x\left(\dfrac{3}{x}\right)$

$5 - 2(8x) = 8(3)$

$5 - 16x = 24$

$-16x = 19$

$x = -\dfrac{19}{16}$

The solution set is $\left\{-\dfrac{19}{16}\right\}$.

17. $n + \dfrac{4}{n} = \dfrac{13}{3}$

$3n\left(n + \dfrac{4}{n}\right) = 3n\left(\dfrac{13}{3}\right)$

$3n^2 + 12 = 13n$

$3n^2 - 13n + 12 = 0$

$(3n-4)(n-3) = 0$

$3n - 4 = 0$ or $n - 3 = 0$

$3n = 4$ or $n = 3$

$n = \dfrac{4}{3}$ or $n = 3$

The solution set is $\left\{\dfrac{4}{3}, 3\right\}$.

18. $\dfrac{x}{8} = \dfrac{6}{x-2},\ x \neq 2$

Cross products are equal.

$x(x-2) = 6(8)$

$x^2 - 2x = 48$

$x^2 - 2x - 48 = 0$

$(x-8)(x+6) = 0$

$x - 8 = 0$ or $x + 6 = 0$

$x = 8$ or $x = -6$

The solution set is $\{-6, 8\}$.

19.
$$\begin{aligned}
\frac{x}{x-1}+\frac{2}{x+1}&=\frac{8}{3},\ x\neq -1 \text{ and } x\neq 1\\
3(x-1)(x+1)\left(\frac{x}{x-1}+\frac{2}{x+1}\right)&=3(x-1)(x+1)\left(\frac{8}{3}\right)\\
3x(x+1)+6(x-1)&=8\left(x^2-1\right)\\
3x^2+3x+6x-6&=8x^2-8\\
3x^2+9x-6&=8x^2-8\\
-5x^2+9x+2&=0\\
5x^2-9x-2&=0\\
(5x+1)(x-2)&=0
\end{aligned}$$
$$5x+1=0 \quad\text{or}\quad x-2=0$$
$$5x=-1 \quad\text{or}\quad x=2$$
$$x=-\frac{1}{5} \quad\text{or}\quad x=2$$
The solution set is $\left\{-\frac{1}{5},2\right\}$.

20. $\frac{3}{2x+1}=\frac{5}{3x-6},\ x\neq -\frac{1}{2}$ and $x\neq 2$

Cross products are equal.
$$\begin{aligned}
3(3x-6)&=5(2x+1)\\
9x-18&=10x+5\\
-x&=23\\
x&=-23
\end{aligned}$$
The solution set is $\{-23\}$.

21.
$$\begin{aligned}
\frac{4}{n^2-n}-\frac{3}{n-1}&=-1,\ n\neq 0 \text{ and } n\neq 1\\
\frac{4}{n(n-1)}-\frac{3}{n-1}&=-1\\
n(n-1)\left[\frac{4}{n(n-1)}-\frac{3}{n-1}\right]&=n(n-1)(-1)\\
4-3n&=-n(n-1)\\
4-3n&=-n^2+n\\
0&=-n^2+4n-4\\
0&=n^2-4n+4\\
0&=(n-2)(n-2)
\end{aligned}$$
$$n-2=0 \quad\text{or}\quad n-2=0$$
$$n=2 \quad\text{or}\quad n=2$$
The solution set is $\{2\}$.

22.
$$\frac{3n-1}{3}+\frac{2n+5}{4}=\frac{4n-6}{9}$$
$$36\left(\frac{3n-1}{3}+\frac{2n+5}{4}\right)=36\left(\frac{4n-6}{9}\right)$$
$$12(3n-1)+9(2n+5)=4(4n-6)$$
$$36n-12+18n+45=16n-24$$
$$54n+33=16n-24$$
$$38n+33=-24$$
$$38x=-57$$
$$x=-\frac{57}{38}=-\frac{3}{2}$$
The solution set is $\left\{-\frac{3}{2}\right\}$.

23. Let n represent the number, then $\frac{1}{n}$ represents the reciprocal.
$$n+(2)\frac{1}{n}=\frac{11}{3}$$
$$3n\left(n+\frac{2}{n}\right)=3n\left(\frac{11}{3}\right)$$
$$3n^2+6=11n$$
$$3n^2-11n+6=0$$
$$(3n-2)(n-3)=0$$
$$3n-2=0 \quad \text{or} \quad n-3=0$$
$$3n=2 \quad \text{or} \quad n=3$$
$$n=\frac{2}{3} \quad \text{or} \quad n=3$$
The number is either $\frac{2}{3}$ or 3.

24. Let r represent Betty's rate, then $r+2$ represents Wendy's rate.

	Distance	Rate	Time $\left(t=\frac{d}{r}\right)$
Betty	36	r	$\frac{36}{r}$
Wendy	42	$r+2$	$\frac{42}{r+2}$

Betty's time = Wendy's time

$$\frac{36}{r} = \frac{42}{r+2},\ r \neq -2 \text{ and } r \neq 0$$

$$r(r+2)\left(\frac{36}{r}\right) = r(r+2)\left(\frac{42}{r+2}\right)$$

$$36(r+2) = 42r$$

$$36r + 72 = 42r$$

$$72 = 6r$$

$$12 = r$$

Wendy's rate is 14 miles per hour.

25. Let t represent the time working together. The sum of the individual rates equals the rate working together.

$$\frac{1}{20} + \frac{1}{30} = \frac{1}{t}$$

$$60t\left(\frac{1}{20} + \frac{1}{30}\right) = 60t\left(\frac{1}{t}\right)$$

$$3t + 2t = 60$$

$$5t = 60$$

$$t = 12$$

It takes them 12 minutes working together.

CHAPTERS 1-7 Cumulative Review

1. $3x - 2xy - 7y + 5xy = 3x + 3xy - 7y$

when $x = \frac{1}{2},\ y = 3:\ 3\left(\frac{1}{2}\right) + 3\left(\frac{1}{2}\right)(3) - 7(3)$

$$= \frac{3}{2} + \frac{9}{2} - 21$$

$$= \frac{12}{2} - 21$$

$$= 6 - 21 = -15$$

2. $a = -3,\ b = -5:$

$7(a-b) - 3(a-b) - (a-b) =$

$3(a-b) = 3[(-3) - (-5)] =$

$3[2] = 6$

3. $x = \frac{2}{3},\ y = \frac{5}{6},\ z = \frac{3}{4}:$

$\frac{xy + yz}{y} = \frac{xy}{y} + \frac{yz}{y} = x + z =$

$\frac{2}{3} + \frac{3}{4} = \frac{8}{12} + \frac{9}{12} = \frac{17}{12}$

4. $x = 0.4,\ b = 0.6:\ ab + b^2 =$

$(0.4)(0.6) + (0.6)^2 = 0.24 + 0.36 = 0.6$

5. $x = -6,\ y = 4:\ x^2 - y^2 =$

$(-6)^2 - (4)^2 = 36 - 16 = 20$

6. $x = -9:\ x^2 + 5x - 36 =$

$(x+9)(x-4) = (-9+9)(-9-4) =$

$0(-13) = 0$

7. $x = -6:\ \frac{x^2 + 2x}{x^2 + 5x + 6} =$

$\frac{x\cancel{(x+2)}}{(x+3)\cancel{(x+2)}} = \frac{x}{x+3} =$

$\frac{-6}{-6+3} = \frac{-6}{-3} = 2$

8. $x = 4: \dfrac{x^2+3x-10}{x^2-9x+14} = \dfrac{(x+5)\cancel{(x-2)}}{(x-7)\cancel{(x-2)}} =$

$\dfrac{x+5}{x-7} = \dfrac{4+5}{4-7} = \dfrac{9}{-3} = -3$

9. $3^{-3} = \dfrac{1}{3^3} = \dfrac{1}{27}$

10. $\left(\dfrac{2}{3}\right)^{-1} = \dfrac{3}{2}$

11. $\left(\dfrac{1}{2}+\dfrac{1}{3}\right)^{0} = 1$

12. $\left(\dfrac{1}{3}+\dfrac{1}{4}\right)^{-1} = \left(\dfrac{1\cdot 4}{3\cdot 4}+\dfrac{1\cdot 3}{4\cdot 3}\right)^{-1} =$

$\left(\dfrac{4}{12}+\dfrac{3}{12}\right)^{-1} = \left(\dfrac{7}{12}\right)^{-1} = \dfrac{12}{7}$

13. $-4^{-2} = -\dfrac{1}{4^2} = -\dfrac{1}{16}$

14. $\left(\dfrac{2}{3}\right)^{-2} = \left(\dfrac{3}{2}\right)^{2} = \dfrac{3^2}{2^2} = \dfrac{9}{4}$

15. $\dfrac{1}{\dfrac{2}{5}^{-2}} = \left(\dfrac{2}{5}\right)^{2} = \dfrac{4}{25}$

16. $(-3)^{-3} = \dfrac{1}{(-3)^3} = \dfrac{1}{-27} = -\dfrac{1}{27}$

17. $\dfrac{7}{5x}+\dfrac{2}{x}-\dfrac{3}{2x} =$

$\dfrac{7}{5x}\cdot\dfrac{2}{2}+\dfrac{2}{x}\cdot\dfrac{10}{10}-\dfrac{3}{2x}\cdot\dfrac{5}{5} =$

$\dfrac{14+20-15}{10x} = \dfrac{19}{10x}$

18. $\dfrac{4x}{5y} \div \dfrac{12x^2}{10y^2} = \dfrac{4x}{5y}\cdot\dfrac{10y^2}{12x^2} =$

$\dfrac{\cancel{4}\cdot\overset{2}{\cancel{10}}\cdot\cancel{x}\cdot\overset{y}{\cancel{y^2}}}{\cancel{5}\cdot\underset{3}{\cancel{12}}\cdot\underset{x}{\cancel{x^2}}\cdot\cancel{y}} = \dfrac{2y}{3x}$

19. $\dfrac{4}{x-6}+\dfrac{3}{x+4} =$

$\dfrac{4}{x-6}\left(\dfrac{x+4}{x+4}\right)+\dfrac{3}{x+4}\left(\dfrac{x-6}{x-6}\right) =$

$\dfrac{4(x+4)+3(x-6)}{(x+4)(x-6)} =$

$\dfrac{4x+16+3x-18}{(x+4)(x-6)} = \dfrac{7x-2}{(x+4)(x-6)}$

20. $\dfrac{2}{x^2-4x}-\dfrac{3}{x^2} = \dfrac{2}{x(x-4)}-\dfrac{3}{x^2} =$

$\left[\dfrac{2}{x(x-4)}\right]\left(\dfrac{x}{x}\right)-\left(\dfrac{3}{x^2}\right)\left(\dfrac{x-4}{x-4}\right) =$

$\dfrac{2x-3(x-4)}{x^2(x-4)} = \dfrac{2x-3x+12}{x^2(x-4)} =$

$\dfrac{-x+12}{x^2(x-4)}$

21. $\dfrac{x^2-8x}{x^2-x-56}\cdot\dfrac{x^2-49}{3xy} =$

$\dfrac{\cancel{x}\cancel{(x-8)}}{\cancel{(x-8)}\cancel{(x+7)}}\cdot\dfrac{(x-7)\cancel{(x+7)}}{3\cdot\cancel{x}\cdot y} =$

$\dfrac{x-7}{3y}$

22. $\dfrac{5}{x^2-x-12}-\dfrac{3}{x-4}=$

$\dfrac{5}{(x-4)(x+3)}-\dfrac{3}{x-4}=$

$\dfrac{5}{(x-4)(x+3)}-\left(\dfrac{3}{x-4}\right)\left(\dfrac{x+3}{x+3}\right)=$

$\dfrac{5-3(x+3)}{(x-4)(x+3)}=\dfrac{5-3x-9}{(x-4)(x+3)}=$

$\dfrac{-3x-4}{(x-4)(x+3)}$

23. $(-5x^2y)(7x^3y^4)=$
$(-5)(7)(x^{2+3})(y^{1+4})=-35x^5y^5$

24. $(9ab^3)^2=9^2a^2(b^3)^2=81a^2b^6$

25. $(-3n^2)(5n^2+6n-2)=$
$(-3n^2)(5n^2)+(-3n^2)(6n)-(-3n^2)(2)=$
$-15n^4-18n^3+6n^2$

26. $(5x-1)(3x+4)=$
$(5x)(3x)+(20-3)x-4=$
$15x^2+17x-4$

27. $(2x+5)^2=(2x)^2+2(2x)(5)+(5)^2=$
$4x^2+20x+25$

28. $(x+2)(2x^2-3x-1)=$
$x(2x^2-3x-1)+2(2x^2-3x-1)=$
$2x^3-3x^2-x+4x^2-6x-2=$
$2x^3+x^2-7x-2$

29. $(x^2-x-1)(x^2+2x-3)=$
$x^2(x^2+2x-3)-x(x^2+2x-3)$
$-1(x^2+2x-3)=$
$x^4+2x^3-3x^2-x^3-2x^2+3x$
$-x^2-2x+3=$
$x^4+x^3-6x^2+x+3$

30. $(-2x-1)(3x-7)=$
$(-2x)(3x)+(14-3)x+7=$
$-6x^2+11x+7$

31. $\dfrac{24x^2y^3-48x^4y^5}{8xy^2}=$

$\dfrac{24x^2y^3}{8xy^2}-\dfrac{48x^4y^5}{8xy^2}=$

$3xy-6x^3y^3$

32.

$$\begin{array}{r} 7x+4 \\ 4x-5\,\overline{)\,28x^2-19x-20} \\ \underline{28x^2-35x} \\ 16x-20 \\ \underline{16x-20} \end{array}$$

33. $3x^3+15x+27x=3x(x^2+5x+9)$
(x^2+5x+9 is not factorable.)

34. $x^2-100=(x-10)(x+10)$

35. $5x^2-22x+8=$
$5x^2-20x-2x+8=$
$5x(x-4)-2(x-4)=$
$(x-4)(5x-2)$

36. $8x^2-22x-63=$
$8x^2+14x-36x-63=$
$2x(4x+7)-9(4x+7)=$
$(4x+7)(2x-9)$

37. $n^2+25n+144=$
$n^2+16n+9n+144=$
$n(n+16)+9(n+16)=$
$(n+16)(n+9)$

38. $nx+ny-2x-2y=$
$n(x+y)-2(x+y)=$
$(x+y)(n-2)$

39. $3x^3-3x=3x(x^2-1)=$
$3x(x-1)(x+1)$

40. $2x^3-6x^2-108x=$
$2x(x^2-3x-54)=$
$2x(x-9)(x+6)$

41. $36x^2-60x+25=$
$(6x)^2-2(6x)(5)+(5)^2=$
$(6x-5)^2$

42. $3x^2 - 5xy - 2y^2 =$
$3x^2 - 6xy + xy - 2y^2 =$
$3x(x - 2y) + y(x - 2y) =$
$(x - 2y)(3x + y)$

43. $3(x-2) - 2(x+6) = -2(x+1)$
$3x - 6 - 2x - 12 = -2x - 2$
$x - 18 = -2x - 2$
$3x - 18 = -2$
$3x = 16$
$x = \frac{16}{3}$
The solution set is $\left\{\frac{16}{3}\right\}$.

44. $x^2 = -11x$
$x^2 + 11x = 0$
$x(x + 11) = 0$
$x = 0$ or $x + 11 = 0$
$x = 0$ or $x = -11$
The solution set is $\{-11, 0\}$.

45. $0.2x - 3(x - 0.4) = 1$
Multiplied by 10.
$2x - 30(x - 0.4) = 10$
$2x - 30x + 12 = 10$
$-28x + 12 = 10$
$-28x = -2$
$x = \frac{2}{28} = \frac{1}{14}$ Reduced.
The solution set is $\left\{\frac{1}{14}\right\}$.

46. $\frac{3n-1}{4} = \frac{5n+2}{7}$
Cross products are equal.
$7(3n - 1) = 4(5n + 2)$
$21n - 7 = 20n + 8$
$n - 7 = 8$
$n = 15$
The solution set is $\{15\}$.

47. $5n^2 - 5 = 0$
Divided by 5.
$n^2 - 1 = 0$
$(n-1)(n+1) = 0$
$n - 1 = 0$ or $n + 1 = 0$
$n = 1$ or $n = -1$
The solution set is $\{-1, 1\}$.

48. $x^2 + 5x - 6 = 0$
$(x-1)(x+6) = 0$
$x - 1 = 0$ or $x + 6 = 0$
$x = 1$ or $x = -6$
The solution set is $\{-6, 1\}$.

49. $n + \frac{4}{n} = 4, n \neq 0$
$n\left(n + \frac{4}{n}\right) = n(4)$
$n^2 + 4 = 4n$
$n^2 - 4n + 4 = 0$
$(n-2)(n-2) = 0$
$n - 2 = 0$ or $n - 2 = 0$
$n = 2$ or $n = 2$
The solution set is $\{2\}$.

50. $\frac{2x+1}{2} + \frac{3x-4}{3} = 1$
$6\left(\frac{2x+1}{2} + \frac{3x-4}{3}\right) = 6(1)$
$3(2x+1) + 2(3x-4) = 6$
$6x + 3 + 6x - 8 = 6$
$12x - 5 = 6$
$12x = 11$
$x = \frac{11}{12}$
The solution set is $\left\{\frac{11}{12}\right\}$.

51. $2(x-1) - x(x-1) = 0$
$(x-1)(2-x) = 0$
$x - 1 = 0$ or $2 - x = 0$
$x = 1$ or $2 = x$
The solution set is $\{1, 2\}$.

52. $$\frac{3}{2x} - 1 = \frac{5}{3x} + 2,\ x \neq 0$$
$$6x\left(\frac{3}{2x} - 1\right) = 6x\left(\frac{5}{3x} + 2\right)$$
$$3(3) - 6x = 2(5) + 12x$$
$$9 - 6x = 10 + 12x$$
$$9 - 18x = 10$$
$$-18x = 1$$
$$x = -\frac{1}{18}$$
The solution set is $\left\{-\frac{1}{18}\right\}$.

53. $$6t^2 + 19t - 7 = 0$$
$$(3t - 1)(2t + 7) = 0$$
$3t - 1 = 0$ or $2t + 7 = 0$

$3t = 1$ or $2t = -7$

$t = \frac{1}{3}$ or $t = -\frac{7}{2}$

The solution set is $\left\{-\frac{7}{2}, \frac{1}{3}\right\}$.

54. $$(2x - 1)(x - 8) = 0$$
$2x - 1 = 0$ or $x - 8 = 0$

$2x = 1$ or $x = 8$

$x = \frac{1}{2}$ or $x = 8$

The solution set is $\left\{\frac{1}{2}, 8\right\}$.

55. $$(x + 1)(x + 6) = 24$$
$$x^2 + 7x + 6 = 24$$
$$x^2 + 7x - 18 = 0$$
$$(x + 9)(x - 2) = 0$$
$x + 9 = 0$ or $x - 2 = 0$

$x = -9$ or $x = 2$

The solution set is $\{-9, 2\}$.

56. $$\frac{x}{x - 2} - \frac{7}{x + 1} = 1,\ x \neq -1 \text{ and } x \neq 2$$
$$(x - 2)(x + 1)\left(\frac{x}{x - 2} - \frac{7}{x + 1}\right) = (x - 2)(x + 1)(1)$$
$$x(x + 1) - 7(x - 2) = x^2 - x - 2$$
$$x^2 + x - 7x + 14 = x^2 - x - 2$$
$$x^2 - 6x + 14 = x^2 - x - 2$$
$$-5x + 14 = -2$$
$$-5x = -16$$
$$x = \frac{16}{5}$$
The solution set is $\left\{\frac{16}{5}\right\}$.

57.

$$\frac{1}{n} - \frac{2}{n-1} = \frac{3}{n},\ n \neq 0 \text{ and } n \neq 1$$
$$n(n-1)\left(\frac{1}{n} - \frac{2}{n-1}\right) = n(n-1)\left(\frac{3}{n}\right)$$
$$1(n-1) - 2n = 3(n-1)$$
$$n - 1 - 2n = 3n - 3$$
$$-1 - n = 3n - 3$$
$$-1 - 4n = -3$$
$$-4n = -2$$
$$n = \frac{2}{4} = \frac{1}{2}$$

The solution set is $\left\{\frac{1}{2}\right\}$.

58. $y = -3x + 5$

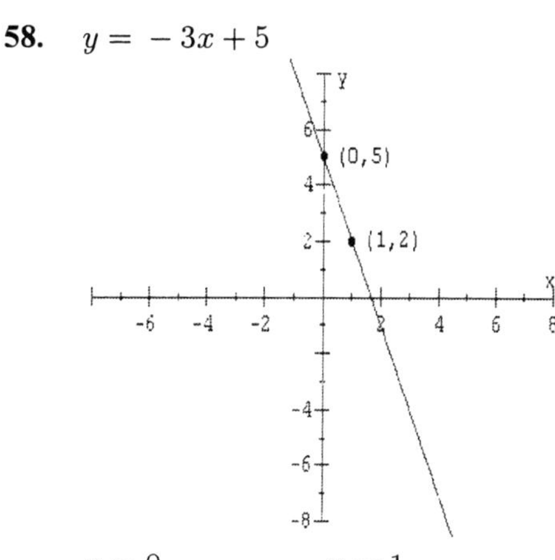

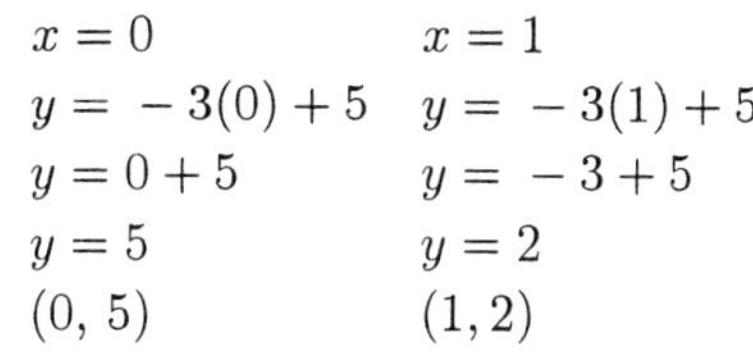

$x = 0$	$x = 1$
$y = -3(0) + 5$	$y = -3(1) + 5$
$y = 0 + 5$	$y = -3 + 5$
$y = 5$	$y = 2$
$(0,\ 5)$	$(1, 2)$

59. $y = \frac{1}{4}x + 2$

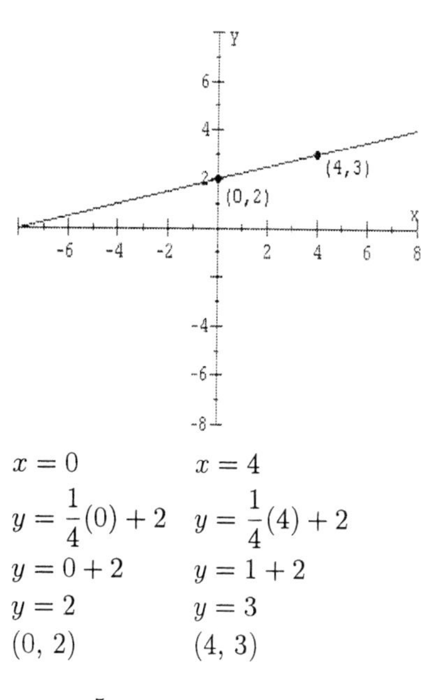

$x = 0$	$x = 4$
$y = \frac{1}{4}(0) + 2$	$y = \frac{1}{4}(4) + 2$
$y = 0 + 2$	$y = 1 + 2$
$y = 2$	$y = 3$
$(0,\ 2)$	$(4,\ 3)$

60. $y = -\frac{5}{2}x + 4$

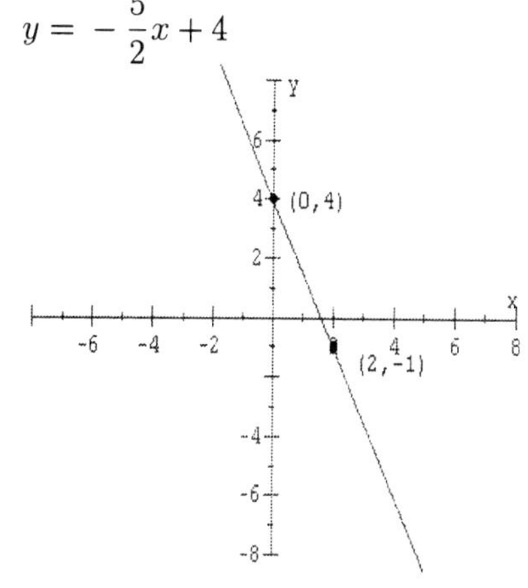

$x = 0$
$y = -\frac{5}{2}(0) + 4$
$y = 0 + 4$
$y = 4$
$(0,\ 4)$

$x = 2$
$y = -\frac{5}{2}(2) + 4$
$y = -5 + 4$
$y = -1$
$(2,\ -1)$

61. $3x - y = 6$

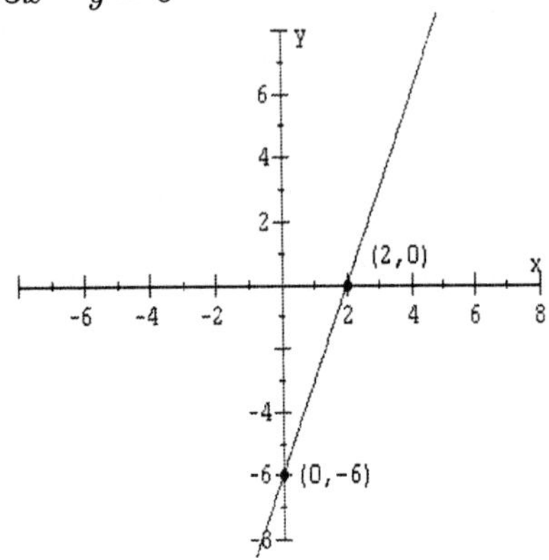

$x = 0$
$3(0) - y = 6$
$-y = 6$
$y = -6$
$(0,\ -6)$

$y = 0$
$3x - 0 = 6$
$3x = 6$
$x = 2$
$(2,\ 0)$

62. $x + 2y = 0$

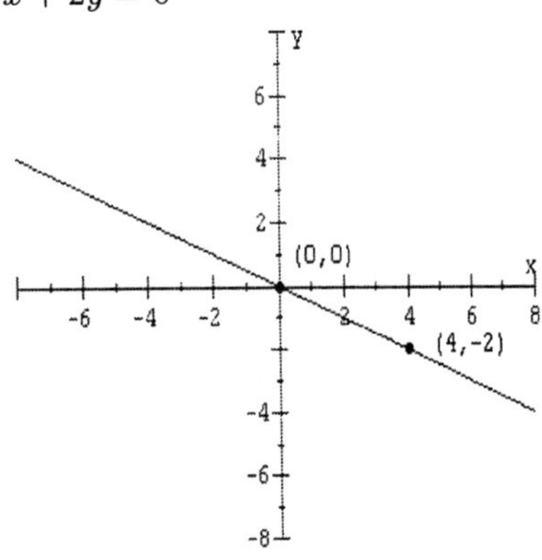

$x = 0$
$0 + 2y = 0$
$2y = 0$
$y = \frac{0}{2} = 0$
$(0,\ 0)$

$x = 4$
$4 + 2y = 0$
$2y = -4$
$y = -2$
$(4,\ -2)$

63. $y = 2x^2 - 4$

let $x = 0$
$y = 2(0)^2 - 4 = -4 \quad (0,\ -4)$

let $x = 1$
$y = 2(1)^2 - 4 = -2 \quad (1,\ -2)$

let $x = 2$
$y = 2(2)^2 - 4 = 4 \quad (2,\ 4)$

let $x = -1$
$y = 2(-1)^2 - 4 = -2 \quad (-1,\ -2)$

let $x = -2$
$y = 2(-2)^2 - 4 = 4 \quad (-2,\ 4)$

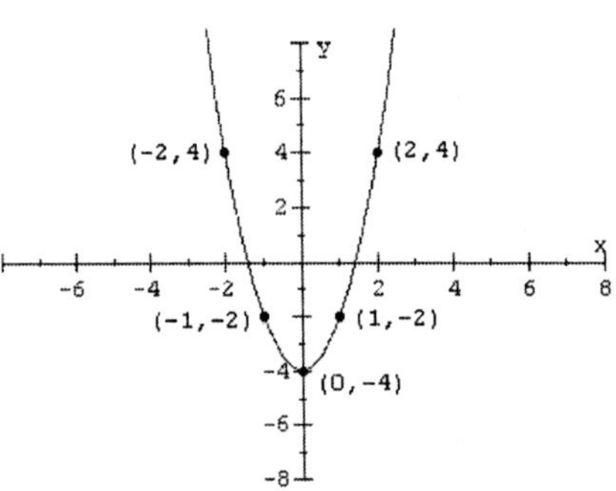

64. The denominator can not be zero.
So set $2x - 6 = 0$
$2x = 6$
$x = 3$
Therefore 3 is excluded from the domain.
The domain is all real numbers except 3.

65. All real numbers

66. Domain $\{-1,\ 0,\ 1\}$
Range $\{2,\ 4,\ 6\}$
It is a function.

67. Domain $\{-1,\ 0,\ 1\}$
Range $\{4\}$
It is a function.

68. Domain $\{-1\}$
Range $\{2, 4, 6\}$
It is not a function.
Since $(-1, 2)$ and $(-1, 4)$ assigns the same domain element to two different range elements it is not a function.

69.

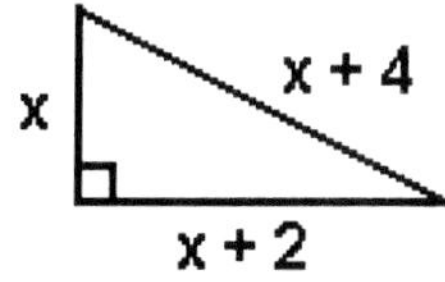

$$x^2 + (x+2)^2 = (x+4)^2$$
$$x^2 + x^2 + 4x + 4 = x^2 + 8x + 16$$
$$2x^2 + 4x + 4 = x^2 + 8x + 16$$
$$x^2 - 4x - 12 = 0$$
$$(x-6)(x+2) = 0$$
$$x - 6 = 0 \quad \text{or} \quad x + 2 = 0$$
$$x = 6 \quad \text{or} \quad x = -2$$

Since length can not be negative, discard the root $x = -2$. The length of the legs are 6 inches and 8 inches. The hypotenuse is 10 inches.

70. Let $x =$ the number
$$0.02(x) = 15$$
$$x = \frac{15}{0.20} = 75$$
The number is 75.

71. Let $x =$ milliliters of 65% HCl acid
$$0.65x + 0.30(40) = 0.55(x+40)$$
$$0.65x + 12 = 0.55x + 22$$
$$0.10x = 10$$
$$x = \frac{10}{0.10} = 100$$
100 ml of 65% hydrochloric acid must be added.

72. Let $x =$ width of rectangle
then $x + 2 =$ length of rectangle
$$P = 2L + 2W$$
$$28 = 2(x+2) + 2x$$
$$28 = 2x + 4 + 2x$$
$$28 = 4x + 4$$
$$24 = 4x$$
$$6 = x$$
The dimensions will be 6 feet by 8 feet.

73. Let $t =$ time traveling
$$55t + 65t = 300$$
$$120t = 300$$
$$t = \frac{300}{120} = 2.5$$
It will take 2.5 hours to be 300 miles apart.

74.
$$A = \frac{1}{2}h(b_1 + b_2)$$
$$120 = \frac{1}{2}h(10 + 22)$$
$$120 = \frac{1}{2}h(32)$$
$$120 = 16h$$
$$\frac{120}{16} = h$$
$$7.5 = h$$
The altitude is 7.5 cm.

75.
$$\frac{16}{352} = \frac{x}{594}$$
$$352x = 16(594)$$
$$352x = 9504$$
$$x = \frac{9504}{352} = 27$$
The consumption would be 27 gallons.

Chapter 8 Coordinate Geometry and Linear Systems

PROBLEM SET 8.1 The Slope of a Line

1. Let $(7, 5)$ be P_1 and $(3, 2)$ be P_2.

$$m = \frac{y_2 - y_1}{x_2 - x_1} = \frac{2-5}{3-7} = \frac{-3}{-4} = \frac{3}{4}$$

3. Let $(-1, 3)$ be P_1 and $(-6, -4)$ be P_2.

$$m = \frac{y_2 - y_1}{x_2 - x_1} = \frac{-4-3}{-6-(-1)}$$

$$m = \frac{-7}{-5} = \frac{7}{5}$$

5. Let $(2, 8)$ be P_1 and $(7, 2)$ be P_2.

$$m = \frac{y_2 - y_1}{x_2 - x_1} = \frac{2-8}{7-2} = \frac{-6}{5} = -\frac{6}{5}$$

7. Let $(-2, 5)$ be P_1 and $(1, -5)$ be P_2.

$$m = \frac{y_2 - y_1}{x_2 - x_1} = \frac{-5-5}{1-(-2)}$$

$$m = \frac{-10}{3} = -\frac{10}{3}$$

9. Let $(4, -1)$ be P_1 and $(-4, -7)$ be P_2.

$$m = \frac{y_2 - y_1}{x_2 - x_1} = \frac{-7-(-1)}{-4-4}$$

$$m = \frac{-6}{-8} = \frac{3}{4}$$

11. Let $(3, -4)$ be P_1 and $(2, -4)$ be P_2.

$$m = \frac{y_2 - y_1}{x_2 - x_1} = \frac{-4-(-4)}{2-3} = \frac{0}{-1} = 0$$

13. Let $(-6, -1)$ be P_1 and $(-2, -7)$ be P_2.

$$m = \frac{y_2 - y_1}{x_2 - x_1} = \frac{-7-(-1)}{-2-(-6)}$$

$$m = \frac{-6}{4} = -\frac{3}{2}$$

15. Let $(-2, 4)$ be P_1 and $(-2, -6)$ be P_2.
The slope is undefined because $x_1 = x_2$.

17. Let $(-1, 10)$ be P_1 and $(-9, 2)$ be P_2.

$$m = \frac{y_2 - y_1}{x_2 - x_1} = \frac{2-10}{-9-(-1)}$$

$$m = \frac{-8}{-8} = 1$$

19. Let (a, b) be P_1 and (c, d) be P_2.

$$m = \frac{y_2 - y_1}{x_2 - x_1} = \frac{d-b}{c-a}$$

21. Let $(7, 8)$ be P_1 and $(2, y)$ be P_2.

$$\frac{y-8}{2-7} = \frac{4}{5}$$
$$\frac{y-8}{-5} = \frac{4}{5}$$
$$5(y-8) = 4(-5)$$
$$5y - 40 = -20$$
$$5y = 20$$
$$y = 4$$

23. Let $(-2, -4)$ be P_1 and $(x, 2)$ be P_2.

$$\frac{2-(-4)}{x-(-2)} = -\frac{3}{2}$$
$$\frac{6}{x+2} = -\frac{3}{2}$$
$$-3(x+2) = 6(2)$$
$$-3x - 6 = 12$$
$$-3x = 18$$
$$x = -6$$

In Problems 25 – 32, the answers will vary but sample points are given.

25. From $(3, 2)$ move 2 units up and 3 units right to $(6, 4)$.
From $(6, 4)$ move 2 units up and 3 units right to $(9, 6)$.
From $(9, 6)$ move 2 units up and 3 units right to $(12, 8)$.

27. From $(-2, -4)$ move 1 unit up and 2 units right to $(0, -3)$.
From $(0, -3)$ move 1 unit up and 2 units right to $(2, -2)$.
From $(2, -2)$ move 1 unit up and 2 units right to $(4, -1)$.

29. From $(-3, 4)$ move 3 units down and 4 units right to $(1, 1)$.
From $(1, 1)$ move 3 units down and 4 units right to $(5, -2)$.
From $(5, -2)$ move 3 units down and 4 units right to $(9, -5)$.

31. From $(4, -5)$ move 2 units down and 1 unit right to $(5, -7)$.
From $(5, -7)$ move 2 units down and 1 unit right to $(6, -9)$.
From $(6, -9)$ move 2 units down and 1 unit right to $(7, -11)$.

33. The line falls from left to right, so slope is negative.
See back of textbook for graph.

35. The line rises from left to right, so the slope is positive.
See back of textbook for graph.

37. The line is horizontal, so the slope is zero.
See back of textbook for graph.

39. The line falls from left to right, so the slope is negative.
See back of textbook for graph.

41.

$$\begin{aligned} & 3x + 2y = 6 \\ x = 0: \quad & 0 + 2y = 6 \\ & 2y = 6 \\ & y = 3 \\ y = 0: \quad & 3x + 0 = 6 \\ & x = 2 \end{aligned}$$

Point
$(0, 3)$
$(2, 0)$

$$m = \frac{0-3}{2-0} = \frac{-3}{2} = -\frac{3}{2}$$

43.

$$\begin{aligned} & 5x - 4y = 20 \\ x = 0: \quad & 0 - 4y = 20 \\ & -4y = 20 \\ & y = -5 \\ y = 0: \quad & 5x + 0 = 20 \\ & 5x = 20 \\ & x = 4 \end{aligned}$$

Point
$(0, -5)$
$(4, 0)$

$$m = \frac{0-(-5)}{4-0} = \frac{5}{4}$$

45.

$$\begin{aligned} & x + 5y = 6 \\ y = 0: \quad & x + 0 = 6 \\ & x = 6 \\ y = 1: \quad & x + 5 = 6 \\ & x = 1 \end{aligned}$$

Point
$(6, 0)$
$(1, 1)$

$$m = \frac{1-0}{1-6} = \frac{1}{-5} = -\frac{1}{5}$$

47.

$$\begin{aligned} & 2x - y = -7 \\ x = 0: \quad & 0 - y = -7 \\ & -y = -7 \\ & y = 7 \\ x = 1: \quad & 2 - y = -7 \\ & -y = -9 \\ & y = 9 \end{aligned}$$

Point
$(0, 7)$
$(1, 9)$

$$m = \frac{9-7}{1-0} = \frac{2}{1} = 2$$

49.

		Point
	$y = 3$	
$x = 0:$	$y = 3$	$(0, 3)$
$x = 2:$	$y = 3$	$(2, 3)$

$$m = \frac{3-3}{2-0} = \frac{0}{2} = 0$$

51.

$$-2x+5y=9$$
$$x=0:\quad 0+5y=9$$
$$5y=9$$
$$y=\frac{9}{5}$$
$$y=0:\quad -2x+0=9$$
$$-2x=9$$
$$x=-\frac{9}{2}$$

Point
$\left(0,\ \frac{9}{5}\right)$
$\left(-\frac{9}{2},\ 0\right)$

$$m=\frac{0-\frac{9}{5}}{-\frac{9}{2}-0}=\frac{-\frac{9}{5}}{-\frac{9}{2}}$$
$$m=\left(-\frac{9}{5}\right)\left(-\frac{2}{9}\right)=\frac{2}{5}$$

53.

$$6x-5y=-30$$
$$x=0:\quad 0-5y=-30$$
$$-5y=-30$$
$$y=6$$
$$y=0:\quad 6x+0=-30$$
$$6x=-30$$
$$x=-5$$

Point
(0, 6)
(– 5, 0)

$$m=\frac{0-6}{-5-0}=\frac{-6}{-5}=\frac{6}{5}$$

55.

$$y=-3x-1$$
$$x=0:\quad y=0-1$$
$$y=-1$$
$$x=1:\quad y=-3-1$$
$$y=-4$$

Point
(0, – 1)
(1, – 4)

$$m=\frac{-4-(-1)}{1-0}=\frac{-3}{1}=-3$$

57.

$$y=4x$$
$$x=0:\quad y=0$$
$$x=1:\quad y=4$$

Point
(0, 0)
(1, 4)

$$m=\frac{4-0}{1-0}=\frac{4}{1}=4$$

59.

$$y=\frac{2}{3}x-\frac{1}{2}$$
$$x=0:\quad y=0-\frac{1}{2}$$
$$y=-\frac{1}{2}$$
$$x=3:\quad y=\frac{2}{3}(3)-\frac{1}{2}$$
$$y=2-\frac{1}{2}$$
$$y=\frac{4}{2}-\frac{1}{2}$$
$$y=\frac{3}{2}$$

Point
$\left(0,\ -\frac{1}{2}\right)$
$\left(3,\ \frac{3}{2}\right)$

$$m=\frac{\frac{3}{2}-\left(-\frac{1}{2}\right)}{3-0}=\frac{\frac{4}{2}}{3}=\frac{2}{3}$$

61. The grade of the highway can be calculated as a ratio as follows:

$$\frac{\text{distance highway rises}}{\text{horizontal distance}}=\frac{135}{2640}=0.051136...$$

The grade of the highway is 5.1%.

63. Let x represent the measure of the run of the stairs. Solve a proportion comparing the rise to the run of the stairs.

$$\frac{\text{rise}}{\text{run}}\quad \frac{3}{5}=\frac{19}{x}$$
$$3x=19(5)$$
$$3x=95$$
$$x=31.667 \text{ or } 32 \text{ to the nearest whole number.}$$

The measure of the run of the stairs is 32 centimeters.

65. Let x represent the vertical drop for the sewage pipe.

$$\frac{\text{fall}}{100 \text{ feet}} \quad \frac{2\frac{1}{4}}{100} = \frac{x}{45}$$

$$100x = 45\left(\frac{9}{4}\right)$$

$$400x = 45(9)$$

$$400x = 405$$

$$x = 1.0125 \text{ or } 1.0 \text{ to the nearest tenth.}$$

The vertical drop must be 1.0 feet for 45 feet.

PROBLEM SET 8.2 Writing Equations of Lines

Problems 1 – 11 can be done by using the general approach demonstrated in Example 1 in the text or by using the point-slope form.

1. The general approach is shown for this problem. The slope for the points (2, 3) and (x, y) is $\frac{2}{3}$.

$$\frac{y-3}{x-2} = \frac{2}{3}$$

$$2(x-2) = 3(y-3)$$

$$2x - 4 = 3y - 9$$

$$2x - 3y = -5$$

3. The general approach is shown for this problem. The slope for the points $(-3, -5)$ and (x, y) is $\frac{1}{2}$.

$$\frac{y-(-5)}{x-(-3)} = \frac{1}{2}$$

$$\frac{y+5}{x+3} = \frac{1}{2}$$

$$1(x+3) = 2(y+5)$$

$$x + 3 = 2y + 10$$

$$x - 2y = 7$$

5. The point-slope form is shown for this problem.

$$y - y_1 = m(x - x_1)$$

$$y - 8 = -\frac{1}{3}[x - (-4)]$$

$$3(y-8) = -1(x+4)$$

$$3y - 24 = -x - 4$$

$$x + 3y = 20$$

7. The point-slope form is shown for this problem.

$$y - y_1 = m(x - x_1)$$

$$y - (-7) = 0(x - 3)$$

$$y + 7 = 0$$

$$y = -7$$

9. The point-slope form is shown for this problem.

$$y - y_1 = m(x - x_1)$$

$$y - 0 = -\frac{4}{9}(x - 0)$$

$$9(y-0) = -4(x-0)$$

$$9y = -4x$$

$$4x + 9y = 0$$

11. The point-slope form is shown for this problem.

$$y - y_1 = m(x - x_1)$$

$$y - (-2) = 3[x - (-6)]$$

$$y + 2 = 3(x+6)$$

$$y + 2 = 3x + 18$$

$$-3x + y = 16$$

$$3x - y = -16$$

Problem Set 8.2

13. Find the slope.

$$m = \frac{y_2 - y_1}{x_2 - x_1} = \frac{10-3}{7-2} = \frac{7}{5}$$

Use the slope and either point in the point-slope form.

$$y - y_1 = m(x - x_1)$$
$$y - 3 = \frac{7}{5}(x-2)$$
$$5(y-3) = 7(x-2)$$
$$5y - 15 = 7x - 14$$
$$-7x + 5y = 1$$
$$7x - 5y = -1$$

15. Find the slope.

$$m = \frac{y_2 - y_1}{x_2 - x_1} = \frac{4-(-2)}{-1-3} = \frac{6}{-4} = -\frac{3}{2}$$

Use the slope and either point in the point-slope form.

$$y - y_1 = m(x - x_1)$$
$$y - 4 = -\frac{3}{2}[x-(-1)]$$
$$2(y-4) = -3(x+1)$$
$$2y - 8 = -3x - 3$$
$$3x + 2y = 5$$

17. Find the slope.

$$m = \frac{y_2 - y_1}{x_2 - x_1} = \frac{-7-(-2)}{-6-(-1)} = \frac{-5}{-5} = 1$$

Use the slope and either point in the point-slope form.

$$y - y_1 = m(x - x_1)$$
$$y - (-2) = 1[x-(-1)]$$
$$y + 2 = x + 1$$
$$-x + y = -1$$
$$x - y = 1$$

19. Find the slope.

$$m = \frac{y_2 - y_1}{x_2 - x_1} = \frac{-5-0}{-3-0} = \frac{-5}{-3} = \frac{5}{3}$$

Use the slope and either point in the point-slope form.

$$y - y_1 = m(x - x_1)$$
$$y - 0 = \frac{5}{3}(x-0)$$
$$3y = 5x$$
$$-5x + 3y = 0$$
$$5x - 3y = 0$$

21. Find the slope.

$$m = \frac{y_2 - y_1}{x_2 - x_1} = \frac{0-4}{7-0} = \frac{-4}{7} = -\frac{4}{7}$$

Use the slope and either point in the point-slope form.

$$y - y_1 = m(x - x_1)$$
$$y - 0 = -\frac{4}{7}(x-7)$$
$$7y = -4(x-7)$$
$$7y = -4x + 28$$
$$4x + 7y = 28$$

23. Use the slope-intercept form.

$$y = mx + b$$
$$y = \frac{3}{5}x + 2$$

25. Use the slope-intercept form.

$$y = mx + b$$
$$y = 2x - 1$$

27. Use the slope-intercept form.

$$y = mx + b$$
$$y = -\frac{1}{6}x - 4$$

29. Use the slope-intercept form.

$$y = mx + b$$
$$y = -x + \frac{5}{2}$$

31. Use the slope-intercept form.

$$y = mx + b$$
$$y = -\frac{5}{9}x - \frac{1}{2}$$

For Problems 33 – 43 see back of textbook for graphs.

33. $m = -2,\ b = -5$

35.
$$\begin{aligned} 3x - 5y &= 15 \\ -5y &= -3x + 15 \\ y &= \frac{3}{5}x - 3 \\ m = \frac{3}{5},\ b &= -3 \end{aligned}$$

37.
$$\begin{aligned} -4x + 9y &= 18 \\ 9y &= 4x + 18 \\ y &= \frac{4}{9}x + 2 \\ m = \frac{4}{9},\ b &= 2 \end{aligned}$$

39.
$$\begin{aligned} -y &= -\frac{3}{4}x + 4 \\ y &= \frac{3}{4}x - 4 \\ m = \frac{3}{4},\ b &= -4 \end{aligned}$$

41.
$$\begin{aligned} -2x - 11y &= 11 \\ -11y &= 2x + 11 \\ y &= -\frac{2}{11}x - 1 \\ m = -\frac{2}{11},\ b &= -1 \end{aligned}$$

43.
$$\begin{aligned} 9x + 7y &= 0 \\ 7y &= -9x + 0 \\ y &= -\frac{9}{7}x + 0 \\ m = -\frac{9}{7},\ b &= 0 \end{aligned}$$

PROBLEM SET 8.3 Solving Linear Systems by Graphing

1.
$$\begin{aligned} 5x + y &= 9 & 3x - 2y &= 4 \\ 5(1) + (4) &= 9 & 3(1) - 2(4) &= 4 \\ 9 &= 9 & 3 - 8 &= 4 \\ & & -5 &\neq 4 \end{aligned}$$
Therefore $(1,\ 4)$ is not a solution of the system.

3.
$$\begin{aligned} x - 3y &= 17 & 2x + 5y &= -21 \\ (2) - 3(-5) &= 17 & 2(2) + 5(-5) &= -21 \\ 2 + 15 &= 17 & 4 - 25 &= -21 \\ 17 &= 17 & -21 &= -21 \end{aligned}$$
Therefore, $(2,\ -5)$ is a solution of the system.

5.
$$\begin{aligned} y &= 2x & 3x - 4y &= 5 \\ -2 &= 2(-1) & 3(-1) - 4(-2) &= 5 \\ -2 &= -2 & -3 + 8 &= 5 \\ & & 5 &= 5 \end{aligned}$$
Therefore, $(-1,\ -2)$ is a solution of the system.

7.
$$\begin{aligned} 6x - 5y &= 5 & 3x + 4y &= -4 \\ 6(0) - 5(-1) &= 5 & 3(0) + 4(-1) &= -4 \\ 0 + 5 &= 5 & 0 - 4 &= -4 \\ 5 &= 5 & -4 &= -4 \end{aligned}$$
Therefore $(0,\ -1)$ is not a solution of the system.

9.
$$\begin{aligned} -3x - y &= 4 & -2x + 3y &= -23 \\ -3(4) - (-5) &= 4 & -2(4) + 3(-5) &= -23 \\ -12 + 5 &= 4 & -8 - 15 &= -23 \\ -7 &\neq 4 & -23 &= -23 \end{aligned}$$
Therefore $(4,\ -5)$ is a solution of the system.

11. See back of textbook for graph. The solution set is $\{(2,\ -1)\}$.

13. See back of textbook for graph. The solution set is $\{(2,\ 1)\}$.

15. See back of textbook for graph. The solution is $\emptyset$.

Problem Set 8.3

17. See back of textbook for graph.
The solution set is $\{(0,\ 0)\}$.

19. See back of textbook for graph.
The solution set is $\{(1,\ -1)\}$.

21. See back of textbook for graph.
The system has infinitely many solutions.

23. See back of textbook for graph.
The solution set is $\{(1,\ 3)\}$.

25. See back of textbook for graph.
The solution set is $\{(3,\ -2)\}$.

27. See back of textbook for graph.
The solution set is $\{(2,\ 4)\}$.

29. See back of textbook for graph.
The solution set is $\{(-2,\ -3)\}$.

PROBLEM SET 8.4 The Elimination-by-Addition Method

1.

$$\begin{array}{rcl} x + y &=& 14 \\ x - y &=& -2 \\ \hline 2x &=& 12 \end{array}$$ Add the two equations.

$x = 6$

Substitute 6 for x in $x + y = 14$.
$6 + y = 14$
$y = 8$
The solution set is $\{(6,\ 8)\}$.

3.

$$\begin{array}{rcl} x + 4y &=& -21 \\ 3x - 4y &=& 1 \\ \hline -4x &=& -20 \end{array}$$ Add the two equations.

$x = -5$

Substitute -5 for x in $x + 4y = -21$.
$-5 + 4y = -21$
$4y = -16$
$y = -4$
The solution set is $\{(-5,\ -4)\}$.

5. $y = 6 - x$ Add x to both sides.
$x - y = -18$ Leave alone.

$$\begin{array}{rcl} x + y &=& 6 \\ x - y &=& -18 \\ \hline 2x &=& -12 \\ x &=& -6 \end{array}$$

Substitute -6 for x in $y = 6 - x$.
$y = 6 - (-6)$
$y = 12$
The solution set is $\{(-6,\ 12)\}$.

7.

$$\begin{array}{lll} 5x+y=23 & \text{Multiply by 2.} & 10x+2y=46 \\ 3x-2y=19 & \text{Leave alone.} & \underline{3x-2y=19} \\ & & 13x \quad = 65 \\ & & x=5 \end{array}$$

Substitute 5 for x in $5x + y = 23$.

$5(5) + y = 23$

$25 + y = 23$

$y = -2$

The solution set is $\{(5, -2)\}$.

9.

$$\begin{array}{ll} x+2y=5 & \\ \underline{3x-2y=6} & \\ 4x \quad = 11 & \text{Add the two equations.} \\ x=\frac{11}{4} & \end{array}$$

$$\begin{array}{lll} x+2y=5 & \text{Multiply by } -3. & -3x-6y=-15 \\ 3x-2y=6 & \text{Leave alone.} & \underline{3x-2y=\quad 6} \\ & & -8y=-9 \\ & & y=\frac{9}{8} \end{array}$$

The solution set is $\left\{\left(\frac{11}{4}, \frac{9}{8}\right)\right\}$.

11.

$$\begin{array}{lll} y=-x & \text{Add } x \text{ to both sides.} & x+y=0 \\ 2x-y=-2 & \text{Leave alone.} & \underline{2x-y=-2} \\ & & 3x \quad =-2 \\ & & x=-\frac{2}{3} \end{array}$$

Substitute $-\frac{2}{3}$ for y in $-x$.

$y = -\left(-\frac{2}{3}\right) = \frac{2}{3}$

The solution set is $\left\{\left(-\frac{2}{3}, \frac{2}{3}\right)\right\}$.

13.

$$\begin{array}{lll} 4x+5y=9 & \text{Multiply by 5.} & 20x+25y=45 \\ 5x-6y=-50 & \text{Multiply by } -4. & \underline{-20x+24y=200} \\ & & 49y=245 \\ & & y=5 \end{array}$$

Substitute 5 for y in $4x + 4y = 9$.

$$4x + 5(5) = 9$$
$$4x + 25 = 9$$
$$4x = -16$$
$$x = -4$$

The solution set is $\{(-4, 5)\}$.

15. $9x - 7y = 29$ Multiply by 5. $\quad 45x - 35y = 145$

$5x - 3y = 17$ Multiply by -9. $\quad -45x + 27y = -153$

$$-8y = -8$$
$$y = 1$$

Substitute 1 for y in $9x - 7y = 29$.

$$9x - 7(1) = 29$$
$$9x - 7 = 29$$
$$9x = 36$$
$$x = 4$$

The solution set is $\{(4, 1)\}$.

17. $6x + 5y = -6$ Multiply by 3. $\quad 18x + 15y = -18$

$8x - 3y = 21$ Multiply by 5. $\quad 40x - 15y = 105$

$$58x = 87$$
$$x = \frac{87}{58} = \frac{3}{2}$$

Substitute $\frac{3}{2}$ for x in $6x + 5y = -6$.

$$6\left(\frac{3}{2}\right) + 5y = -6$$
$$9 + 5y = -6$$
$$5y = -15$$
$$y = -3$$

The solution set is $\left\{\left(\frac{3}{2}, -3\right)\right\}$.

19. $2x - 7y = -1$ Multiply by 4. $\quad 8x - 28y = -4$

$9x + 4y = -2$ Multiply by 7. $\quad 63x + 28y = -14$

$$71x = -18$$
$$x = -\frac{18}{71}$$

$2x - 7y = -1$ Multiply by 9. $\quad 18x - 63y = -9$

$9x + 4y = -2$ Multiply by -2. $\quad -18x - 8y = 4$

$$-71y = -5$$
$$y = \frac{5}{71}$$

The solution set is $\left\{\left(-\frac{18}{71}, \frac{5}{71}\right)\right\}$.

21.

$$\begin{array}{rll} x + y = 750 & \text{Multiply by } -7. & -7x - 7y = -5250 \\ 0.07x + 0.08y = 57.5 & \text{Multiply by } 100. & 7x + 8y = 5750 \\ \hline & & y = 500 \end{array}$$

Substitute 500 for y in $x + y = 750$.

$x + 500 = 750$

$x = 250$

The solution set is $\{(250, 500)\}$.

23.

$$\begin{array}{rll} 0.09x + 0.11y = 31 & \text{Multiply by } 100. & 9x + 11y = 3100 \\ y = x + 100 & \text{Subtract } x \text{ from both sides.} & -x + y = 100 \end{array}$$

$$\begin{array}{rll} 9x + 11y = 3100 & \text{Leave alone.} & 9x + 11y = 3100 \\ -x + y = 100 & \text{Multiply by } 9. & -9x + 9y = 900 \\ \hline & & 20y = 4000 \\ & & y = 200 \end{array}$$

Substitute 200 for y in $y = x + 100$.

$200 = x + 100$

$100 = x$

The solution set is $\{(100, 200)\}$.

25. Let x and y represent the two numbers.

$$\begin{array}{rl} x + y = 30 & \text{The sum of the numbers is 30.} \\ x - y = 12 & \text{The difference of the numbers is 12.} \\ \hline 2x = 42 & \\ x = 21 & \end{array}$$

Substitute 21 for x in $x + y = 30$.

$21 + y = 30$

$y = 9$

The two numbers are 21 and 9.

27. Let x represent the smaller number and y the larger number.

$y - x = 7$ Their difference is 7.

$2y - 3x = 6$ Three times the smaller subtracted from twice the larger is 6.

$$\begin{array}{rll} y - x = 7 & \text{Multiply by } -2. & -2y + 2x = -14 \\ 2y - 3x = 6 & \text{Leave alone.} & 2y - 3x = 6 \\ \hline & & -x = -8 \\ & & x = 8 \end{array}$$

Substitute 8 for x in $y - x = 7$.

$y - 8 = 7$

$y = 15$

The numbers are 8 and 15.

29. Let x represent the smaller number and y the larger number.

$y = 2x$ One number is twice the other.

$3x + 5y = 78$ The sum of three times the smaller and five times the larger is 78.

$y = 2x$ Subtract y from both sides. $2x - y = 0$

$3x + 5y = 78$ Leave alone. $3x + 5y = 78$

$2x - y = 0$ Multiply by 5. $10x - 5y = 0$

$3x + 5y = 78$ Leave alone. $3x + 5y = 78$

$$13x = 78$$
$$x = 6$$

Substitute 6 for x in $y = 2x$.
$y = 2(6) = 12$
The numbers are 6 and 12.

31. Let b represent the cost of one lemon and a represent the cost of one apple.

$3b + 2a = 105$ The cost of 3 lemons and 2 apples is $1.05.

$2b + 3a = 120$ The cost of 2 lemons and 3 apples is $1.20.

$3b + 2a = 105$ Multiply by 2. $6b + 4b = 210$

$2b + 3a = 120$ Multiply by -3. $-6b - 9a = -360$

$$-5a = -150$$
$$a = 30$$

Substitute 30 for a in $3b + 2a = 105$.
$3b + 2(30) = 105$
$3b + 60 = 105$
$3b = 45$
$b = 15$
The cost of a lemon is $0.15 and the cost of an apple is $0.30.

33. Let d represent the number of dimes and q the number of quarters.

$d + q = 10$ The number of coins is 10.

$10d + 25q = 145$ The total value of the coins is $1.45.

$d + q = 10$ Multiply by -10. $-10d - 10q = -100$

$10d + 25q = 145$ Leave alone. $10d + 25q = 145$

$$15q = 45$$
$$q = 3$$

Substitute 3 for q in $d + q = 10$.
$d + 3 = 10$
$d = 7$
He has 7 dimes and 3 quarters.

35. Let x represent the \$12 book and y represent the \$14 book.

$x + y = 35$ The total number of books is 35.
$12x + 14y = 462$ The total value of the books is \$462.

$x + y = 35$ Multiply by -12. $\quad -12x - 12y = -420$
$12x + 14y = 462$ Leave alone. $\quad 12x + 14y = 462$

$$2y = 42$$
$$y = 21$$

Substitute 21 for y in $x + y = 35$.
$x + 21 = 35$
$x = 14$
They bought 14 books at \$12 per book and 21 books at \$14 per book.

37. Let x represent the number of gallons of 10% salt solution and y represent the number of gallons of 15% salt solution.

$x + y = 10$ The total of gallons is 10.
$0.10x + 0.15y = 0.13(10)$ The quantity of salt in the solution.

$x + y = 10$ Multiply by -10. $\quad -10x - 10y = -100$
$0.10x + 0.15y = 1.30$ Multiply by 100. $\quad 10x + 15y = 130$

$$5y = 30$$
$$y = 6$$

Substitute 6 for y in $x + y = 10$.
$x + 6 = 10$
$x = 4$
The quantity needed would be 4 gallons of the 10% solution and 6 gallons of the 15% solution.

39. Let x represent the investment at 10% and y represent the investment at 12%.

$x + y = 1300$ The total investment was \$1300.
$0.10x + 0.12y = 146$ The total yearly interest was \$146.

$x + y = 1300$ Multiply by -10. $\quad -10x - 10y = -13,000$
$0.10x + 0.12y = 146$ Multiply by 100. $\quad 10x + 12y = 14,600$

$$y = 800$$

Substitute 800 for y in $x + y = 1300$.
$x + 800 = 1300$
$x = 500$
The investments were \$500 at 10% and \$800 at 12%.

PROBLEM SET 8.5 The Substitution Method

1. Substitute $2x - 1$ for y in $x + y = 14$.
$x + (2x - 1) = 14$
$3x - 1 = 14$
$3x = 15$
$x = 5$
Substitute 5 for x in $y = 2x - 1$.
$y = 2(5) - 1$
$y = 10 - 1$
$y = 9$
The solution set is $\{(5, 9)\}$.

3. Substitute $-3x - 2$ for y in $x - y = -14$.
$x - (-3x - 2) = -14$
$x + 3x + 2 = -14$
$4x = -16$
$x = -4$
Substitute -4 for x in $y = -3x - 2$.
$y = -3(-4) - 2$
$y = 12 - 2$
$y = 10$
The solution set is $\{(-4, 10)\}$.

5. Substitute $-2x + 7$ for y in $4x - 3y = -6$.
$4x - 3(-2x + 7) = -6$
$4x + 6x - 21 = -6$
$10x - 21 = -6$
$10x = 15$
$x = \frac{15}{10} = \frac{3}{2}$
Substitute $\frac{3}{2}$ for x in $y = -2x + 7$.
$y = -2\left(\frac{3}{2}\right) + 7$
$y = -3 + 7$
$y = 4$
The solution set is $\left\{\left(\frac{3}{2}, 4\right)\right\}$.

7. Solve $x + y = 1$ for x.
$x + y = 1$
$x = 1 - y$
Substitute $1 - y$ for x in $3x + 6y = 7$.
$3(1 - y) + 6y = 7$
$3 - 3y + 6y = 7$
$3 + 3y = 7$
$3y = 4$
$y = \frac{4}{3}$
Substitute $\frac{4}{3}$ for y in $x = 1 - y$.
$x = 1 - \frac{4}{3}$
$x = \frac{3}{3} - \frac{4}{3} = -\frac{1}{3}$
The solution set is $\left\{\left(-\frac{1}{3}, \frac{4}{3}\right)\right\}$.

9. Substitute $\frac{3}{4}y$ for x in $2x - y = 12$.
$2\left(\frac{3}{4}y\right) - y = 12$
$\frac{3}{2}y - \frac{2}{2}y = 12$
$\frac{1}{2}y = 12$
$y = 24$
Substitute 24 for y in $x = \frac{3}{4}y$.
$x = \frac{3}{4}(24) = 18$
The solution set is $\{(18, 24)\}$.

11. Substitute $\frac{3}{2}x$ for y in $6x - 5y = 15$.
$6x - 5\left(\frac{3}{2}x\right) = 15$
$6x - \frac{15}{2}x = 15$
$12x - 15x = 30$
$-3x = 30$
$x = -10$
Substitute -10 for x in $y = \frac{3}{2}x$.
$y = \frac{3}{2}(-10) = -15$
The solution set is $\{(-10, -15)\}$.

13. Substitute $4y - 1$ for x in $2x - 8y = 3$.

$2(4y - 1) - 8y = 3$

$8y - 2 - 8y = 3$

$-2 = 3$ This is not possible.

The solution set is $\emptyset$.

15. Solve $7x + 2y = -2$ for y.

$7x + 2y = -2$

$2y = -7x - 2$

$y = \frac{-7x - 2}{2}$

Substitute $\frac{-7x-2}{2}$ for y in $6x + 5y = 18$.

$6x + 5\left(\frac{-7x - 2}{2}\right) = 18$

Multiplied by 2.

$12x + 5(-7x - 2) = 36$

$12x - 35x - 10 = 36$

$-13x - 10 = 36$

$-23x = 46$

$x = -2$

Substitute -2 for x in $7x + 2y = -2$.

$7(-2) + 2y = -2$

$-14 + 2y = -2$

$2y = 12$

$y = 6$

The solution set is $\{(-2, 6)\}$.

17. Solve $x + 5y = -71$ for x.

$x + 5y = -71$

$x = -5y - 71$

Substitute $-5y - 71$ for x in $8x - 3y = -9$.

$8(-5y - 71) - 3y = -9$

$-40y - 568 - 3y = -9$

$-43y - 568 = -9$

$-43y = 559$

$y = -13$

Substitute -13 for y in $x + 5y = -71$.

$x + 5(-13) = -71$

$x - 65 = -71$

$x = -6$

The solution set is $\{(-6, -13)\}$.

19. Solve $4x - 6y = 1$ for y.

$4x - 6y = 1$

$-6y = -4x + 1$

Multiplied by -1.

$6y = 4x - 1$

$y = \frac{4x - 1}{6}$

Substitute $\frac{4x-1}{6}$ for y in $2x + 3y = 4$.

$2x + 3\left(\frac{4x - 1}{6}\right) = 4$

Multiplied by 2.

$4x + 1(4x - 1) = 8$

$4x + 4x - 1 = 8$

$8x - 1 = 8$

$8x = 9$

$x = \frac{9}{8}$

Substitute $\frac{9}{8}$ for x in $4x - 6y = 1$.

$4\left(\frac{9}{8}\right) - 6y = 1$

$\frac{9}{2} - 6y = 1$

$9 - 12y = 2$

$-12y = -7$

$y = \frac{7}{12}$

The solution set is $\left\{\left(\frac{9}{8}, \frac{7}{12}\right)\right\}$.

21. Solve $3x - 2y = 0$ for y.

$3x - 2y = 0$

$-2y = -3x$

$y = \frac{3x}{2}$

Substitute $\frac{3x}{2}$ for y in $5x + 7y = 3$.

$5x+7\left(\frac{3x}{2}\right)=3$

Multiplied by 2.

$10x+7(3x)=6$

$10x+21x=6$

$31x=6$

$x=\frac{6}{31}$

Substitute $\frac{6}{31}$ for x in $3x-2y=0$.

$3\left(\frac{6}{31}\right)-2y=0$

$3(6)-62y=0$

$18-62y=0$

$-62y=-18$

$y=\frac{18}{62}=\frac{9}{31}$

The solution set is $\left\{\left(\frac{6}{31}, \frac{9}{31}\right)\right\}$.

23. Substitute $x+300$ for y in $0.05x+0.07y=33$.

$0.05x+0.07(x+300)=33$

Multiplied by 100.

$5x+7(x+300)=3300$

$5x+7x+2100=3300$

$12x+2100=3300$

$12x=1200$

$x=100$

Substitute 100 for x in $y=x+300$.

$y=100+300=400$

The solution set is $\{(100, 400)\}$.

25. Solve $x+y=13$ for y.

$x+y=13$

$y=13-x$

Substitute $13-x$ for y in $0.05x+0.1y=1.15$.

$0.05x+0.1(13-x)=1.15$

Multiplied by 100.

$5x+10(13-x)=115$

$5x+130-10x=115$

$130-5x=115$

$-5x=-15$

$x=3$

Substitute 3 for x in $x+y=13$.

$3+y=13$

$y=10$

The solution set is $\{(3, 10)\}$.

27. Use the elimination-by-addition method.

$5x-4y=14$ Multiply by 3.

$7x+3y=-32$ Multiply by 4.

$15x-12y=42$

$28x+12y=-128$

$43x=-86$

$x=-2$

Substitute -2 for x in $5x-4y=14$.

$5(-2)-4y=14$

$-10-4y=14$

$-4y=24$

$y=-6$

The solution set is $\{(-2, -6)\}$.

29. Use the substitution method.

Substitute $-x$ for y in $2x+9y=6$.

$2x+9(-x)=6$

$2x-9x=6$

$-7x=6$

$x=-\frac{6}{7}$

Substitute $-\frac{6}{7}$ for x in $y=-x$.

$y=-\left(-\frac{6}{7}\right)=\frac{6}{7}$

The solution set is $\left\{\left(-\frac{6}{7}, \frac{6}{7}\right)\right\}$.

31. Use the elimination-by-addition method.

$x+y=22$ Multiply by -5.

$0.6x+0.5y=12$ Multiply by 10.

$-5x-5y=-110$

$6x+5y=120$

$x=10$

Substitute 10 for x in $x+y=22$.

$10+y=22$

$y=12$

The solution set is $\{(10, 12)\}$.

33. Use the elimination-by-addition method.
$4x - y = 0$ Multiply by 2.
$7x + 2y = 9$ Leave alone.

$$\begin{array}{l} 8x - 2y = 0 \\ \underline{7x + 2y = 9} \\ 15x \quad\;\; = 9 \\ \quad x = \frac{9}{15} = \frac{3}{5} \end{array}$$

Substitute $\frac{3}{5}$ for x in $4x - y = 0$.

$$4\left(\frac{3}{5}\right) - y = 0$$
$$\frac{12}{5} - y = 0$$
$$\frac{12}{5} = y$$

The solution set is $\left\{\left(\frac{3}{5}, \frac{12}{5}\right)\right\}$.

35. Use the elimination-by-addition method.
$2x + y = 1$ Multiply by -3.
$6x - 7y = -57$ Leave alone.

$$\begin{array}{l} -6x - 3y = -3 \\ \underline{\;\;6x - 7y = -57} \\ \quad -10y = -60 \\ \qquad y = 6 \end{array}$$

Substitute 6 for y in $2x + y = 1$.

$$2x + 6 = 1$$
$$2x = -5$$
$$x = -\frac{5}{2}$$

The solution set is $\left\{\left(-\frac{5}{2}, 6\right)\right\}$.

37. Use the elimination-by-addition method.
$6x - y = -1$ Multiply by 2.
$10x + 2y = 13$ Leave alone.

$$\begin{array}{l} 12x - 2y = -2 \\ \underline{10x + 2y = 13} \\ 22x \quad\;\; = 11 \\ \quad x = \frac{11}{22} = \frac{1}{2} \end{array}$$

Substitute $\frac{1}{2}$ for x in $6x - y = -1$.

$$6\left(\frac{1}{2}\right) - y = -1$$
$$3 - y = -1$$
$$-y = -4$$
$$y = 4$$

The solution set is $\left\{\left(\frac{1}{2}, 4\right)\right\}$.

39. Use the elimination-by-addition method.
$4x + 8y = 20$ Divide by -4.
$x + 2y = 5$ Leave alone.

$$\begin{array}{l} -x - 2y = -5 \\ \underline{\;\;x + 2y = \;\;5} \\ \qquad 0 = 0 \quad \text{(Always true.)} \end{array}$$

The system has infinitely many solutions.

41. Use the substitution method.
Substitute $2y$ for x in $3x - 8y = -5$.

$$3(2y) - 8y = -5$$
$$6y - 8y = -5$$
$$-2y = -5$$
$$y = \frac{5}{2}$$

Substitute $\frac{5}{2}$ for y in $x = 2y$.

$$x = 2\left(\frac{5}{2}\right) = 5$$

The solution set is $\left\{\left(5, \frac{5}{2}\right)\right\}$.

43. Use the elimination-by-addition method.
$5y - 2x = -4$ Multiply by -2.
$10y = 3x + 4$ Subtract $3x$ from both sides.

$$\begin{array}{l} -10y + 4x = 8 \\ \underline{\;\;10y - 3x = 4} \\ \qquad\quad x = 12 \end{array}$$

Substitute 12 for x in $10y = 3x + 4$.

$$10y = 3(12) + 4$$
$$10y = 36 + 4$$
$$10y = 40$$
$$y = 4$$

The solution set is $\{(12, 4)\}$.

45. Use the substitution method.
Substitute $-y-1$ for x
in $6x-5y=4$.
$6(-y-1)-5y=4$
$-6y-6-5y=4$
$-11y-6=4$
$-11y=10$
$y=-\frac{10}{11}$
Substitute $-\frac{10}{11}$ for y in $x=-y-1$.
$x=-\left(-\frac{10}{11}\right)-1$
$x=\frac{10}{11}-1$
$x=\frac{10}{11}-\frac{11}{11}$
$x=-\frac{1}{11}$
The solution set is $\left\{\left(-\frac{1}{11}, -\frac{10}{11}\right)\right\}$.

47. Let x and y represent the two numbers.
$x+y=46$ Their sum is 46.
$x-y=22$ Their difference is 22.
$2x \quad =68$
$x=34$
Substitute 34 for x in $x+y=46$.
$34+y=46$
$y=12$
The numbers are 12 and 34.

49. Let x represent the number of double rooms at \$28. Let y represent the number of single rooms at \$19.
$x+y=50$ The total number of rooms is 50.
$28x+19y=1265$ The total revenue was \$1265.

$x+y=50$ Multiply by -19.
$28x+19y=1265$ Leave alone.

$-19x-19y=-950$
$28x+19y=1265$
$9x \quad = 315$
$x=35$
Substitute 35 for x in $x+y=50$.
$35+y=50$
$y=15$
There were 35 double rooms at \$28 and 15 single rooms at \$19.

51. Let t represent the tens digit.
Let u represent the units digit.
$t+u=9$ Their sum is 9.
$10t+u=9u$ The two-digit number is nine times its units digit.

$t+u=9$ Multiply by -10.
$10t+u=9u$ Subtract $9u$ from both sides..

$-10t-10u=-90$
$10t-\ 8u=\ 0$
$-18u=-90$
$u=5$
Substitute 5 for u in $t+u=9$.
$t+5=9$
$t=4$
The tens digit is 4 and the units digit is 5. Thus, the number is 45.

53. Let d represent the number of dimes.
Let q represent the number of quarters.
$10d+25q=1205$ The value of the coins was \$12.05.
$q=2d+5$ The number of quarters is five more than twice the number of dimes.

Substitute $2d+5$ for q in $10d+25q=1205$.
$10d+25(2d+5)=1205$
$10d+50d+125=1205$
$60d+125=1205$
$60d=1080$
$d=18$

Substitute 18 for d in $q = 2d + 5$.
$q = 2(18) + 5$
$q = 36 + 5$
$q = 41$
There were 18 dimes and 41 quarters.

55. Let t represent the tens digit.
Let u represent the units digit.
$t + u = 12$ Their sum is 12.
$t = 3u$ The tens digit is three times the units digit.

Substitute $3u$ for t in $t + u = 12$.
$3u + u = 12$
$4u = 12$
$u = 3$
Substitute 3 for u in $t + u = 12$.
$t + 3 = 12$
$t = 9$
The tens digit is 9 and the units digit is 3.
Thus, the number is 93.

57. Let x represent the money invested at 8%.
Let y represent the money invested at 9%.
$y = x + 250$ $250 more was invested at 9% than at 8%.
$0.08x + 0.09y = 48$ The total yearly interest was $48.

Substitute $x + 250$ for y in
$0.08x + 0.09y = 48$.
$0.08x + 0.09(x + 250) = 48$
Multiplied by 100.
$8x + 9(x + 250) = 4800$
$8x + 9x + 2250 = 4800$
$17x + 2250 = 4800$
$17x = 2550$
$x = 150$

Substitute 150 for x in $y = x + 250$.
$y = 150 + 250 = 400$
The investment was $150 at 8%
and $400 at 9%.

59. Let x represent the quantity of 30% alcohol. Let y represent the quantity of 70% alcohol.

$x + y = 10$ The total quantity of solution.
$0.30x + 0.70y = 0.40(10)$ The total quantity of alcohol.

$x + y = 10$ Multiply by -3.
$0.30x + 0.70y = 4$ Multiply by 10.

$$\begin{aligned} -3x - 3y &= -30 \\ 3x + 7y &= 40 \\ \hline 4y &= 10 \\ y &= 2.5 \end{aligned}$$

Substitute 2.5 for y in $x + y = 10$.
$x + 2.5 = 10$
$x = 7.5$
The quantity of each solution to be used would be 7.5 liters of 30% alcohol and 2.5 liters of 70% alcohol.

PROBLEM SET 8.6 Graphing Linear Inequalities and Systems of Linear Inequalities

1. Use the points (0, 1) and (1, 0) to graph a dashed line for $x + y = 1$. Use (0, 0) as a test point. The given inequality $x + y > 1$ becomes $0 + 0 > 1$ which is a false statement. Therefore, the solution set is the half-plane that does not contain the origin.

3. Use the points $(0, 3)$ and $(2, 0)$ to graph a dashed line for $3x + 2y = 6$. Use $(0, 0)$ as a test point. The given inequality $3x + 2y < 1$ becomes $0 + 0 < 1$ which is a true statement. Therefore, the solution set is the half-plane that does contain the origin.

5. Use the points $(0, 1)$ and $(1, 0)$ to graph a solid line for $2x - y = 4$. Use $(0, 0)$ as a test point. The given inequality $2x - y \geq 4$ becomes $0 - 0 \geq 4$ which is a false statement. Therefore, the solution set is the half-plane that does not contain the origin.

7. Use the points $(0, -4)$ and $(3, 0)$ to graph a solid line for $4x - 3y = 12$. Use $(0, 0)$ as a test point. The given inequality $4x - 3y \leq 12$ becomes $0 - 0 \leq 12$ which is a true statement. Therefore, the solution set is the half-plane that does contain the origin.

9. Use the points $(0, 0)$ and $(1, 2)$ to graph a dashed line for $y = -x$. Since the origin is on the line use $(2, 1)$ as a test point. The given inequality $y > -x$ becomes $1 > -2$ which is a true statement. Therefore, the solution set is the half-plane that contains the point $(2, 1)$.

11. Use the points $(0, 0)$ and $(1, -1)$ to graph a solid line for $2x - y = 0$. Since the origin is on the line use $(2, -1)$ as a test point. The given inequality $2x - y \geq 0$ becomes $4 + 1 \geq 0$ which is a true statement. Therefore, the solution set is the half-plane that contains the point $(2, -1)$.

13. Use the points $(0, -1)$ and $(2, 0)$ to graph a dashed line for $-x + 2y = -2$. Use $(0, 0)$ as a test point. The given inequality $-x + 2y < -2$ becomes $0 + 0 < -2$ which is a false statement. Therefore, the solution set is the half-plane that does not contain the origin.

15. Use the points $(0, -2)$ and $(4, 0)$ to graph a solid line for $y = \frac{1}{2}x - 2$. Use $(0, 0)$ as a test point. The given inequality $y \leq \frac{1}{2}x - 2$ becomes $0 \leq 0 - 2$ which is a false statement. Therefore, the solution set is the half-plane that does not contain the origin.

17. Use the points $(0, 4)$ and $(4, 0)$ to graph a solid line for $y = -x + 4$. Use $(0, 0)$ as a test point. The given inequality $y \geq -x + 4$ becomes $0 \geq 0 + 4$ which is a false statement. Therefore, the solution set is the half-plane that does not contain the origin.

19. Use the points $(0, -3)$ and $(-4, 0)$ to graph a dashed line for $3x + 4y = -12$. Use $(0, 0)$ as a test point. The given inequality $3x + 4y > -12$ becomes $0 + 0 > -12$ which is a true statement. Therefore, the solution set is the half-plane that does contain the origin.

21. Use the points $(0, 2)$ and $(3, 0)$ to graph a dashed line for $2x + 3y = 6$. Use $(0, 0)$ as a test point. The inequality $2x + 3y > 6$ becomes $0 + 0 > 6$ which is a false statement. Therefore, $2x + 3y > 6$ is satisfied by the points in the half-plane above the line $2x + 3y = 6$.

Use the points $(0, -2)$ and $(2, 0)$ to graph a dashed line for $x - y = 2$. Use $(0, 0)$ as a test point. The inequality $x - y < 2$ becomes $0 + 0 < 2$ which is a true statement. Therefore, $x - y < 2$ is satisfied by the points in the half-plane above the line $x - y = 2$.

The solution set for the system is the intersection of the individual solution sets.

23. Use the points $(0, -1)$ and $(3, 0)$ to graph a solid line for $x - 3y = 3$. Use $(0, 0)$ as a test point. The inequality $x - 3y \geq 3$ becomes $0 + 0 \geq 3$ which is a false statement. Therefore, $x - 3y \geq 3$ is satisfied by the points in the half-plane on or below the line $x - 3y = 3$.

Use the points $(0, 3)$ and $(1, 0)$ to graph a solid line for $3x + y = 3$. Use $(0, 0)$ as a test point. The inequality $3x + y \leq 3$ becomes $0 + 0 \leq 3$ which is a true statement. Therefore, $3x + y \leq 3$ is satisfied by the points in the half-plane on or below the line $3x + y = 3$.

The solution set for the system is the intersection of the individual solution sets.

25. Use the points $(0, 0)$ and $(1, 2)$ to graph a solid line for $y = 2x$. Since $(0, 0)$ is on the line use $(3, -1)$as a test point. The inequality $y \geq 2x$ becomes $-1 \geq 6$ which is a false statement. Therefore, $y \geq 2x$ is satisfied by the points in the half-plane on or above the line $y = 2x$.

Use the points $(0, 0)$ and $(2, 2)$ to graph a dashed line for $y = x$. Since $(0, 0)$ is on the line use $(3, -1)$as a test point. The inequality $y < x$ becomes $-1 < 3$ which is a true statement. Therefore, $y < x$ is satisfied by the points in the half-plane below the line $y = x$.

The solution set for the system is the intersection of the individual solution sets.

27. Use the points $(0, 1)$ and $(1, 0)$ to graph a dashed line for $y = -x + 1$. Use $(0, 0)$ as a test point. The inequality $y < -x + 1$ becomes $0 < 0 + 1$ which is a true statement. Therefore, $y < -x + 1$ is satisfied by the points in the half-plane below the line $y = -x + 1$.

Use the points $(0, -1)$ and $(-1, 0)$ to graph a dashed line for $y = -x - 1$. Use $(0, 0)$ as a test point. The inequality $y > -x - 1$ becomes $0 > 0 - 1$ which is a true statement. Therefore, $y > -x - 1$ is satisfied by the points in the half-plane above the line $y = -x - 1$.

The solution set for the system is the intersection of the individual solution sets.

29. Use the points $(0, 2)$ and $(-4, 0)$ to graph a dashed line for $y = \frac{1}{2}x + 2$. Use $(0, 0)$ as a test point. The inequality $y < \frac{1}{2}x + 2$ becomes $0 < 0 + 2$ which is a true statement. Therefore, $y < \frac{1}{2}x + 2$ is satisfied by the points in the half-plane below the line $y = \frac{1}{2}x + 2$.

Use the points $(0, -1)$ and $(2, 0)$ to graph a dashed line for $y = \frac{1}{2}x - 1$. Use $(0, 0)$ as a test point. The inequality $y < \frac{1}{2}x - 1$ becomes $0 < 0 - 1$ which is a false statement. Therefore, $y < \frac{1}{2}x - 1$ is satisfied by the points in the half-plane below the line $y = \frac{1}{2}x - 1$.

The solution set for the system is the intersection of the individual solution sets.

Problem Set 8.7

PROBLEM SET 8.7

1. $\begin{pmatrix} 3x+y+2z=6 \\ 6y+5z=-4 \\ -4z=8 \end{pmatrix}$

Solve equation (3).

$-4z=8$

$z=-2$

Substitute $z=-2$ into equation (2).

$6y+5(-2)=-4$

$6y-10=-4$

$6y=6$

$y=1$

Substitute $z=-2$ and $y=1$ into equation (1).

$3x+(1)+2(-2)=6$

$3x+1-4=6$

$3x-3=6$

$3x=9$

$x=3$

The solution set is $\{(3,\ 1,\ -2)\}$.

3. $\begin{pmatrix} x+2y-z=1 \\ y+2z=11 \\ 2y-z=2 \end{pmatrix}$

Multiply equation (2) by -2 and add the result to equation (3).

$-2y-4z=-22$

$2y-z=2$

$-5z=-20$

$z=4$

Substitute $z=4$ into equation (2).

$y+2(4)=11$

$y+8=11$

$y=3$

Substitute $y=3$ and $z=4$ into equation (1).

$x+2(3)-(4)=1$

$x+6-4=1$

$x+2=1$

$x=-1$

The solution set is $\{(-1,\ 3,\ 4)\}$.

5. $\begin{pmatrix} 4x+3y-2z=9 \\ 2x+y=7 \\ 3x-2y=21 \end{pmatrix}$

Multiply equation (2) by 2 and add the result to equation (3).

$4x+2y=14$

$3x-2y=21$

$7x \quad\quad =35$

$x=5$

Substitute $x=5$ into equation (2).

$2(5)+y=7$

$10+y=7$

$y=-3$

Substitute $x=5$ and $y=-3$ into equation (1).

$4(5)+3(-3)-2z=9$

$20-9-2z=9$

$11-2z=9$

$-2z=-2$

$z=1$

The solution set is $\{(5,\ -3,\ 1)\}$.

7. $\begin{pmatrix} x+2y-3z=-11 \\ 2x-y+2z=3 \\ 4x+3y+z=6 \end{pmatrix}$

Multiply equation (1) by -2 and add the result to equation (2).

$-2x-4y+6z=22$

$2x-\ \ y+2z=3$

$-5y+8z=25 \quad (4)$

Multiply equation (1) by -4 and add the result to equation (3).

$$\begin{array}{r} -4x-8y+12z=44 \\ 4x+3y+\quad z=\ 6 \\ \hline -5y+13z=50 \quad (5) \end{array}$$

Multiply equation (4) by -1 and add the result to equation (5).

$$\begin{array}{r} 5y-\ 8z=-25 \\ -5y+13z=\ 50 \\ \hline 5z=\ 25 \end{array}$$

$z=5$

Substitute $z=5$ into equation (4).

$-5y+8(5)=25$

$-5y+40=25$

$-5y=-15$

$y=3$

Substitute $y=3$ and $z=5$ into equation (1).

$x+2(3)-3(5)=-11$

$x+6-15=-11$

$x-9=-11$

$x=-2$

The solution set is $\{(-2, 3, 5)\}$.

9. $\begin{pmatrix} 4x-3y+z=14 \\ 2x+y-3z=16 \\ 3x-4y+2z=9 \end{pmatrix}$

Multiply equation (1) by 3 and add the result to equation (2).

$$\begin{array}{r} 12x-9y+3z=42 \\ 2x+\ y-3z=16 \\ \hline 14x-8y \quad =58 \quad (4) \end{array}$$

Multiply equation (1) by -2 and add the result to equation (3).

$$\begin{array}{r} -8x+6y-2z=-28 \\ 3x-4y+2z=\ 9 \\ \hline -5x+2y \quad =-19 \quad (5) \end{array}$$

Multiply equation (5) by 4 and add the result to equation (4).

$$\begin{array}{r} -20x+8y=-76 \\ 14x-8y=\ 58 \\ \hline -6x \quad =-18 \end{array}$$

$x=3$

Substitute $x=3$ into equation (5).

$-5(3)+2y=-19$

$-15+2y=-19$

$2y=-4$

$y=-2$

Substitute $x=3$ and $y=-2$ into equation (1).

$4(3)-3(-2)+z=14$

$12+6+z=14$

$18+z=14$

$z=-4$

The solution set is $\{(3, -2, -4)\}$.

11. $\begin{pmatrix} 2x+y+4z=5 \\ 5x-2y+z=-10 \\ 3x+3y-2z=4 \end{pmatrix}$

Multiply equation (1) by 2 and add the result to equation (2).

$$\begin{array}{r} 4x+2y+8z=\ 10 \\ 5x-2y+\ z=-10 \\ \hline 9x \quad +9z=\ 0 \quad (4) \end{array}$$

Multiply equation (1) by -3 and add the result to equation (3).

$$\begin{array}{r} -6x-3y-12z=-15 \\ 3x+3y-\ 2z=\ 4 \\ \hline -3x \quad -14z=-11 \quad (5) \end{array}$$

Multiply equation (5) by 3 and add the result to equation (4).

$$\begin{array}{r} -9x-42z=-33 \\ 9x+\ 9z=\ 0 \\ \hline -33z=-33 \end{array}$$

$z=1$

Substitute $z=1$ into equation (4).

$9x+9(1)=0$

$9x+9=0$

$9x=-9$

$x=-1$

Substitute $x=-1$ and $z=1$ into equation (1).

$2(-1)+y+4(1)=5$

$-2+y+4=5$

$y+2=5$

$y=3$

The solution set is $\{(-1, 3, 1)\}$.

13. $\begin{pmatrix} x+3y-4z=11 \\ 3x-y+2z=5 \\ 2x+5y-z=8 \end{pmatrix}$

Multiply equation (1) by -3 and add the result to equation (2).

$$\begin{array}{r} -3x-9y+12z=-33 \\ 3x-y+2z=5 \\ \hline -10y+14z=-28 \quad (4) \end{array}$$

Multiply equation (1) by -2 and add the result to equation (3).

$$\begin{array}{r} -2x-6y+8z=-22 \\ 2x+5y-z=8 \\ \hline -y+7z=-14 \quad (5) \end{array}$$

Multiply equation (5) by -10 and add the result to equation (4).

$$\begin{array}{r} 10y-70z=140 \\ 10y+14z=-28 \\ \hline -56z=112 \\ z=-2 \end{array}$$

Substitute $z=-2$ into equation (4).

$$-10y+14(-2)=-28$$
$$-10y-28=-28$$
$$-10y=0$$
$$y=\frac{0}{-10}=0$$

Substitute $y=0$ and $z=-2$ into equation (1).

$$x+3(0)-4(-2)=11$$
$$x+0+8=11$$
$$x+8=11$$
$$x=3$$

The solution set is $\{(3, 0, -2)\}$.

15. $\begin{pmatrix} 3x+y-2z=3 \\ 2x-3y+4z=-2 \\ 4x+z=6 \end{pmatrix}$

Multiply equation (1) by 3 and add the result to equation (2).

$$\begin{array}{r} 9x+3y-6z=9 \\ 2x-3y+4z=-2 \\ \hline 11x-2z=7 \quad (4) \end{array}$$

Multiply equation (3) by 2 and add the result to equation (4).

$$\begin{array}{r} 8x+2z=12 \\ 11x-2z=7 \\ \hline 19x=19 \\ x=1 \end{array}$$

Substitute $x=1$ into equation (3).

$$4(1)+z=6$$
$$4+z=6$$
$$z=2$$

Substitute $x=1$ and $z=2$ into equation (1).

$$3(1)+y-2(2)=3$$
$$3+y-4=3$$
$$y-1=3$$
$$y=4$$

The solution set is $\{(1, 4, 2)\}$.

17. Let $x=$ number of quarters

$y=$ number of dimes

$z=$ number of nickels

$$x+y+z=20$$
$$y+z=x \quad \Rightarrow \quad -x+y+z=0$$
$$25x+10y+5z=340$$

$$\begin{pmatrix} x+y+z=20 \\ -x+y+z=0 \\ 25x+10y+5z=340 \end{pmatrix}$$

Add equation (1) and equation (2).

$$\begin{array}{r} x+y+z=20 \\ -x+y+z=0 \\ \hline 2y+2z=20 \quad (4) \end{array}$$

Multiply equation (1) by -25 and add the result to equation (3).

$$\begin{array}{r} -25x-25y-25z=-500 \\ 25x+10y+5z=340 \\ \hline -15y-20z=-160 \quad (5) \end{array}$$

Multiply equation (4) by 10 and add the result to equation (5).

$$\begin{array}{r} 20y+20z=200 \\ -15y-20z=-160 \\ \hline 5y=40 \\ y=8 \end{array}$$

Substitute $y = 8$ into equation (4).
$2(8) + 2z = 20$
$16 + 2z = 20$
$2z = 4$
$z = 2$
Substitute $y = 8$ and $z = 2$ into equation (1).
$x + 8 + 2 = 20$
$x + 10 = 20$
$x = 10$
There are 10 quarters, 8 dimes, and 2 nickels.

19. Let a = measure of $\angle$A
b = measure of $\angle$B
c = measure of $\angle$C

$a = 5b \quad \Rightarrow \quad a - 5b = 0$
$b + c = a - 60 \quad \Rightarrow \quad -a + b + c = -60$
$a + b + c = 180$

$$\begin{pmatrix} a + b + c = 180 \\ -a + b + c = -60 \\ a - 5b = 0 \end{pmatrix}$$

Multiply equation (1) by -1 and add the result to equation (2).
$-a - b - c = -180$
$-a + b + c = -60$
$-2a = -240$
$a = 120$
Substitute $a = 120$ into equation (3).
$120 - 5b = 0$
$-5b = -120$
$b = 24$
Substitute $a = 120$ and $b = 24$ into equation (1).
$120 + 24 + c = 180$
$144 + c = 180$
$c = 36$
The measure of $\angle$A = 120°, the measure of $\angle$B = 24°, and the measure of $\angle$C = 36°.

21. Let x = wages per hour of plumber
y = wages per hour of apprentice
z = wages per hour of laborer

$x + y + z = 80$
$x = y + z + 20 \quad \Rightarrow \quad x - y - z = 20$
$x = 5z \quad \Rightarrow \quad x - 5z = 0$

$$\begin{pmatrix} x + y + z = 80 \\ x - y - z = 20 \\ x - 5z = 0 \end{pmatrix}$$

Add equation (1) and equation (2).
$x + y + z = 80$
$x - y - z = 20$
$2x = 100$
$x = 50$
Substitute $x = 50$ into equation (3).
$50 - 5z = 0$
$-5z = -50$
$z = 10$
Substitute $x = 50$ and $z = 10$ into equation (1).
$50 + y + 10 = 80$
$y + 60 = 80$
$y = 20$
The plumber's wages are \$50 per hour, the apprentice's wages are \$20 per hour, and the laborer's wages are \$10 per hour.

23. Let x = price per pound of peaches
y = price per pound of cherries
z = price per pound of pears

$$\begin{pmatrix} 2x + y + 3z = 5.64 \\ x + 2y + 2z = 4.65 \\ 2x + 4y + z = 7.23 \end{pmatrix}$$

Multiply equation (1) by -2 and add the result to equation (2).
$-4x - 2y - 6z = -11.28$
$x + 2y + 2z = 4.65$
$-3x - 4z = -6.63 \quad (4)$
Multiply equation (1) by -4 and add the result to equation (3).

$$\begin{array}{rl} -8x - 4y - 12z = & -22.56 \\ 2x + 4y + \quad z = & 7.23 \\ \hline -6x \qquad - 11z = & -15.33 \quad (5) \end{array}$$

Multiply equation (4) by -2 and add the result to equation (5).

$$\begin{array}{rl} 6x + 8z = & 13.26 \\ -6x - 11z = & -15.33 \\ \hline -3z = & -2.07 \end{array}$$

$$z = 0.69$$

Substitute $z = 0.69$ into equation (4).

$$-3x - 4(0.69) = -6.63$$
$$-3x - 2.76 = -6.63$$
$$-3x = -3.87$$
$$x = 1.29$$

Substitute $x = 1.29$ and $z = 0.69$ into equation (1).

$$2(1.29) + y + 3(0.69) = 5.64$$
$$2.58 + y + 2.07 = 5.64$$
$$y + 4.65 = 5.64$$
$$y = 0.99$$

Peaches cost \$1.29 per pound, cherries cost \$0.99 per pound, and pears cost \$0.69 per pound.

25. Let $x =$ cost of helmet
$y =$ cost of jacket
$z =$ cost of gloves

$$x + y + z = 650$$
$$y = x + 100 \quad \Rightarrow \quad -x + y = 100$$
$$x + z = y - 50 \quad \Rightarrow \quad x - y + z = -50$$

$$\begin{pmatrix} x + y + z = 650 \\ -x + y = 100 \\ x - y + z = -50 \end{pmatrix}$$

Multiply equation (1) by -1 and add the result to equation (3).

$$\begin{array}{rl} -x - y - z = & -650 \\ x - y + z = & -50 \\ \hline -2y \qquad = & -700 \end{array}$$

$$y = 350$$

Substitute $y = 350$ into equation (2).

$$-x + 350 = 100$$
$$-x = -250$$
$$x = 250$$

Substitute $x = 250$ and $y = 350$ into equation (1).

$$250 + 350 + z = 650$$
$$600 + z = 650$$
$$z = 50$$

The helmet cost \$250,
the jacket cost \$350,
and the gloves cost \$50.

CHAPTER 8 Review Problem Set

1. $2x - 5y = 10$

$$-5y = -2x + 10$$
$$y = \frac{-2}{-5}x + \frac{10}{-5}$$
$$y = \frac{2}{5}x - 2$$

$m = \frac{2}{5} \qquad b = -2$

See back of textbook for graph.

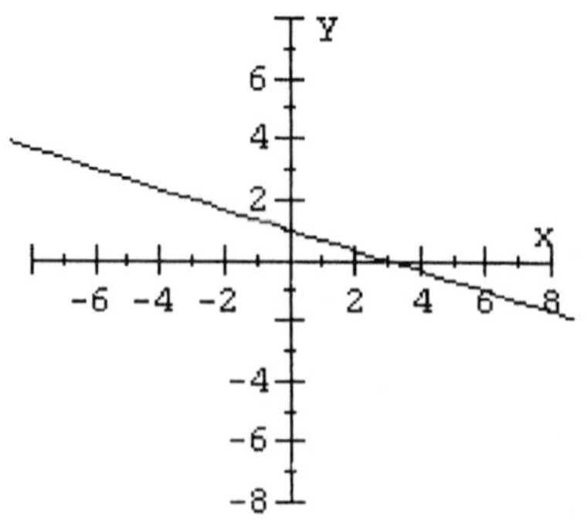

2. $y = -\frac{1}{3}x + 1$

$m = -\frac{1}{3} \qquad b = 1$

3. $x + 2y = 2$

$$2y = -x + 2$$
$$y = \frac{-1}{2}x + \frac{2}{2}$$
$$y = -\frac{1}{2}x + 1$$
$$m = -\frac{1}{2} \qquad b = 1$$

See back of textbook for graph.

4. $3x + y = -2$

$$y = -3x - 2$$
$$m = -2 \qquad b = -2$$

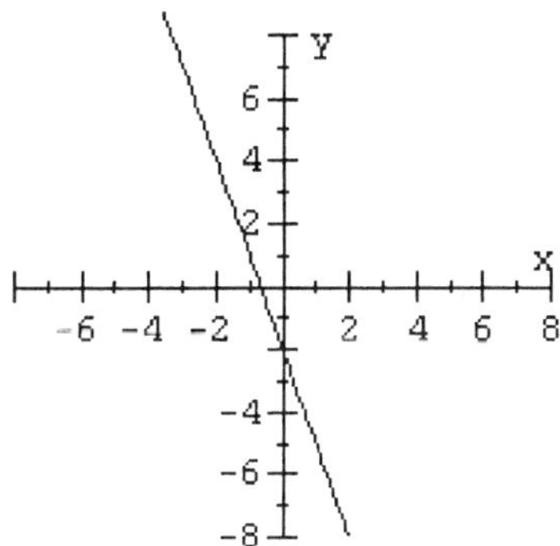

5. $2x - y = 4$

$$-y = -2x + 4$$
$$y = 2x - 4$$
$$m = 2 \qquad b = -4$$

See back of textbook for graph.

6. $3x - 4y = 12$

$$-4y = -3x + 12$$
$$y = \frac{-3}{-4}x + \frac{12}{-4}$$
$$y = \frac{3}{4}x - 3$$
$$m = \frac{3}{4} \qquad b = -3$$

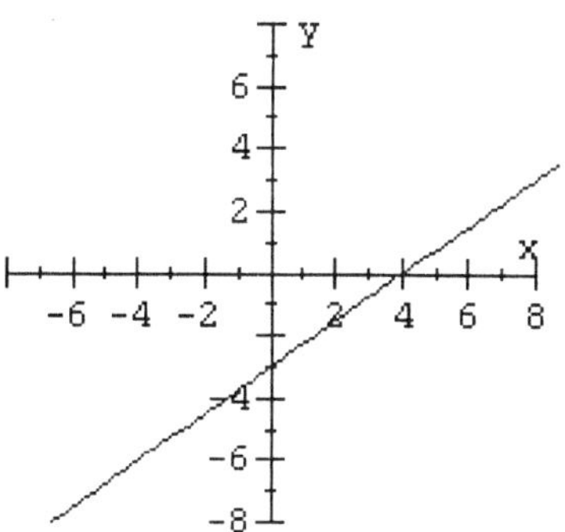

7. Let $(3, -4)$ be P_1 and $(-2, 5)$ be P_2.

$$m = \frac{y_2 - y_1}{x_2 - x_1} = \frac{5-(-4)}{-2-3} = \frac{9}{-5} = -\frac{9}{5}$$

8. $5x - 6y = 30$

$$-6y = -5x + 30$$
$$y = \frac{-5}{-6}x + \frac{30}{-6}$$
$$y = \frac{5}{6}x - 5$$
$$m = \frac{5}{6}$$

9. The point-slope form is shown for this problem.

$$y - y_1 = m(x - x_1)$$
$$y - (-3) = -\frac{5}{7}(x - 2)$$
$$7(y + 3) = -5(x - 2)$$
$$7y + 21 = -5x + 10$$
$$5x + 7y = -11$$

10. $m = \dfrac{-3-5}{-1-2} = \dfrac{-8}{-3} = \dfrac{8}{3}$

$$y - y_1 = m(x - x_1)$$
$$y - 5 = \frac{8}{3}(x - 2)$$
$$3(y - 5) = 3\left[\frac{8}{3}(x - 2)\right]$$
$$3(y - 5) = 8(x - 2)$$
$$3y - 15 = 8x - 16$$
$$-15 = 8x - 3y - 16$$
$$1 = 8x - 3y$$
$$8x - 3y = 1$$

11. Use the slope-intercept form.

$$y = mx + b$$
$$y = \frac{2}{9}x - 1$$
$$9y = 2x - 9$$
$$-2x + 9y = -9$$
$$2x - 9y = 9$$

12. A line perpendicular to the x-axis will be a vertical line. The equation of a vertical line through $(2, 4)$ is $x = 2$.

13.

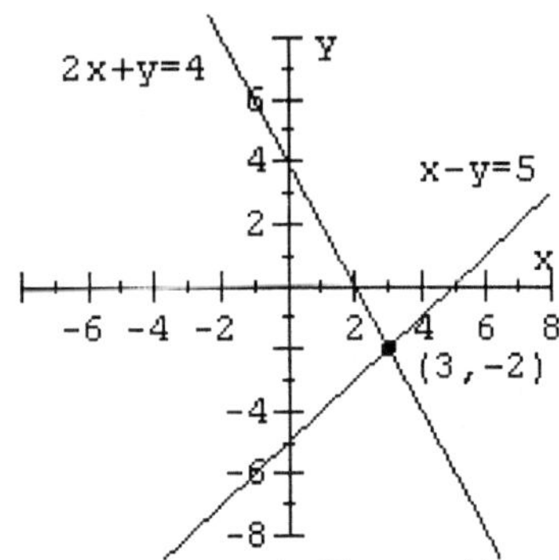

The solution set is $\{(3, -2)\}$.

14. $\begin{pmatrix} 2x - y = 1 \\ 3x - 2y = -5 \end{pmatrix}$

Multiply equation (1) by -2 and add the result to equation (2).

$$\begin{aligned} -4x + 2y &= -2 \\ 3x - 2y &= -5 \\ \hline -x \quad &= -7 \end{aligned}$$
$$x = 7$$

Substitute $x = 7$ into equation (1).

$$2(7) - y = 1$$
$$14 - y = 1$$
$$-y = -13$$
$$y = 13$$

The solution set is $\{(7, 13)\}$.

15. Use the substitution method.

Substitute $-3y + 1$ for x in $2x + 5y = 7$.

$$2(-3y + 1) + 5y = 7$$
$$-6y + 2 + 5y = 7$$
$$-y + 2 = 7$$
$$-y = 5$$
$$y = -5$$

Substitute -5 for y in $x = -3y + 1$.

$$x = -3(-5) + 1$$
$$x = 15 + 1$$
$$x = 16$$

The solution set is $\{(16, -5)\}$.

16. $\begin{pmatrix} 3x + 2y = 7 \\ 4x - 5y = 3 \end{pmatrix}$

Multiply equation (1) by -4.

$-12x - 8y = -28 \quad (3)$

Multiply equiation (2) by 3.

$12x - 15y = 9 \quad (4)$

Add equation (3) and equation (4).

$$\begin{aligned} -12x - 8y &= -28 \\ 12x - 15y &= 9 \\ \hline -23y &= -19 \end{aligned}$$
$$y = \frac{-19}{-23} = \frac{19}{23}$$

Substitute $y = \frac{19}{23}$ into equation (1).

$$3x + 2\left(\frac{19}{23}\right) = 7$$
$$3x + \frac{38}{23} = 7$$
$$3x = 7 - \frac{38}{23}$$
$$3x = \frac{161}{23} - \frac{38}{23}$$
$$3x = \frac{123}{23}$$
$$\frac{1}{3}(3x) = \frac{1}{3}\left(\frac{123}{23}\right)$$
$$x = \frac{41}{23}$$

The solution set is $\left\{\left(\frac{41}{23}, \frac{19}{23}\right)\right\}$.

17. Use the elimination-by-addition method.

$9x + 2y = 140$ Leave alone.

$x + 5y = 135$ Multiply by -9.

$$\begin{array}{r} 9x + 2y = 140 \\ -9x - 45y = -1215 \\ \hline -43y = -1075 \end{array}$$
$y = 25$
Substitute 25 for y in $x + 5y = 135$.
$x + 5(25) = 135$
$x + 125 = 135$
$x = 10$
The solution set is $\{(10, 25)\}$.

18. $$\begin{pmatrix} \frac{1}{2}x + \frac{1}{4}y = -5 \\ \frac{2}{3}x - \frac{1}{2}y = 0 \end{pmatrix}$$

Multiply equation (1) by 4.
$2x + y = -20$
Multiply equation (2) by 6.
$4x - 3y = 0$
The equivalent system is

$$\begin{pmatrix} 2x + y = -20 \\ 4x - 3y = 0 \end{pmatrix} \begin{matrix} (3) \\ (4) \end{matrix}$$

Multiply equation (3) by 3 and add the result to equation (4).
$$\begin{array}{r} 6x - 3y = -60 \\ 4x - 3y = 0 \\ \hline 10x \quad = -60 \end{array}$$
$x = -6$
Substitute $x = -6$ into equation (3).
$2(-6) + y = -20$
$-12 + y = -20$
$y = -8$
The solution set is $\{(-6, -8)\}$.

19. Use the elimination-by-addition method.
$x + y = 1000$ Multiply by -7.
$0.07x + 0.09y = 82$ Multiply by 100.

$$\begin{array}{r} -7x - 7y = -7000 \\ 7x + 9y = 8200 \\ \hline 2y = 1200 \end{array}$$
$y = 600$
Substitute 600 for y in $x + y = 1000$.
$x + 600 = 1000$
$x = 400$
The solution set is $\{(400, 600)\}$.

20. $$\begin{pmatrix} y = 5x + 2 \\ 10x - 2y = 1 \end{pmatrix}$$

Substitute in equation (2) for y.
$10x - 2(5x + 2) = 1$
$10x - 10x - 4 = 1$
$-4 = 1$
Since $-4 \neq 1$ the system is inconsistent and the solution set is $\emptyset$.

21. Use the substitution method.
Substitute $3x - 2$ for y in $5x - 7y = 9$.
$5x - 7(3x - 2) = 9$
$5x - 21x + 14 = 9$
$-16x + 14 = 9$
$-16x = -5$
$x = \frac{5}{16}$
Substitute $\frac{5}{16}$ for x in $y = 3x - 2$.
$y = 3\left(\frac{5}{16}\right) - 2$
$y = \frac{15}{16} - \frac{32}{16}$
$y = -\frac{17}{16}$
The solution set is $\left\{\left(\frac{5}{16}, -\frac{17}{16}\right)\right\}$.

22. $$\begin{pmatrix} 10t + u = 6u \\ t + u = 12 \end{pmatrix} \Rightarrow \begin{pmatrix} 10t - 5u = 0 \\ t + u = 12 \end{pmatrix} \begin{matrix} (1) \\ (2) \end{matrix}$$

Multiply equation (2) by 5 and add the result to equation (1).
$$\begin{array}{r} 5t + 5u = 60 \\ 10t - 5u = 0 \\ \hline 10t \quad = 60 \end{array}$$
$t = 4$
Substitute $t = 4$ into equation (2).
$4 + u = 12$
$u = 8$
The solution set is $t = 4$ and $u = 8$.

23. Use the substitution method.
Substitute $2u$ for t in
$10t + u - 36 = 10u + t.$

$$10(2u) + u - 36 = 10u + (2u)$$
$$20u + u - 36 = 10u + 2u$$
$$21u - 36 = 12u$$
$$-36 = -9u$$
$$4 = u$$

Substitute 4 for u in $t = 2u$.
$t = 2(4) = 8$
The solution set is $t = 8$ and $u = 4$.

24. $\begin{pmatrix} u = 2t + 1 \\ 10t + u + 10u + t = 100 \end{pmatrix} \Rightarrow$

$\begin{pmatrix} u = 2t + 1 \\ 11t + 11u = 110 \end{pmatrix}$

Substitute for u into equation (2).

$$11t + 11(2t + 1) = 110$$
$$11t + 22t + 11 = 110$$
$$33t + 11 = 110$$
$$33t = 99$$
$$t = 3$$

Substitute $t = 3$ into equation (1).
$u = 2(3) + 1$
$u = 6 + 1$
$u = 7$
The solution set is $t = 3$ and $u = 7$.

25. Use the substitution method.
Substitute $-\frac{2}{3}x$ for y
in $\frac{1}{3}x - y = -9.$

$$\frac{1}{3}x - \left(-\frac{2}{3}x\right) = -9$$
$$\frac{1}{3}x + \frac{2}{3}x = -9$$
$$x = -9$$

Substitute -9 for x in $y = -\frac{2}{3}x.$
$y = -\frac{2}{3}(-9) = 6$
The solution set is $\{(-9, 6)\}$.

26. $y > \frac{2}{3}x - 1$

Use $(0, -1)$ and $(3, 1)$ to graph a dashed line for $y = \frac{2}{3}x - 1.$
Use $(0, 0)$ for a test point.
$y > \frac{2}{3}x - 1$
$0 > \frac{2}{3}(0) - 1$
$0 > -1$
Since $0 > -1$ is a true statement, the solution set is the half-plane that contains $(0, 0)$.

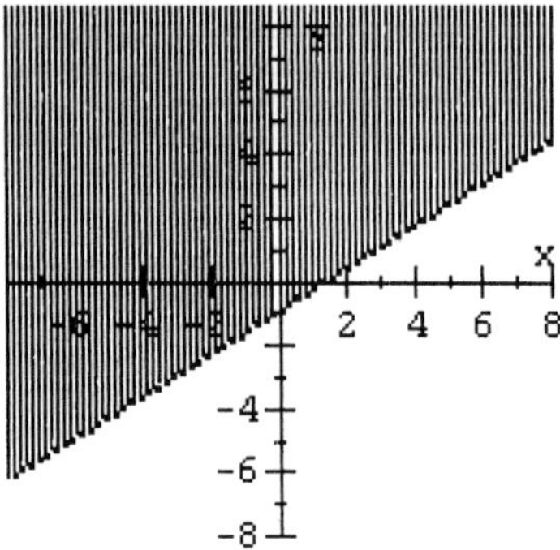

27. $x - 2y \leq 4$
Use $(0, -2)$ and $(4, 0)$ to graph a solid line for $x - 2y = 4$.
Use $(0, 0)$ for a test point.

$$x - 2y \leq 4$$
$$0 - 2(0) \leq 4$$
$$0 \leq 4$$

Since $0 \leq 4$ is a true statement, the solution set is the half-plane that contains $(0, 0)$.
See back of textbook for graph.

28. $y \leq -2x$
Use $(0, 0)$ and $(1, -2)$ to graph a solid line for $y = -2x$.
Use $(3, 3)$ for a test point.
$y \leq -2x$
$3 \leq -2(3)$
$3 \leq -6$
Since $3 \leq -6$ is a false statement, the solution set is the half-plane that does **not** contains $(3, 3)$.

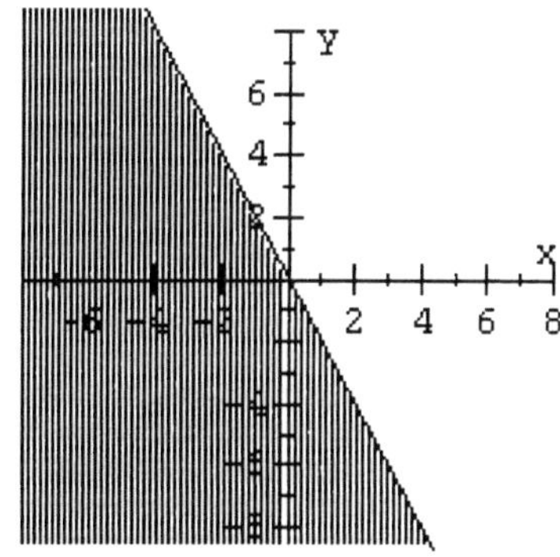

29. $3x + 2y > -6$
Use $(0, -3)$ and $(-2, 0)$ to graph
a dashed line for $3x + 2y = -6$.
Use $(0, 0)$ for a test point.

$$3x + 2y > -6$$
$$3(0) + 2(0) > -6$$
$$0 > -6$$

Since $0 > -6$ is a true statement,
the solution set is the half-plane
that contains $(0, 0)$.
See back of textbook for graph.

30. $\begin{pmatrix} x + y < 4 \\ 2x - y > 2 \end{pmatrix}$

Use $(0, 4)$ and $(4, 0)$ to graph
a dashed line for $x + y = 4$.
Use $(0, 0)$ for a test point.

$$x + y < 4$$
$$0 + 0 < 4$$
$$0 < 4$$

Since $0 < 4$ is a true statement,
the solution set is the half-plane
that contains $(0, 0)$.

Use $(0, -2)$ and $(1, 0)$ to graph
a dashed line for $2x - y = 2$.
Use $(0, 0)$ for a test point.

$$2x - y > 2$$
$$2(0) - 0 > 2$$
$$0 > 2$$

Since $0 > 2$ is a false statement,
the solution set is the half-plane
that does **not** contains $(0, 0)$.

The solution set for the system is the
intersection of the individual solution sets.

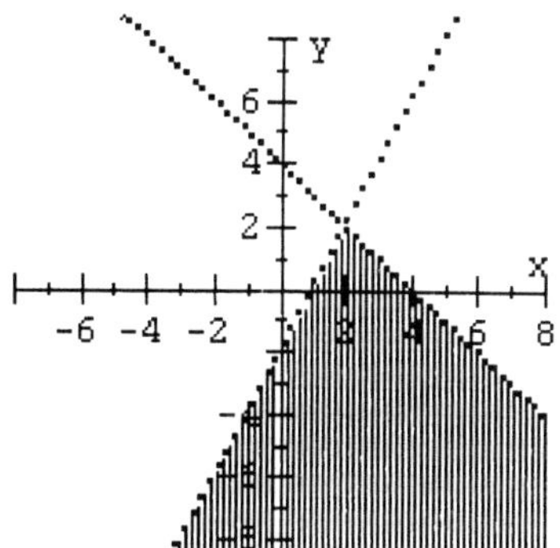

31. $\begin{pmatrix} x + 2y < 4 \\ 3x - y < -3 \end{pmatrix}$

Use $(0, 2)$ and $(4, 0)$ to graph
a dashed line for $x + 2y = 4$.
Use $(0, 0)$ for a test point.

$$x + 2y < 4$$
$$0 + 2(0) < 4$$
$$0 < 4$$

Since $0 < 4$ is a true statement,
the solution set is the half-plane
that contains $(0, 0)$.

Use $(0, 3)$ and $(-1, 0)$ to graph
a dashed line for $3x - y = -3$.
Use $(0, 0)$ for a test point.

$$3x - y < -3$$
$$2(0) - 0 < -3$$
$$0 < -3$$

Since $0 < -3$ is a false statement,
the solution set is the half-plane
that does **not** contain $(0, 0)$.

The solution set for the system is the
intersection of the individual solution sets.
See back of textbook for graph.

32. $\begin{pmatrix} y \geq -\frac{1}{4}x + 3 \\ y < x + 1 \end{pmatrix}$

Use $(0, 3)$ and $(4, 2)$ to graph a solid line for $y = -\frac{1}{4}x + 3$.
Use $(0, 0)$ for a test point.

$y \geq -\frac{1}{4}x + 3$

$0 \geq -\frac{1}{4}(0) + 3$

$0 \geq 3$

Since $0 \geq 3$ is a false statement, the solution set is the half-plane that does **not** contain $(0, 0)$.

Use $(0, 1)$ and $(1, 2)$ to graph a dashed line for $y = x + 1$.
Use $(0, 0)$ for a test point.

$y < x + 1$

$0 < 0 + 1$

$0 < 1$

Since $0 < 1$ is a true statement, the solution set is the half-plane that contains $(0, 0)$.

The solution set for the system is the intersection of the individual solution sets.

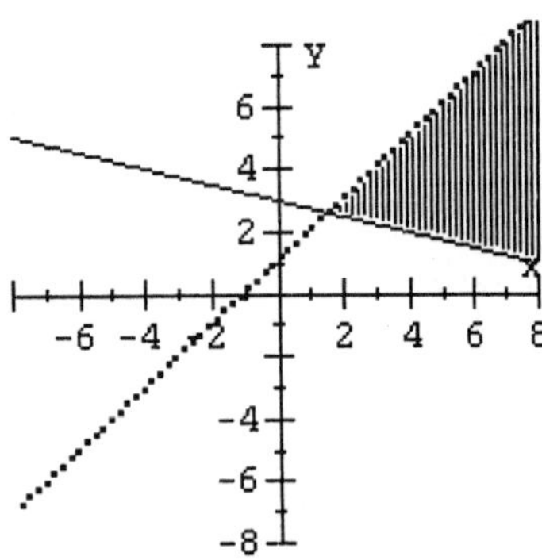

33. $\begin{pmatrix} y > 2x - 3 \\ y \geq -\frac{1}{3}x + 1 \end{pmatrix}$

Use $(0, -3)$ and $(1, -1)$ to graph a dashed line for $y = 2x - 3$.
Use $(0, 0)$ for a test point.

$y > 2x - 3$

$0 > 2(0) - 3$

$0 > -3$

Since $0 > -3$ is a true statement, the solution set is the half-plane that contains $(0, 0)$.

Use $(0, 1)$ and $(3, 0)$ to graph a solid line for $y = -\frac{1}{3}x + 1$.
Use $(0, 0)$ for a test point.

$y \geq -\frac{1}{3}x + 1$

$0 \geq -\frac{1}{3}(0) + 1$

$0 \geq 1$

Since $0 \geq 1$ is a false statement, the solution set is the half-plane that does not contain $(0, 0)$.

The solution set for the system is the intersection of the individual solution sets.
See back of textbook for graph.

34. $\begin{pmatrix} x + 3y - z = 1 \\ 2x - y + z = 3 \\ 3x + y + 2z = 12 \end{pmatrix}$

Multiply equation (1) by -2 and add the result to equation (2).

$$\begin{array}{r} -2x - 6y + 2z = -2 \\ 2x - y + z = 3 \\ \hline -7y + 3z = 1 \quad (4) \end{array}$$

Multiply equation (1) by -3 and add the result to equation (3).

$$\begin{array}{r} -3x - 9y + 3z = -3 \\ 3x + y + 2z = 12 \\ \hline -8y + 5z = 9 \quad (5) \end{array}$$

Multiply equation (4) by -5 and multiply equation (5) by 3. Then add the resulting equations.

$$\begin{array}{r} 35y - 15z = -5 \\ -24y + 15z = 27 \\ \hline 11y = 22 \\ y = 2 \end{array}$$

Substitute for $y = 2$ into equation (4).

$-7(2) + 3z = 1$
$-14 + 3z = 1$
$3z = 15$
$z = 5$

Substitute $y = 2$ and $z = 5$ into equation (1).

$x + 3(2) - 5 = 1$
$x + 6 - 5 = 1$
$x + 1 = 1$
$x = 0$

The solution set is $\{(0, 2, 5)\}$.

34. $\begin{pmatrix} x + 3y - z = 1 \\ 2x - y + z = 3 \\ 3x + y + 2z = 12 \end{pmatrix}$

Add equation (1) and equation (2).

$$\begin{array}{l} x + 3y - z = 1 \\ 2x - y + z = 3 \\ \hline 3x + 2y = 4 \quad (4) \end{array}$$

Multiply equation (1) by 2 and add the result to equation (3).

$$\begin{array}{l} 2x + 6y - 2z = 2 \\ 3x + y + 2z = 12 \\ \hline 5x + 7y = 14 \quad (5) \end{array}$$

Multiply equation (4) by 5 and multiply equation (5) by -3. Then add the resulting equations.

$$\begin{array}{r} 15x + 10y = 20 \\ -15x - 21y = -42 \\ \hline -11y = -22 \\ y = 2 \end{array}$$

Substitute for $y = 2$ into equation (4).

$3x + 2(2) = 4$
$3x + 4 = 4$
$3x = 0$
$x = 0$

Substitute $x = 0$ and $y = 2$ into equation (1).

$x + 3y - z = 1$
$0 + 3(2) - z = 1$
$0 + 6 - z = 1$
$6 - z = 1$
$-z = -5$
$z = 5$

The solution set is $\{(0, 2, 5)\}$.

35. $\begin{pmatrix} 2x + 3y - z = 4 \\ x + 2y + z = 7 \\ 3x + y + 2z = 13 \end{pmatrix}$

Add equation (1) and equation (2).

$$\begin{array}{l} 2x + 3y - z = 4 \\ x + 2y + z = 7 \\ \hline 3x + 5y = 11 \quad (4) \end{array}$$

Multiply equation (1) by 2 and add the result to equation (3).

$$\begin{array}{l} 4x + 6y - 2z = 8 \\ 3x + y + 2z = 13 \\ \hline 7x + 7y = 21 \quad (5) \end{array}$$

Multiply equation (4) by -7 and multiply equation (5) by 3. Then add the resulting equations.

$$\begin{array}{r} -21x - 35y = -77 \\ 21x + 21y = 63 \\ \hline -14y = -14 \\ y = 1 \end{array}$$

Substitute for $y = 1$ into equation (4).

$3x + 5(1) = 11$
$3x + 5 = 11$
$3x = 6$
$x = 2$

Substitute $x = 2$ and $y = 1$ into equation (1).

$2x + 3y - z = 4$
$2(2) + 3(1) - z = 4$
$4 + 3 - z = 4$
$7 - z = 4$
$-z = -3$
$z = 3$

The solution set is $\{(2, 1, 3)\}$.

36. Let $x =$ one number
$y =$ other number

$$\begin{pmatrix} x + y = 113 \\ x = 2y - 1 \end{pmatrix}$$

Use substitution to solve.

$$\begin{aligned} x + y &= 113 \\ (2y - 1) + y &= 113 \\ 3y - 1 &= 113 \\ 3y &= 114 \\ y &= 38 \end{aligned}$$

Substitute $y = 38$ into equation (2).
$x = 2(38) - 1$
$x = 76 - 1$
$x = 75$
The numbers are 75 and 38.

37. Let x represent the money invested at 9%.
Let y represent the money invested at 11%.

$y = x + 50$
The money invested at 11% is \$50 more than the money invested at 9%.

$0.09x + 0.11y = 55.50$
The yearly interest was \$55.50.

Substitute $x + 50$ for y in
$0.09x + 0.11y = 55.50$

$0.09x + 0.11(x + 50) = 55.50$
Multiplied by 100.

$$\begin{aligned} 9x + 11(x + 50) &= 5550 \\ 9x + 11x + 550 &= 5550 \\ 20x + 550 &= 5550 \\ 20x &= 5000 \\ x &= 250 \end{aligned}$$

Substitute 250 for x in $y = x + 50$.
$y = 250 + 50 = 300$
The money invested was \$250 at 9% and \$300 at 11%.

38. Let n represent the number of nickels and d the number of dimes.
$n + d = 43$
The number of coins is 43.
$5n + 10d = 340$
The total value of the coins is 340 cents.

$n + d = 43$	Multiply by -5.
$5n + 10d = 340$	Leave alone.

$$\begin{array}{r} -5n - 5d = -215 \\ 5n + 10d = 340 \\ \hline 5d = 125 \\ d = 25 \end{array}$$

Substitute 25 for d in $n + d = 43$.
$n + 25 = 43$
$n = 18$
She has 18 nickels and 25 dimes.

39. Let l represent the length of the rectangle.
Let w represent the width of the rectangle.
$l = 3w + 1$
The length is one more than three times the width.
$2l + 2w = 50$
The perimeter is 50 inches.

Substitute $3w + 1$ for l in $2l + 2w = 50$.

$$\begin{aligned} 2(3w + 1) + 2w &= 50 \\ 6w + 2 + 2w &= 50 \\ 8w + 2 &= 50 \\ 8w &= 48 \\ w &= 6 \end{aligned}$$

Substitute 6 for w in $l = 3w + 1$.
$l = 3(6) + 1$
$l = 18 + 1$
$l = 19$
The rectangle has a length of 19 inches and a width of 6 inches.

40. Let $x =$ length of rectangle
$y =$ width of rectangle

$$\begin{pmatrix} y = x - 5 \\ 2x + 2y = 38 \end{pmatrix}$$

Substitute for y in equation (2).
$2x + 2(x - 5) = 38$
$2x + 2x - 10 = 38$
$4x - 10 = 38$
$4x = 48$
$x = 12$
Substitute $x = 12$ into equation (1).
$y = 12 - 5$
$y = 7$
The length is 12 inches and the width is 7 inches.

41. Let $x =$ number of quarters
$y =$ number of dimes

$$\begin{pmatrix} x + y = 32 \\ 25x + 10y = 485 \end{pmatrix}$$

Multiply equation (1) by -25 and add the resulting equation to equation (2).
$-25x - 25y = -800$
$25x + 10y = 485$
$-15y = -315$
$y = 21$
Substitute $y = 21$ in equation (1).
$x + 21 = 32$
$x = 11$
There are 11 quarters and 21 dimes.

42. Let a and b represent the angles.
$a + b = 90$
The angles are complementary.
$a = 2b - 6$
One angle is 6° less than twice the other angle.

Substitute $2b - 6$ for a in $a + b = 90$.
$(2b - 6) + b = 90$
$3b - 6 = 90$
$3b = 96$
$b = 32$
Substitute 32 for b in $a + b = 90$.
$a + 32 = 90$
$a = 58$
The two angles are 32° and 58°.

43. Let a represent the larger angle.
Let b represent the smaller angle.
$a + b = 180$
The angles are supplementary.
$a = 3b - 20$
The larger angle is 20° less than three times the smaller angle.
Substitute $3b - 20$ for a in $a + b = 180$.
$(3b - 20) + b = 180$
$4b - 20 = 180$
$4b = 200$
$b = 50$
Substitute 50 for b in $a + b = 180$.
$a + 50 = 180$
$a = 130$
The two angles are 50° and 130°.

44. Let c represent the cost of a cheeseburger and m the cost of a milkshake.
$4c + 5m = 835$
The cost of four cheeseburgers and five milkshakes is $8.35.
$2m = c + 35$
Two milkshakes cost 35 cents more than one cheeseburger.

$4c + 5m = 835$ Leave alone.
$-c + 2m = 35$ Multiply by 4.

$4c + 5m = 835$
$-4c + 8m = 140$
$13m = 975$
$m = 75$
Substitute 75 for m in $2m = c + 35$.
$2(75) = c + 35$
$150 = c + 35$
$115 = c$
The cost of a cheeseburger is $1.15 and a milkshake is $.75

45. Let p represent the cost of a can of prune juice. Let t represent the cost of a can of tomato juice.
$3p + 2t = 385$
The cost of three prune and 2 tomato cans.
$2p + 3t = 355$
The cost of two prune and 3 tomato cans.

Chapter 8 Review Problem Set

$3p + 2t = 385$ Multiply by -2.
$2p + 3t = 355$ Multiply by 3.

$$\begin{aligned} -6p - 4t &= -770 \\ 6p + 9t &= 1065 \\ \hline 5t &= 295 \\ t &= 59 \end{aligned}$$

Substitute 59 for t in $3p + 2t = 385$.

$$\begin{aligned} 3p + 2(59) &= 385 \\ 3p + 118 &= 385 \\ 3p &= 267 \\ p &= 89 \end{aligned}$$

Prune juice costs $\$0.89$ per can and tomato juice costs $\$0.59$ per can.

CHAPTER 8 Test

1. $5x + 3y = 15$

$$\begin{aligned} 3y &= -5x + 15 \\ y &= \frac{-5}{3}x + 5 \end{aligned}$$

$m = -\frac{5}{3}$ $\quad b = 5$

See back of textbook for graph.

2. $-2x + y = -4$

$$y = 2x - 4$$

$m = 2$ $\quad b = -4$

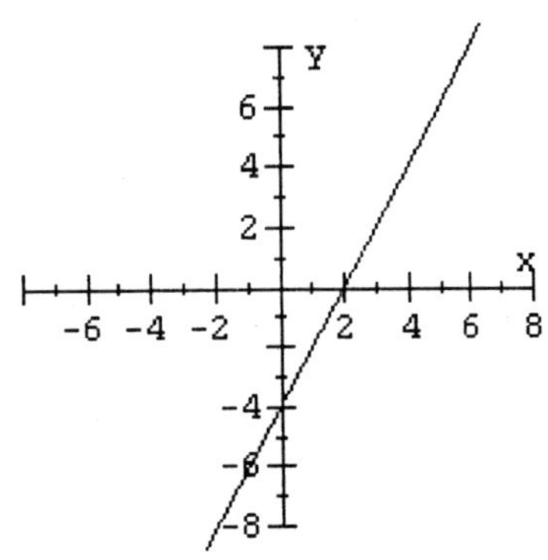

3. $y = -\frac{1}{2}x - 2$

$m = -\frac{1}{2}$ $\quad b = -2$

See back of textbook for graph.

4. $3x + y = 0$

$$y = -3x + 0$$

$m = -3$ $\quad b = 0$

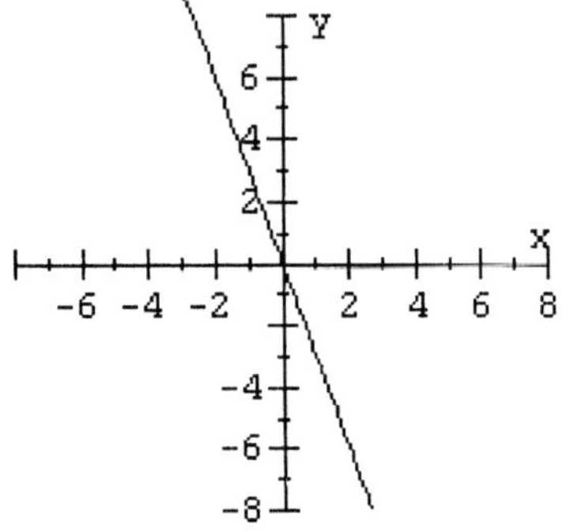

5.

$$\begin{aligned} \frac{3}{2} &= \frac{13-7}{x-4} \\ \frac{3}{2} &= \frac{6}{x-4} \\ 3(x-4) &= 2(6) \\ 3x - 12 &= 12 \\ 3x &= 24 \\ x &= 8 \end{aligned}$$

6.

$$\begin{aligned} -\frac{3}{5} &= \frac{5-y}{6-1} \\ \frac{-3}{5} &= \frac{5-y}{5} \\ -3(5) &= 5(5-y) \\ -15 &= 25 - 5y \\ -40 &= -5y \\ 8 &= y \end{aligned}$$

7. Answers may vary
(7, 6); (11, 7)

8. Answers may vary
$(3, -2)$; $(4, -5)$

9. $m = \frac{\text{rise}}{\text{run}} = \frac{85}{1850}$
$m = 0.046$
$0.046 \times 100\% = 4.6\%$

10. $\begin{pmatrix} y > -\frac{1}{5}x + 3 \\ 2x + y < 4 \end{pmatrix}$

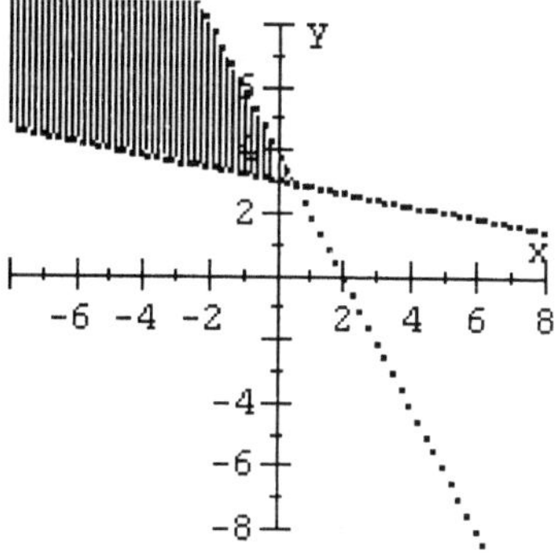

Use $(0, 3)$ and $(5, 2)$ to graph a dashed line for $y = -\frac{1}{5}x + 3$.
Use $(0, 0)$ for a test point.
$y > -\frac{1}{5}x + 3$
$0 > -\frac{1}{5}(0) + 3$
$0 > 3$
Since $0 > 3$ is a false statement, the solution set is the half-plane that does not contain $(0, 0)$.

Use $(0, 4)$ and $(2, 0)$ to graph a dashed line for $2x + y = 4$.
Use $(0, 0)$ for a test point.
$2x + y < 4$
$2(0) + 0 < 4$
$0 < 4$
Since $0 < 4$ is a true statement, the solution set is the half-plane that does contain $(0, 0)$.

The solution set for the system is the intersection of the individual solution sets.

11. Use the slope-intercept form.
$y = mx + b$
$y = -\frac{3}{5}x + 4$
$5y = -3x + 20$
$3x + 5y = 20$

12. Use the point-slope form.
$y - y_1 = m(x - x_1)$
$y - (-2) = \frac{4}{9}(x - 4)$
$9(y + 2) = 4(x - 4)$
$9y + 18 = 4x - 16$
$-4x + 9y = -34$
$4x - 9y = 34$
The solution set is $\{(2, 5)\}$.

13. Find the slope.
$m = \frac{y_2 - y_1}{x_2 - x_1} = \frac{-3 - 6}{-2 - 4} = \frac{-9}{-6} = \frac{3}{2}$
Use the slope and either point in the point-slope form.
$y - y_1 = m(x - x_1)$
$y - 6 = \frac{3}{2}(x - 4)$
$2(y - 6) = 3(x - 4)$
$2y - 12 = 3x - 12$
$-3x + 2y = 0$
$3x - 2y = 0$

14.

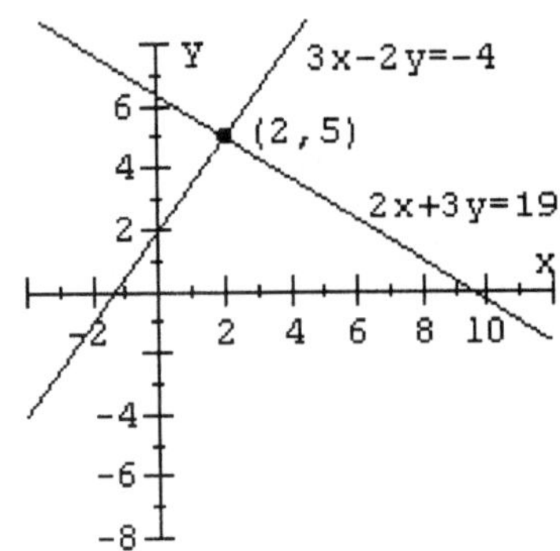

15. $x - 3y = -9$ Multiply by -4.
$4x + 7y = 40$ Leave alone.

$$\begin{aligned} -4x + 12y &= 36 \\ 4x + 7y &= 40 \\ \hline 19y &= 76 \\ y &= 4 \end{aligned}$$

Substitute 4 for y in $x - 3y = -9$.

$$\begin{aligned} x - 3(4) &= -9 \\ x - 12 &= -9 \\ x &= 3 \end{aligned}$$

The solution set is $\{(3, 4)\}$.

16. Solve $5x + y = -14$ for y.

$$\begin{aligned} 5x + y &= -14 \\ y &= -5x - 14 \end{aligned}$$

Substitute $-5x - 14$ for y in
$6x - 7y = -66$.

$$\begin{aligned} 6x - 7(-5x - 14) &= -66 \\ 6x + 35x + 98 &= -66 \\ 41x + 98 &= -66 \\ 41x &= -164 \\ x &= -4 \end{aligned}$$

Substitute -4 for x in $y = -5x - 14$.

$$\begin{aligned} y &= -5(-4) - 14 \\ y &= 20 - 14 \\ y &= 6 \end{aligned}$$

The solution set is $\{(-4, 6)\}$.

17. Use the elimination-by-addition method.
$2x - 7y = 26$ Multiply by 2.
$3x + 2y = -11$ Multiply by 7.

$$\begin{aligned} 4x - 14y &= 52 \\ 21x + 14y &= -77 \\ \hline 25x &= -25 \\ x &= -1 \end{aligned}$$

Substitute -1 for x in $2x - 7y = 26$.

$$\begin{aligned} 2(-1) - 7y &= 26 \\ -2 - 7y &= 26 \\ -7y &= 28 \\ y &= -4 \end{aligned}$$

The solution set is $\{(-1, -4)\}$.

18. Use the elimination-by-addition method.
$8x + 5y = -6$ Leave alone.
$4x - y = 18$ Multiply by 5.

$$\begin{aligned} 8x + 5y &= -6 \\ 20x - 5y &= 90 \\ \hline 28x &= 84 \\ x &= 3 \end{aligned}$$

Substitute 3 for x in $4x - y = 18$.

$$\begin{aligned} 4(3) - y &= 18 \\ 12 - y &= 18 \\ -y &= 6 \\ y &= -6 \end{aligned}$$

The solution set is $\{(3, -6)\}$.

19. Let $x =$ cost of a ream of paper
$y =$ cost of a notebook

$$\begin{pmatrix} 3x + 4y = 19.63 \\ 4x + y = 16.25 \end{pmatrix}$$

$3x + 4y = 19.63$ Leave alone.
$4x + y = 16.25$ Multiply by -4.

$$\begin{aligned} 3x + 4y &= 19.63 \\ -16x - 4y &= -65.00 \\ \hline -13x &= -45.37 \\ x &= 3.49 \end{aligned}$$

Substitute for $x = 3.49$ in equation (1).

$$\begin{aligned} 3(3.49) + 4y &= 19.63 \\ 10.47 + 4y &= 19.63 \\ 4y &= 9.16 \\ y &= 2.29 \end{aligned}$$

A ream of paper cost \$3.49 and a notebook cost \$2.29.

20. Let l represent the length of the rectangle and w the width.
$l = 2w - 12$
The length is 1 foot (12 inches) less than twice the width.
$2l + 2w = 40$
The perimeter is 40 inches.

Substitute $2w - 12$ for l in $2l + 2w = 40$.

$$\begin{aligned} 2(2w - 12) + 2w &= 40 \\ 4w - 24 + 2w &= 40 \\ 6w - 24 &= 40 \\ 6w &= 64 \\ w &= \frac{64}{6} = \frac{32}{3} \end{aligned}$$

Substitute $\frac{32}{3}$ for w in $l = 2w - 12$.

$$l = 2\left(\frac{32}{3}\right) - 12$$
$$l = \frac{64}{3} - \frac{36}{3}$$
$$l = \frac{28}{3} = 9\frac{1}{3}$$

The length of the rectangle is $9\frac{1}{3}$ inches.

Chapter 9 Square Roots and Radicals

PROBLEM SET 9.1 Square Roots and Radicals

1. $\sqrt{49} = 7$ because $7^2 = 49$.

3. $-\sqrt{64} = -8$ because $-(8^2) = -64$.

5. $\sqrt{121} = 11$ because $11^2 = 121$.

7. $\sqrt{3600} = 60$ because $60^2 = 3600$.

9. $-\sqrt{1600} = -40$ because $-(40^2) = -1600$.

11. $\sqrt{6400} = 80$ because $80^2 = 6400$.

13. $\sqrt{324} = 18$ because $18^2 = 324$.

15. $\sqrt{\frac{25}{9}} = \frac{5}{3}$ because $\left(\frac{5}{3}\right)^2 = \frac{25}{9}$.

17. $\sqrt{0.16} = 0.4$ because $(0.4)^2 = 0.16$.

19. $\sqrt[3]{27} = 3$ because $3^3 = 27$.

21. $\sqrt[4]{16} = 2$ because $2^4 = 16$.

23. $\sqrt[5]{32} = 2$ because $2^5 = 32$.

25. $\sqrt[3]{-216} = -6$ because $(-6)^3 = -216$.

For Problems 27 – 59, use a calculator and check answers in the text.

59. $7\sqrt{2} + 14\sqrt{2} = (7+14)\sqrt{2} = 21\sqrt{2}$

61. $17\sqrt{7} - 9\sqrt{7} = (17-9)\sqrt{7} = 8\sqrt{7}$

63. $4\sqrt[3]{2} + 7\sqrt[3]{2} = 11\sqrt[3]{2}$

65. $9\sqrt[3]{7} + 2\sqrt[3]{5} - 6\sqrt[3]{7} = 3\sqrt[3]{7} + 2\sqrt[3]{5}$

67. $8\sqrt{2} - 4\sqrt{3} - 9\sqrt{2} + 6\sqrt{3} =$
$(8-9)\sqrt{2} + (-4+6)\sqrt{3} =$
$-\sqrt{2} + 2\sqrt{3}$

69. $6\sqrt{7} + 5\sqrt{10} - 8\sqrt{10} - 4\sqrt{7} - 11\sqrt{7} + \sqrt{10} =$
$(6-4-11)\sqrt{7} + (5-8+1)\sqrt{10} =$
$-9\sqrt{7} - 2\sqrt{10}$

71. $9\sqrt{3} + \sqrt{3} = 10\sqrt{3} \approx$
$10(1.732) \approx 17.3$

73. $9\sqrt{5} - 3\sqrt{5} = 6\sqrt{5} \approx$
$6(2.236) \approx 13.4$

75. $14\sqrt{2} - 15\sqrt{2} = -\sqrt{2} \approx$
$-(1.414) \approx -1.4$

77. $8\sqrt{7} - 4\sqrt{7} + 6\sqrt{7} = 10\sqrt{7} \approx$
$10(2.646) \approx 26.5$

79. $4\sqrt{3} - 2\sqrt{2} \approx$
$4(1.732) - 2(1.414) \approx$
$6.928 - 2.828 \approx 4.1$

81. $9\sqrt{6} - 3\sqrt{5} + 2\sqrt{6} - 7\sqrt{5} - \sqrt{6} =$
$10\sqrt{6} - 10\sqrt{5} \approx 10(2.449) - 10(2.236) \approx$
$24.49 - 22.36 \approx 2.1$

83. $4\sqrt{11} - 5\sqrt{11} - 7\sqrt{11} + 2\sqrt{11} - 3\sqrt{11} =$
$-9\sqrt{11} \approx -9(3.317) \approx -29.9$

85. $L = 2:\quad T = 2\pi\sqrt{\frac{L}{32}} = 2(3.14)\sqrt{\frac{2}{32}}$
$= 6.28(0.25) = 1.6$ seconds

$L = 3.5:\quad T = 2\pi\sqrt{\frac{L}{32}} = 2(3.14)\sqrt{\frac{3.5}{32}}$
$= 6.28(0.33) = 2.1$ seconds

$L = 4:\quad T = 2\pi\sqrt{\frac{L}{32}} = 2(3.14)\sqrt{\frac{4}{32}}$
$= 6.28(0.35) = 2.2$ seconds

87. $d = 75:\quad T = \sqrt{\frac{d}{16}} = \sqrt{\frac{75}{16}}$
$= \sqrt{4.6875} = 2.2$ seconds

$d = 125:\quad T = \sqrt{\frac{d}{16}} = \sqrt{\frac{125}{16}}$
$= \sqrt{7.8125} = 2.8$ seconds

$d = 5280:\quad T = \sqrt{\frac{d}{16}} = \sqrt{\frac{5280}{16}}$
$= \sqrt{330} = 18.2$ seconds

PROBLEM SET 9.2 Simplifying Radicals

1. $\sqrt{24} = \sqrt{4}\sqrt{6} = 2\sqrt{6}$

3. $\sqrt{18} = \sqrt{9}\sqrt{2} = 3\sqrt{2}$

5. $\sqrt{27} = \sqrt{9}\sqrt{3} = 3\sqrt{3}$

7. $\sqrt{40} = \sqrt{4}\sqrt{10} = 2\sqrt{10}$

9. $\sqrt[3]{-54} = \sqrt[3]{-27}\sqrt[3]{2} = -3\sqrt[3]{2}$

11. $\sqrt{80} = \sqrt{16}\sqrt{5} = 4\sqrt{5}$

13. $\sqrt{117} = \sqrt{9}\sqrt{13} = 3\sqrt{13}$

15. $4\sqrt{72} = 4\sqrt{36}\sqrt{2} = 4(6)\sqrt{2} = 24\sqrt{2}$

17. $7\sqrt[4]{162} = 7\sqrt[4]{81}\sqrt[4]{2} = 7(3)\sqrt[4]{2} = 21\sqrt[4]{2}$

19. $-5\sqrt{20} = -5\sqrt{4}\sqrt{5} =$
$-5(2)\sqrt{5} = -10\sqrt{5}$

21. $-8\sqrt{96} = -8\sqrt{16}\sqrt{6} =$
$-8(4)\sqrt{6} = -32\sqrt{6}$

23. $\frac{3}{2}\sqrt{8} = \frac{3}{2}\sqrt{4}\sqrt{2} = \frac{3}{2}(2)\sqrt{2} = 3\sqrt{2}$

25. $\frac{3}{4}\sqrt{12} = \frac{3}{4}\sqrt{4}\sqrt{3} = \frac{3}{4}(2)\sqrt{3} = \frac{3}{2}\sqrt{3}$

27. $-\frac{2}{3}\sqrt{45} = -\frac{2}{3}\sqrt{9}\sqrt{5} =$
$-\frac{2}{3}(3)\sqrt{5} = -2\sqrt{5}$

29. $-\sqrt[4]{32} = -\sqrt[4]{16}\sqrt[4]{2} = -2\sqrt[4]{2}$

31. $\sqrt{x^2y^3} = \sqrt{x^2y^2}\sqrt{y} = xy\sqrt{y}$

33. $\sqrt{2x^2y} = \sqrt{x^2}\sqrt{2y} = x\sqrt{2y}$

35. $\sqrt{8x^2} = \sqrt{4x^2}\sqrt{2} = 2x\sqrt{2}$

37. $\sqrt{27a^3b} = \sqrt{9a^2}\sqrt{3ab} = 3a\sqrt{3ab}$

39. $\sqrt[3]{64x^4y^2} = \sqrt[3]{64x^3}\sqrt[3]{xy^2} = 4x\sqrt[3]{xy^2}$

41. $\sqrt{63x^4y^2} = \sqrt{9x^4y^2}\sqrt{7} = 3x^2y\sqrt{7}$

43. $3\sqrt{48x^2} = 3\sqrt{16x^2}\sqrt{3} =$
$3(4x)\sqrt{3} = 12x\sqrt{3}$

45. $-6\sqrt{72x^7} = -6\sqrt{36x^6}\sqrt{2x} =$
$-6(6x^3)\sqrt{2x} = -36x^3\sqrt{2x}$

47. $\frac{2}{9}\sqrt{54xy} = \frac{2}{9}\sqrt{9}\sqrt{6xy} =$
$\frac{2}{9}(3)\sqrt{6xy} = \frac{2}{3}\sqrt{6xy}$

49. $\frac{1}{8}\sqrt[3]{250x^4} = \frac{1}{8}\sqrt[3]{125x^3}\sqrt[3]{2x} =$
$\frac{1}{8}(5x)\sqrt[3]{2x} = \frac{5x\sqrt[3]{2x}}{8}$

51. $-\frac{2}{3}\sqrt{169a^8} = -\frac{2}{3}(13a^4) = -\frac{26}{3}a^4$

53. $7\sqrt{32} + 5\sqrt{2} =$
$7\left(\sqrt{16}\sqrt{2}\right) + 5\sqrt{2} =$
$7(4)\sqrt{2} + 5\sqrt{2} =$
$28\sqrt{2} + 5\sqrt{2} = 33\sqrt{2}$

55. $4\sqrt{45} - 9\sqrt{5} =$
$4\sqrt{9}\sqrt{5} - 9\sqrt{5} =$
$4(3)\sqrt{5} - 9\sqrt{5} =$
$12\sqrt{5} - 9\sqrt{5} = 3\sqrt{5}$

57. $2\sqrt[3]{54} + 6\sqrt[3]{16} =$
$2\sqrt[3]{27}\sqrt[3]{2} + 6\sqrt[3]{8}\sqrt[3]{2} =$
$2(3)\sqrt[3]{2} + 6(2)\sqrt[3]{2} =$
$6\sqrt[3]{2} + 12\sqrt[3]{2} = 18\sqrt[3]{2}$

59. $4\sqrt{63} - 7\sqrt{28} =$
$4\sqrt{9}\sqrt{7} - 7\sqrt{4}\sqrt{7} =$
$4(3)\sqrt{7} - 7(2)\sqrt{7} =$
$12\sqrt{7} - 14\sqrt{7} = -2\sqrt{7}$

61. $5\sqrt{12} + 3\sqrt{27} - 2\sqrt{75} =$
$5\sqrt{4}\sqrt{3} + 3\sqrt{9}\sqrt{3} - 2\sqrt{25}\sqrt{3} =$
$5(2)\sqrt{3} + 3(3)\sqrt{3} - 2(5)\sqrt{3} =$
$10\sqrt{3} + 9\sqrt{3} - 10\sqrt{3} = 9\sqrt{3}$

63. $\frac{1}{2}\sqrt{20} + \frac{2}{3}\sqrt{45} - \frac{1}{4}\sqrt{80} =$
$\frac{1}{2}\sqrt{4}\sqrt{5} + \frac{2}{3}\sqrt{9}\sqrt{5} - \frac{1}{4}\sqrt{16}\sqrt{5} =$
$\frac{1}{2}(2)\sqrt{5} + \frac{2}{3}(3)\sqrt{5} - \frac{1}{4}(4)\sqrt{5} =$
$\sqrt{5} + 2\sqrt{5} - \sqrt{5} = 2\sqrt{5}$

65. $3\sqrt{8} - 5\sqrt{20} - 7\sqrt{18} - 9\sqrt{125} =$
$3\sqrt{4}\sqrt{2} - 5\sqrt{4}\sqrt{5} - 7\sqrt{9}\sqrt{2} - 9\sqrt{25}\sqrt{5} =$
$3(2)\sqrt{2} - 5(2)\sqrt{5} - 7(3)\sqrt{2} - 9(5)\sqrt{5} =$
$6\sqrt{2} - 10\sqrt{5} - 21\sqrt{2} - 45\sqrt{5} =$
$-15\sqrt{2} - 55\sqrt{5}$

PROBLEM SET 9.3 More on Simplifying Radicals

1. $\sqrt{\frac{16}{25}} = \frac{\sqrt{16}}{\sqrt{25}} = \frac{4}{5}$

3. $-\sqrt{\frac{81}{9}} = -\sqrt{9} = -3$

5. $\sqrt{\frac{1}{64}} = \frac{\sqrt{1}}{\sqrt{64}} = \frac{1}{8}$

7. $\sqrt[3]{\frac{125}{64}} = \frac{\sqrt[3]{125}}{\sqrt[3]{64}} = \frac{5}{4}$

9. $-\sqrt{\frac{25}{256}} = -\frac{\sqrt{25}}{\sqrt{256}} = -\frac{5}{16}$

11. $\sqrt{\frac{19}{25}} = \frac{\sqrt{19}}{\sqrt{25}} = \frac{\sqrt{19}}{5}$

13. $\sqrt{\frac{8}{49}} = \frac{\sqrt{8}}{\sqrt{49}} = \frac{\sqrt{4}\sqrt{2}}{7} = \frac{2\sqrt{2}}{7}$

15. $\frac{\sqrt[3]{375}}{\sqrt[3]{216}} = \frac{\sqrt[3]{125}\sqrt[3]{3}}{\sqrt[3]{216}} = \frac{5\sqrt[3]{3}}{6}$

17. $\frac{\sqrt{12}}{\sqrt{36}} = \frac{\sqrt{4}\sqrt{3}}{6} = \frac{2\sqrt{3}}{6} = \frac{\sqrt{3}}{3}$

19. $\sqrt{\frac{3}{2}} = \frac{\sqrt{3}}{\sqrt{2}} \cdot \frac{\sqrt{2}}{\sqrt{2}} = \frac{\sqrt{6}}{2}$

21. $\sqrt{\frac{5}{8}} = \frac{\sqrt{5}}{\sqrt{8}} \cdot \frac{\sqrt{2}}{\sqrt{2}} = \frac{\sqrt{10}}{\sqrt{16}} = \frac{\sqrt{10}}{4}$

23. $\frac{\sqrt{56}}{\sqrt{8}} = \sqrt{\frac{56}{8}} = \sqrt{7}$

25. $\frac{\sqrt{63}}{\sqrt{7}} = \sqrt{\frac{63}{7}} = \sqrt{9} = 3$

27. $\frac{\sqrt{5}}{\sqrt{18}} = \frac{\sqrt{5}}{3\sqrt{2}} \cdot \frac{\sqrt{2}}{\sqrt{2}} = \frac{\sqrt{10}}{6}$

29. $\frac{\sqrt{4}}{\sqrt{27}} = \frac{2}{3\sqrt{3}} \cdot \frac{\sqrt{3}}{\sqrt{3}} = \frac{2\sqrt{3}}{9}$

31. $\sqrt{\frac{1}{24}} = \frac{\sqrt{1}}{\sqrt{24}} = \frac{1}{2\sqrt{6}} \cdot \frac{\sqrt{6}}{\sqrt{6}} = \frac{\sqrt{6}}{12}$

33. $\frac{2\sqrt{3}}{\sqrt{5}} = \frac{2\sqrt{3}}{\sqrt{5}} \cdot \frac{\sqrt{5}}{\sqrt{5}} = \frac{2\sqrt{15}}{5}$

35. $\frac{4\sqrt{2}}{3\sqrt{3}} = \frac{4\sqrt{2}}{3\sqrt{3}} \cdot \frac{\sqrt{3}}{\sqrt{3}} = \frac{4\sqrt{6}}{9}$

37. $\frac{3\sqrt{7}}{4\sqrt{12}} = \frac{3\sqrt{7}}{8\sqrt{3}} \cdot \frac{\sqrt{3}}{\sqrt{3}} = \frac{3\sqrt{21}}{24} = \frac{\sqrt{21}}{8}$

39. $\sqrt{4\frac{1}{9}} = \sqrt{\frac{37}{9}} = \frac{\sqrt{37}}{\sqrt{9}} = \frac{\sqrt{37}}{3}$

41. $\frac{3}{\sqrt{x}} = \frac{3}{\sqrt{x}} \cdot \frac{\sqrt{x}}{\sqrt{x}} = \frac{3\sqrt{x}}{x}$

43. $\frac{5}{\sqrt{2x}} = \frac{5}{\sqrt{2x}} \cdot \frac{\sqrt{2x}}{\sqrt{2x}} = \frac{5\sqrt{2x}}{2x}$

45. $\sqrt{\frac{3}{x}} = \frac{\sqrt{3}}{\sqrt{x}} = \frac{\sqrt{3}}{\sqrt{x}} \cdot \frac{\sqrt{x}}{\sqrt{x}} = \frac{\sqrt{3x}}{x}$

47. $\sqrt{\frac{12}{x^2}} = \frac{2\sqrt{3}}{\sqrt{x^2}} = \frac{2\sqrt{3}}{x}$

49. $\frac{\sqrt{2x}}{\sqrt{5y}} = \frac{\sqrt{2x}}{\sqrt{5y}} \cdot \frac{\sqrt{5y}}{\sqrt{5y}} = \frac{\sqrt{10xy}}{5y}$

51. $\frac{\sqrt{5x}}{\sqrt{27y}} = \frac{\sqrt{5x}}{3\sqrt{3y}} \cdot \frac{\sqrt{3y}}{\sqrt{3y}} = \frac{\sqrt{15xy}}{9y}$

53. $\frac{\sqrt{2x^3}}{\sqrt{8y}} = \frac{x\sqrt{2x}}{2\sqrt{2y}} \cdot \frac{\sqrt{2y}}{\sqrt{2y}} = \frac{x\sqrt{4xy}}{4y} =$
$\frac{2x\sqrt{xy}}{4y} = \frac{x\sqrt{xy}}{2y}$

55. $\sqrt{\frac{9}{x^3}} = \frac{\sqrt{9}}{x\sqrt{x}} \cdot \frac{\sqrt{x}}{\sqrt{x}} = \frac{3\sqrt{x}}{x^2}$

57. $\frac{4}{\sqrt{x^7}} = \frac{4}{x^3\sqrt{x}} \cdot \frac{\sqrt{x}}{\sqrt{x}} = \frac{4\sqrt{x}}{x^4}$

59. $\frac{3\sqrt{x}}{2\sqrt{y^3}} = \frac{3\sqrt{x}}{2y\sqrt{y}} \cdot \frac{\sqrt{y}}{\sqrt{y}} = \frac{3\sqrt{xy}}{2y^2}$

61. $7\sqrt{3} + \sqrt{\frac{1}{3}} = 7\sqrt{3} + \frac{\sqrt{1}}{\sqrt{3}} \cdot \frac{\sqrt{3}}{\sqrt{3}} =$
$7\sqrt{3} + \frac{\sqrt{3}}{3} = \left(7 + \frac{1}{3}\right)\sqrt{3} =$
$\left(\frac{21}{3} + \frac{1}{3}\right)\sqrt{3} = \frac{22}{3}\sqrt{3}$

63. $4\sqrt{10} - \sqrt{\frac{2}{5}} = 4\sqrt{10} - \frac{\sqrt{2}}{\sqrt{5}} \cdot \frac{\sqrt{5}}{\sqrt{5}} =$
$4\sqrt{10} - \frac{\sqrt{10}}{5} = \left(4 - \frac{1}{5}\right)\sqrt{10} =$
$\left(\frac{20}{5} - \frac{1}{5}\right)\sqrt{10} = \frac{19}{5}\sqrt{10}$

65. $-2\sqrt{5} - 5\sqrt{\frac{1}{5}} = -2\sqrt{5} - \frac{5\sqrt{1}}{\sqrt{5}} \cdot \frac{\sqrt{5}}{\sqrt{5}} =$
$-2\sqrt{5} - \frac{5\sqrt{5}}{5} = -2\sqrt{5} - \sqrt{5} =$
$-3\sqrt{5}$

67. $-3\sqrt{6}-\frac{5\sqrt{2}}{\sqrt{3}} = -3\sqrt{6}-\frac{5\sqrt{2}}{\sqrt{3}}\cdot\frac{\sqrt{3}}{\sqrt{3}} =$

$-3\sqrt{6}-\frac{5\sqrt{6}}{3} = -3-\frac{5}{3}\sqrt{6} =$

$-\frac{9}{3}-\frac{5}{3}\sqrt{6} = -\frac{14}{3}\sqrt{6}$

69. $4\sqrt{12}+\frac{3}{\sqrt{3}}-5\sqrt{27} =$

$8\sqrt{3}+\frac{3}{\sqrt{3}}\cdot\frac{\sqrt{3}}{\sqrt{3}}-15\sqrt{3} =$

$8\sqrt{3}+\frac{3\sqrt{3}}{3}-15\sqrt{3} =$

$8\sqrt{3}+\sqrt{3}-15\sqrt{3} =$

$(8+1-15)\sqrt{3} = -6\sqrt{3}$

71. $\frac{9\sqrt{5}}{\sqrt{3}}-6\sqrt{60}+\frac{10\sqrt{3}}{\sqrt{5}} =$

$\frac{9\sqrt{5}}{\sqrt{3}}\cdot\frac{\sqrt{3}}{\sqrt{3}}-12\sqrt{15}+\frac{10\sqrt{3}}{\sqrt{5}}\cdot\frac{\sqrt{5}}{\sqrt{5}} =$

$\frac{9\sqrt{15}}{3}-12\sqrt{15}+\frac{10\sqrt{15}}{5} =$

$3\sqrt{15}-12\sqrt{15}+2\sqrt{15} = -7\sqrt{15}$

PROBLEM SET 9.4 Products and Quotients Involving Radicals

1. $\sqrt{7}\sqrt{5} = \sqrt{35}$

3. $\sqrt{6}\sqrt{8} = \sqrt{48} = \sqrt{16}\sqrt{3} = 4\sqrt{3}$

5. $\sqrt{5}\sqrt{10} = \sqrt{50} = \sqrt{25}\sqrt{2} = 5\sqrt{2}$

7. $\sqrt[3]{9}\sqrt[3]{6} = \sqrt[3]{54} = \sqrt[3]{27}\sqrt[3]{2} = 3\sqrt[3]{2}$

9. $\sqrt{8}\sqrt{12} = \sqrt{96} = \sqrt{16}\sqrt{6} = 4\sqrt{6}$

11. $\left(3\sqrt{3}\right)\left(5\sqrt{7}\right) =$

$3\cdot 5\cdot\sqrt{3}\cdot\sqrt{7} =$

$15\sqrt{21}$

13. $\left(-\sqrt[3]{6}\right)\left(5\sqrt[3]{4}\right) = -5\sqrt[3]{24} =$

$-5\sqrt[3]{8}\sqrt[3]{3} = -5(2)\sqrt[3]{3} = -10\sqrt[3]{3}$

15. $\left(3\sqrt{6}\right)\left(4\sqrt{6}\right) = 3\cdot 4\cdot\sqrt{6}\cdot\sqrt{6} =$

$12\sqrt{36} = 12\cdot 6 = 72$

17. $\left(5\sqrt{2}\right)\left(4\sqrt{12}\right) = 5\cdot 4\cdot\sqrt{2}\cdot\sqrt{12} =$

$20\sqrt{24} = 20\sqrt{4}\sqrt{6} = 40\sqrt{6}$

19. $\left(4\sqrt{3}\right)\left(2\sqrt{15}\right) = 4\cdot 2\cdot\sqrt{3}\cdot\sqrt{15} =$

$8\sqrt{45} = 8\sqrt{9}\sqrt{5} = 8\cdot 3\sqrt{5} = 24\sqrt{5}$

21. $\sqrt{2}\left(\sqrt{3}+\sqrt{5}\right) = \sqrt{2}\sqrt{3}+\sqrt{2}\sqrt{5} =$

$\sqrt{6}+\sqrt{10}$

23. $\sqrt{6}\left(\sqrt{2}-5\right) = \sqrt{6}\sqrt{2}-\sqrt{6}(5) =$

$\sqrt{12}-5\sqrt{6} = \sqrt{4}\sqrt{3}-5\sqrt{6} =$

$2\sqrt{3}-5\sqrt{6}$

25. $\sqrt[3]{2}\left(\sqrt[3]{4}+\sqrt[3]{10}\right) = \sqrt[3]{8}+\sqrt[3]{20} =$

$2+\sqrt[3]{20}$

27. $\sqrt{12}\left(\sqrt{6}-\sqrt{8}\right) =$

$\sqrt{12}\sqrt{6}-\sqrt{12}\sqrt{8} =$

$\sqrt{72}-\sqrt{96} =$

$\sqrt{36}\sqrt{2}-\sqrt{16}\sqrt{6} =$

$6\sqrt{2}-4\sqrt{6}$

29. $4\sqrt{3}\left(\sqrt{2}-2\sqrt{5}\right) =$
$4\sqrt{3}\left(\sqrt{2}\right) - 4\sqrt{3}\left(2\sqrt{5}\right) =$
$4\sqrt{6} - 8\sqrt{15}$

31. $\left(\sqrt{2}+6\right)\left(\sqrt{2}+9\right) =$
$\sqrt{2}\left(\sqrt{2}+9\right) + 6\left(\sqrt{2}+9\right) =$
$2 + 9\sqrt{2} + 6\sqrt{2} + 54 =$
$56 + 15\sqrt{2}$

33. $\left(\sqrt{6}-5\right)\left(\sqrt{6}+3\right) =$
$\sqrt{6}\left(\sqrt{6}+3\right) - 5\left(\sqrt{6}+3\right) =$
$6 + 3\sqrt{6} - 5\sqrt{6} - 15 =$
$-9 - 2\sqrt{6}$

35. $\left(\sqrt{3}+\sqrt{6}\right)\left(\sqrt{6}+\sqrt{8}\right) =$
$\sqrt{3}\left(\sqrt{6}+\sqrt{8}\right) + \sqrt{6}\left(\sqrt{6}+\sqrt{8}\right) =$
$\sqrt{18} + \sqrt{24} + \sqrt{36} + \sqrt{48} =$
$\sqrt{9}\sqrt{2} + \sqrt{4}\sqrt{6} + 6 + \sqrt{16}\sqrt{3} =$
$3\sqrt{2} + 2\sqrt{6} + 6 + 4\sqrt{3}$

37. $\left(5+\sqrt{10}\right)\left(5-\sqrt{10}\right) =$
$5^2 - \left(\sqrt{10}\right)^2 = 25 - 10 = 15$

39. $\left(3\sqrt{2}-\sqrt{3}\right)\left(3\sqrt{2}+\sqrt{3}\right) =$
$\left(3\sqrt{2}\right)^2 - \left(\sqrt{3}\right)^2 = 9(2) - 3 =$
$18 - 3 = 15$

41. $\left(5\sqrt{3}+2\sqrt{6}\right)\left(5\sqrt{3}-2\sqrt{6}\right) =$
$\left(5\sqrt{3}\right)^2 - \left(2\sqrt{6}\right)^2 = 25(3) - 4(6) =$
$75 - 24 = 51$

43. $\sqrt{xy}\sqrt{x} = \sqrt{x^2y} = \sqrt{x^2}\sqrt{y} = x\sqrt{y}$

45. $\sqrt[3]{25x^2}\sqrt[3]{5x} = \sqrt[3]{125x^3} = 5x$

47. $\left(4\sqrt{a}\right)\left(3\sqrt{ab}\right) = 12\sqrt{a^2b} = 12a\sqrt{b}$

49. $\sqrt{2x}\left(\sqrt{3x}-\sqrt{6y}\right) =$
$\sqrt{2x}\left(\sqrt{3x}\right) - \sqrt{2x}\left(\sqrt{6y}\right) =$
$\sqrt{6x^2} - \sqrt{12xy} =$
$x\sqrt{6} - 2\sqrt{3xy}$

51. $\left(\sqrt{x}+5\right)\left(\sqrt{x}-3\right) =$
$\sqrt{x}\left(\sqrt{x}-3\right) + 5\left(\sqrt{x}-3\right) =$
$x - 3\sqrt{x} + 5\sqrt{x} - 15 =$
$x + 2\sqrt{x} - 15$

53. $\left(\sqrt{x}+7\right)\left(\sqrt{x}-7\right) =$
$\left(\sqrt{x}\right)^2 - 7^2 = x - 49$

55. $\dfrac{3}{\sqrt{2}+4} = \dfrac{3}{\sqrt{2}+4} \cdot \dfrac{\sqrt{2}-4}{\sqrt{2}-4} =$
$\dfrac{3\left(\sqrt{2}-4\right)}{2-16} = \dfrac{3\sqrt{2}-12}{-14} =$
$\dfrac{-3\sqrt{2}+12}{14}$

57. $\dfrac{8}{\sqrt{6}-2} = \dfrac{8}{\sqrt{6}-2} \cdot \dfrac{\sqrt{6}+2}{\sqrt{6}+2} =$
$\dfrac{8\left(\sqrt{6}+2\right)}{6-4} = \dfrac{8\left(\sqrt{6}+2\right)}{2} =$
$4\left(\sqrt{6}+2\right) = 4\sqrt{6} + 8$

59. $\dfrac{2}{\sqrt{5}+\sqrt{3}} = \dfrac{2}{\sqrt{5}+\sqrt{3}} \cdot \dfrac{\sqrt{5}-\sqrt{3}}{\sqrt{5}-\sqrt{3}} =$
$\dfrac{2\left(\sqrt{5}-\sqrt{3}\right)}{5-3} = \dfrac{2\left(\sqrt{5}-\sqrt{3}\right)}{2} =$
$\sqrt{5} - \sqrt{3}$

61. $\dfrac{10}{2-3\sqrt{3}} = \dfrac{10}{2-3\sqrt{3}} \cdot \dfrac{2+3\sqrt{3}}{2+3\sqrt{3}} =$
$\dfrac{10\left(2+3\sqrt{3}\right)}{4-27} = \dfrac{20+30\sqrt{3}}{-23} =$
$\dfrac{-20-30\sqrt{3}}{23}$

63. $\dfrac{4}{\sqrt{x}-2}=\dfrac{4}{\sqrt{x}-2}\cdot\dfrac{\sqrt{x}+2}{\sqrt{x}+2}=$
$\dfrac{4(\sqrt{x}+2)}{x-4}=\dfrac{4\sqrt{x}+8}{x-4}$

65. $\dfrac{\sqrt{x}}{\sqrt{x}+3}=\dfrac{\sqrt{x}}{\sqrt{x}+3}\cdot\dfrac{\sqrt{x}-3}{\sqrt{x}-3}=$
$\dfrac{\sqrt{x}(\sqrt{x}-3)}{x-9}=\dfrac{x-3\sqrt{x}}{x-9}$

67. $\dfrac{\sqrt{a}+2}{\sqrt{a}-5}=\dfrac{\sqrt{a}+2}{\sqrt{a}-5}\cdot\dfrac{\sqrt{a}+5}{\sqrt{a}+5}=$
$\dfrac{(\sqrt{a}+2)(\sqrt{a}+5)}{a-25}=$
$\dfrac{a+5\sqrt{a}+2\sqrt{a}+10}{a-25}=$
$\dfrac{a+7\sqrt{a}+10}{a-25}$

69. $\dfrac{2+\sqrt{3}}{3-\sqrt{2}}=\dfrac{2+\sqrt{3}}{3-\sqrt{2}}\cdot\dfrac{3+\sqrt{2}}{3+\sqrt{2}}=$
$\dfrac{\left(2+\sqrt{3}\right)\left(3+\sqrt{2}\right)}{9-2}=$
$\dfrac{6+2\sqrt{2}+3\sqrt{3}+\sqrt{6}}{7}$

PROBLEM SET 9.5 Solving Radical Equations

1. $\sqrt{x}=7$
$(\sqrt{x})^2=7^2$
$x=49$

CHECK
$\sqrt{49}\overset{?}{=}7$
$7=7$
The solution set is $\{49\}$.

3. $\sqrt{2x}=6$
$\left(\sqrt{2x}\right)^2=6^2$
$2x=36$
$x=18$

CHECK
$\sqrt{2(18)}\overset{?}{=}6$
$\sqrt{36}\overset{?}{=}6$
$6=6$
The solution set is $\{18\}$.

5. $\sqrt{3x}=-6$
$\left(\sqrt{3x}\right)^2=(-6)^2$
$3x=36$
$x=12$

CHECK
$\sqrt{2(12)}\overset{?}{=}-6$
$\sqrt{36}\overset{?}{=}-6$
$6\neq-6$
Since 12 does not check, the solution set is $\emptyset$.

7. $\sqrt{4x}=3$
$\left(\sqrt{4x}\right)^2=3^2$
$4x=9$
$x=\dfrac{9}{4}$

CHECK

$$\sqrt{4\left(\frac{9}{4}\right)} \stackrel{?}{=} 3$$
$$\sqrt{9} \stackrel{?}{=} 3$$
$$3 = 3$$

The solution set is $\left\{\frac{9}{4}\right\}$.

9.
$$3\sqrt{x} = 2$$
$$(3\sqrt{x})^2 = 2^2$$
$$9x = 4$$
$$x = \frac{4}{9}$$

CHECK

$$3\sqrt{\frac{4}{9}} \stackrel{?}{=} 2$$
$$3\left(\frac{2}{3}\right) \stackrel{?}{=} 2$$
$$2 = 2$$

The solution set is $\left\{\frac{4}{9}\right\}$.

11.
$$\sqrt{2n-3} = 5$$
$$\left(\sqrt{2n-3}\right)^2 = (5)^2$$
$$2n - 3 = 25$$
$$2n = 28$$
$$n = 14$$

CHECK

$$\sqrt{2(14)-3} \stackrel{?}{=} 5$$
$$\sqrt{25} \stackrel{?}{=} 5$$
$$5 = 5$$

The solution set is $\{14\}$.

13.
$$\sqrt{5y+2} = -1$$
$$\left(\sqrt{5y+2}\right)^2 = (-1)^2$$
$$5y + 2 = 1$$
$$5y = -1$$
$$y = -\frac{1}{5}$$

CHECK

$$\sqrt{5\left(-\frac{1}{5}\right) + 2} \stackrel{?}{=} -1$$
$$\sqrt{1} \stackrel{?}{=} -1$$
$$1 \neq -1$$

Since $-\frac{1}{5}$ does not check, the solution set is $\emptyset$.

15.
$$\sqrt{6x-5} - 3 = 0$$
$$\sqrt{6x-5} = 3$$
$$\left(\sqrt{6x-5}\right)^2 = (3)^2$$
$$6x - 5 = 9$$
$$6x = 14$$
$$x = \frac{14}{6} = \frac{7}{3}$$

CHECK

$$\sqrt{6\left(\frac{7}{3}\right) - 5} - 3 \stackrel{?}{=} 0$$
$$\sqrt{9} - 3 \stackrel{?}{=} 0$$
$$3 - 3 \stackrel{?}{=} 0$$
$$0 = 0$$

The solution set is $\left\{\frac{7}{3}\right\}$.

17.
$$5\sqrt{x} = 30$$
$$\sqrt{x} = 6$$
$$\left(\sqrt{x}\right)^2 = (6)^2$$
$$x = 36$$

CHECK

$$5\sqrt{36} \stackrel{?}{=} 30$$
$$5(6) \stackrel{?}{=} 30$$
$$30 = 30$$

The solution set is $\{36\}$.

19.
$$\sqrt{3a-2} = \sqrt{2a+4}$$
$$\left(\sqrt{3a-2}\right)^2 = \left(\sqrt{2a+4}\right)^2$$
$$3a-2 = 2a+4$$
$$a-2 = 4$$
$$a = 6$$

CHECK
$$\sqrt{3(6)-2} \stackrel{?}{=} \sqrt{2(6)+4}$$
$$\sqrt{16} \stackrel{?}{=} \sqrt{16}$$
$$4 = 4$$
The solution set is $\{6\}$.

21.
$$\sqrt{7x-3} = \sqrt{4x+3}$$
$$\left(\sqrt{7x-3}\right)^2 = \left(\sqrt{4x+3}\right)^2$$
$$7x-3 = 4x+3$$
$$3x-3 = 3$$
$$3x = 6$$
$$x = 2$$

CHECK
$$\sqrt{7(2)-3} \stackrel{?}{=} \sqrt{4(2)+3}$$
$$\sqrt{11} = \sqrt{11}$$
The solution set is $\{2\}$.

23.
$$2\sqrt{y+1} = 5$$
$$\left(2\sqrt{y+1}\right)^2 = (5)^2$$
$$4(y+1) = 25$$
$$4y+4 = 25$$
$$4y = 21$$
$$y = \frac{21}{4}$$

CHECK
$$2\sqrt{\frac{21}{4}+1} \stackrel{?}{=} 5$$
$$2\sqrt{\frac{25}{4}} \stackrel{?}{=} 5$$
$$2\left(\frac{5}{2}\right) \stackrel{?}{=} 5$$
$$5 = 5$$

The solution set is $\left\{\frac{21}{4}\right\}$.

25.
$$\sqrt{x+3} = x+3$$
$$\left(\sqrt{x+3}\right)^2 = (x+3)^2$$
$$x+3 = x^2+6x+9$$
$$0 = x^2+5x+6$$
$$0 = (x+2)(x+3)$$
$$x+2 = 0 \quad \text{or} \quad x+3 = 0$$
$$x = -2 \quad \text{or} \quad x = -3$$

CHECK
$$\sqrt{-2+3} \stackrel{?}{=} -2+3$$
$$\sqrt{1} = 1$$

$$\sqrt{-3+3} \stackrel{?}{=} -3+3$$
$$\sqrt{0} = 0$$
The solution set is $\{-3, -2\}$.

27.
$$\sqrt{-2x+28} = x-2$$
$$\left(\sqrt{-2x+28}\right)^2 = (x-2)^2$$
$$-2x+28 = x^2-4x+4$$
$$0 = x^2-2x-24$$
$$0 = (x+4)(x-6)$$
$$x+4 = 0 \quad \text{or} \quad x-6 = 0$$
$$x = -4 \quad \text{or} \quad x = 6$$

CHECK
$$\sqrt{-2(-4)+28} \stackrel{?}{=} -4-2$$
$$\sqrt{36} \stackrel{?}{=} -6$$
$$6 \neq -6$$

$$\sqrt{-2(6)+28} \stackrel{?}{=} 6-2$$
$$\sqrt{16} \stackrel{?}{=} 4$$
$$4 = 4$$
The solution set is $\{6\}$.

29.
$$\sqrt{3n-4}=\sqrt{n}$$
$$\left(\sqrt{3n-4}\right)^2=\left(\sqrt{n}\right)^2$$
$$3n-4=n$$
$$-4=-2n$$
$$2=n$$

CHECK
$$\sqrt{3(2)-4}\stackrel{?}{=}\sqrt{2}$$
$$\sqrt{2}=\sqrt{2}$$
The solution set is $\{2\}$.

31.
$$\sqrt{3x}=x-6$$
$$\left(\sqrt{3x}\right)^2=(x-6)^2$$
$$3x=x^2-12x+36$$
$$0=x^2-15x+36$$
$$0=(x-12)(x-3)$$
$$x-12=0 \quad \text{or} \quad x-3=0$$
$$x=12 \quad \text{or} \quad x=3$$

CHECK
$$\sqrt{3(12)}\stackrel{?}{=}12-6$$
$$\sqrt{36}\stackrel{?}{=}6$$
$$6=6$$

$$\sqrt{3(3)}\stackrel{?}{=}3-6$$
$$\sqrt{9}\stackrel{?}{=}-3$$
$$3\neq-3$$
The solution set is $\{12\}$.

33.
$$4\sqrt{x}+5=x$$
$$4\sqrt{x}=x-5$$
$$\left(4\sqrt{x}\right)^2=(x-5)^2$$
$$16x=x^2-10x+25$$
$$0=x^2-26x+25$$
$$0=(x-25)(x-1)$$
$$x-25=0 \quad \text{or} \quad x-1=0$$
$$x=25 \quad \text{or} \quad x=1$$

CHECK
$$4\sqrt{25}+5\stackrel{?}{=}25$$
$$4(5)+5\stackrel{?}{=}25$$
$$25=25$$

$$4\sqrt{1}+5\stackrel{?}{=}1$$
$$9\neq1$$
The solution set is $\{25\}$.

35.
$$\sqrt{x^2+27}=x+3$$
$$\left(\sqrt{x^2+27}\right)^2=(x+3)^2$$
$$x^2+27=x^2+6x+9$$
$$18=6x$$
$$3=x$$

CHECK
$$\sqrt{(3)^2+27}\stackrel{?}{=}3+3$$
$$\sqrt{36}\stackrel{?}{=}6$$
$$6=6$$
The solution set is $\{3\}$.

37.
$$\sqrt{x^2+2x+3}=x+2$$
$$\left(\sqrt{x^2+2x+3}\right)^2=(x+2)^2$$
$$x^2+2x+3=x^2+4x+4$$
$$-2x=1$$
$$x=-\frac{1}{2}$$

CHECK
$$\sqrt{\left(-\frac{1}{2}\right)^2+2\left(-\frac{1}{2}\right)+3}\stackrel{?}{=}-\frac{1}{2}+2$$
$$\sqrt{\frac{1}{4}-1+3}\stackrel{?}{=}\frac{3}{2}$$
$$\sqrt{\frac{9}{4}}\stackrel{?}{=}\frac{3}{2}$$
$$\frac{3}{2}=\frac{3}{2}$$
The solution set is $\left\{-\frac{1}{2}\right\}$.

39. $\sqrt{8x} - 2 = x$

$$\sqrt{8x} = x + 2$$
$$\left(\sqrt{8x}\right)^2 = (x+2)^2$$
$$8x = x^2 + 4x + 4$$
$$0 = x^2 - 4x + 4$$
$$0 = (x-2)(x-2)$$
$$x - 2 = 0 \quad \text{or} \quad x - 2 = 0$$
$$x = 2 \quad \text{or} \quad x = 2$$

CHECK

$$\sqrt{8(2)} - 2 \stackrel{?}{=} 2$$
$$\sqrt{16} - 2 \stackrel{?}{=} 2$$
$$4 - 2 \stackrel{?}{=} 2$$
$$2 = 2$$

The solution set is $\{2\}$.

41. $S = 40 : D = \dfrac{S^2}{30f} = \dfrac{40^2}{30(0.95)} =$

$\dfrac{1600}{28.5} = 56$ to the nearest foot.

The car will skid approximately 56 feet.

$S = 55 : D = \dfrac{S^2}{30f} = \dfrac{55^2}{30(0.95)} =$

$\dfrac{3025}{28.5} = 106$ to the nearest foot.

The car will skid approximately 106 feet.

$S = 65 : D = \dfrac{S^2}{30f} = \dfrac{65^2}{30(0.95)} =$

$\dfrac{4225}{28.5} = 148$ to the nearest foot.

The car will skid approximately 148 feet.

43. $T = 2 : L = \dfrac{32T^2}{4\pi^2} = \dfrac{32(2)^2}{4(3.14)^2} =$

$\dfrac{128}{39.4384} = 3.2$ to the nearest tenth.

The length of the pendulum would be approximately 3.2 feet.

$T = 2.5 : L = \dfrac{32T^2}{4\pi^2} = \dfrac{32(2.5)^2}{4(3.14)^2} =$

$\dfrac{200}{39.4384} = 5.1$ to the nearest tenth.

The length of the pendulum would be approximately 5.1 feet.

$T = 3 : L = \dfrac{32T^2}{4\pi^2} = \dfrac{32(3)^2}{4(3.14)^2} =$

$\dfrac{288}{39.4384} = 7.3$ to the nearest tenth.

The length of the pendulum would be approximately 7.3 feet.

CHAPTER 9 Review Problem Set

1. $\sqrt{64} = 8$ because $8^2 = 64$.

2. $-\sqrt{49} = -7$ because $-(7)^2 = -49$.

3. $\sqrt{1600} = 40$ because $40^2 = 1600$.

4. $\sqrt{\dfrac{81}{25}} = \dfrac{9}{5}$ because $\left(\dfrac{9}{5}\right)^2 = \dfrac{81}{25}$.

5. $-\sqrt{\dfrac{4}{9}} = -\dfrac{2}{3}$ because $\left(\dfrac{2}{3}\right)^2 = \dfrac{4}{9}$.

6. $\sqrt{\dfrac{49}{36}} = \dfrac{7}{6}$ because $\left(\dfrac{7}{6}\right)^2 = \dfrac{49}{36}$.

7. $\sqrt{20} = \sqrt{4}\sqrt{5} = 2\sqrt{5}$

8. $\sqrt{32} = \sqrt{16}\sqrt{2} = 4\sqrt{2}$

9. $5\sqrt{8} = 5\sqrt{4}\sqrt{2} = 5 \cdot 2\sqrt{2} = 10\sqrt{2}$

10. $\sqrt{80} = \sqrt{16}\sqrt{5} = 4\sqrt{5}$

11. $2\sqrt[3]{-125} = 2(-5) = -10$

12. $\frac{\sqrt[3]{40}}{\sqrt[3]{8}} = \sqrt[3]{\frac{40}{8}} = \sqrt[3]{5}$

13. $\frac{\sqrt{36}}{\sqrt{7}} = \frac{6}{\sqrt{7}} \cdot \frac{\sqrt{7}}{\sqrt{7}} = \frac{6\sqrt{7}}{7}$

14. $\sqrt{\frac{7}{8}} = \frac{\sqrt{7}}{2\sqrt{2}} = \frac{\sqrt{7}}{2\sqrt{2}} \cdot \frac{\sqrt{2}}{\sqrt{2}} = \frac{\sqrt{14}}{4}$

15. $\sqrt{\frac{8}{24}} = \sqrt{\frac{1}{3}} = \frac{\sqrt{1}}{\sqrt{3}} \cdot \frac{\sqrt{3}}{\sqrt{3}} = \frac{\sqrt{3}}{3}$

16. $\frac{3\sqrt{2}}{\sqrt{5}} = \frac{3\sqrt{2}}{\sqrt{5}} \cdot \frac{\sqrt{5}}{\sqrt{5}} = \frac{3\sqrt{10}}{5}$

17. $\frac{4\sqrt{3}}{\sqrt{12}} = \frac{4\sqrt{3}}{\sqrt{4}\sqrt{3}} = \frac{4}{\sqrt{4}} = \frac{4}{2} = 2$

18. $\frac{5\sqrt{2}}{2\sqrt{3}} = \frac{5\sqrt{2}}{2\sqrt{3}} \cdot \frac{\sqrt{3}}{\sqrt{3}} = \frac{5\sqrt{6}}{6}$

19. $\frac{-3\sqrt{2}}{\sqrt{27}} = \frac{-3\sqrt{2}}{3\sqrt{3}} \cdot \frac{\sqrt{3}}{\sqrt{3}} = \frac{-3\sqrt{6}}{3\sqrt{9}} =$
$-\frac{\sqrt{6}}{\sqrt{9}} = -\frac{\sqrt{6}}{3}$

20. $\frac{4\sqrt{6}}{3\sqrt{12}} = \frac{4\sqrt{1}}{3\sqrt{2}} \cdot \frac{\sqrt{2}}{\sqrt{2}} = \frac{4\sqrt{2}}{6} = \frac{2\sqrt{2}}{3}$

21. $\sqrt{27} = \sqrt{9}\sqrt{3} = 3\sqrt{3} \approx 3(1.73) \approx$
5.19 or 5.2 to the nearest tenth.

22. $\frac{2}{\sqrt{3}} = \frac{2}{\sqrt{3}} \cdot \frac{\sqrt{3}}{\sqrt{3}} = \frac{2\sqrt{3}}{3} \approx \frac{2(1.73)}{3} \approx 1.2$

23. $3\sqrt{12} + \sqrt{48} = 3\sqrt{4}\sqrt{3} + \sqrt{16}\sqrt{3} =$
$3 \cdot 2\sqrt{3} + 4\sqrt{3} = 6\sqrt{3} + 4\sqrt{3} =$
$10\sqrt{3} \approx 10(1.73) \approx 17.3$

24. $2\sqrt{27} - 2\sqrt{75} = 2\sqrt{9}\sqrt{3} - 2\sqrt{25}\sqrt{3} =$
$2 \cdot 3\sqrt{3} - 2 \cdot 5\sqrt{3} = 6\sqrt{3} - 10\sqrt{3} =$
$-4\sqrt{3} = -4(1.73) = -6.9$

25. $\sqrt{12a^2b^3} = \sqrt{4a^2b^2}\sqrt{3b} = 2ab\sqrt{3b}$

26. $\sqrt{50xy^4} = \sqrt{25y^4}\sqrt{2x} = 5y^2\sqrt{2x}$

27. $\sqrt[4]{32x^4y^8} = \sqrt[4]{16x^4y^8}\sqrt[4]{2} = 2xy^2\sqrt[4]{2}$

28. $\sqrt[3]{125a^2b} = \sqrt[3]{125}\sqrt[3]{a^2b} = 5\sqrt[3]{a^2b}$

29. $\frac{4}{3}\sqrt{27xy^2} = \frac{4}{3}\sqrt{9y^2}\sqrt{3x} =$
$\frac{4}{3}(3y)\sqrt{3x} = 4y\sqrt{3x}$

30. $\frac{3}{4}\left(\sqrt[3]{24x^3}\right) = \frac{3\left(\sqrt[3]{8x^3}\sqrt[3]{3}\right)}{4} =$
$\frac{3\left(2x\sqrt[3]{3}\right)}{4} = \frac{6x\sqrt[3]{3}}{4} = \frac{3x\sqrt[3]{3}}{2}$

31. $\frac{\sqrt{2x}}{\sqrt{5y}} = \frac{\sqrt{2x}}{\sqrt{5y}} \cdot \frac{\sqrt{5y}}{\sqrt{5y}} = \frac{\sqrt{10xy}}{5y}$

32. $\frac{\sqrt{72x}}{\sqrt{16y}} = \frac{6\sqrt{2x}}{4\sqrt{y}} \cdot \frac{\sqrt{y}}{\sqrt{y}} =$
$\frac{6\sqrt{2xy}}{4y} = \frac{3\sqrt{2xy}}{2y}$

33. $\sqrt{\frac{4}{x}} = \frac{\sqrt{4}}{\sqrt{x}} \cdot \frac{\sqrt{x}}{\sqrt{x}} = \frac{\sqrt{4x}}{x} = \frac{2\sqrt{x}}{x}$

34. $\sqrt{\frac{2x^3}{9}} = \frac{x\sqrt{2x}}{3}$

35. $\frac{3\sqrt{x}}{4\sqrt{y^3}} = \frac{3\sqrt{x}}{4y\sqrt{y}} \cdot \frac{\sqrt{y}}{\sqrt{y}} = \frac{3\sqrt{xy}}{4y^2}$

36. $\frac{-2\sqrt{x^2y}}{5\sqrt{xy}} = \frac{-2\sqrt{x}}{5}$

37. $\left(\sqrt{6}\right)\left(\sqrt{12}\right) = \sqrt{72} = \sqrt{36}\sqrt{2} = 6\sqrt{2}$

38. $\left(2\sqrt{3}\right)\left(3\sqrt{6}\right) = 6\sqrt{18} = 6\sqrt{9}\sqrt{2} =$
$6 \cdot 3\sqrt{2} = 18\sqrt{2}$

39. $\left(-5\sqrt{8}\right)\left(2\sqrt{2}\right) = -10\sqrt{16} =$
$-10(4) = -40$

40. $\left(2\sqrt[3]{7}\right)\left(5\sqrt[3]{4}\right) = 10\sqrt[3]{28}$

41. $\sqrt[3]{2}\left(\sqrt[3]{3} + \sqrt[3]{4}\right) = \sqrt[3]{6} + \sqrt[3]{8} = \sqrt[3]{6} + 2$

42. $3\sqrt{5}\left(\sqrt{8} - 2\sqrt{12}\right) =$
$3\sqrt{5}\sqrt{8} - 3\sqrt{5}\left(2\sqrt{12}\right) =$
$3\sqrt{40} - 6\sqrt{60} =$
$3\sqrt{4}\sqrt{10} - 6\sqrt{4}\sqrt{15} =$
$6\sqrt{10} - 12\sqrt{15}$

43. $\left(\sqrt{3} + \sqrt{5}\right)\left(\sqrt{3} + \sqrt{7}\right) =$
$\sqrt{3}\left(\sqrt{3} + \sqrt{7}\right) + \sqrt{5}\left(\sqrt{3} + \sqrt{7}\right) =$
$\sqrt{9} + \sqrt{21} + \sqrt{15} + \sqrt{35} =$
$3 + \sqrt{21} + \sqrt{15} + \sqrt{35}$

44. $\left(2\sqrt{3} + 3\sqrt{2}\right)\left(\sqrt{3} - 5\sqrt{2}\right) =$
$2\sqrt{3}\left(\sqrt{3} - 5\sqrt{2}\right) + 3\sqrt{2}\left(\sqrt{3} - 5\sqrt{2}\right) =$
$2 \cdot 3 - 10\sqrt{6} + 3\sqrt{6} - 15 \cdot 2 =$
$6 - 7\sqrt{6} - 30 = -24 - 7\sqrt{6}$

45. $\left(\sqrt{6} + 2\sqrt{7}\right)\left(3\sqrt{6} - \sqrt{7}\right) =$
$\sqrt{6}\left(3\sqrt{6} - \sqrt{7}\right) + 2\sqrt{7}\left(3\sqrt{6} - \sqrt{7}\right) =$
$3\sqrt{36} - \sqrt{42} + 6\sqrt{42} - 2\sqrt{49} =$
$18 - \sqrt{42} + 6\sqrt{42} - 14 = 4 + 5\sqrt{42}$

46. $\left(3 + 2\sqrt{5}\right)\left(4 - 3\sqrt{5}\right) =$
$3\left(4 - 3\sqrt{5}\right) + 2\sqrt{5}\left(4 - 3\sqrt{5}\right) =$
$12 - 9\sqrt{5} + 8\sqrt{5} - 6 \cdot 5 =$
$12 - \sqrt{5} - 30 = -18 - \sqrt{5}$

47. $\dfrac{5}{\sqrt{7} - \sqrt{5}} = \dfrac{5}{\sqrt{7} - \sqrt{5}} \cdot \dfrac{\sqrt{7} + \sqrt{5}}{\sqrt{7} + \sqrt{5}} =$
$\dfrac{5\left(\sqrt{7} + \sqrt{5}\right)}{7 - 5} = \dfrac{5\sqrt{7} + 5\sqrt{5}}{2}$

48. $\dfrac{\sqrt{6}}{\sqrt{3} - \sqrt{2}} = \dfrac{\sqrt{6}}{\sqrt{3} - \sqrt{2}} \cdot \dfrac{\sqrt{3} + \sqrt{2}}{\sqrt{3} + \sqrt{2}} =$
$\dfrac{\sqrt{6}\left(\sqrt{3} + \sqrt{2}\right)}{3 - 2} = \dfrac{\sqrt{18} + \sqrt{12}}{1} =$
$3\sqrt{2} + 2\sqrt{3}$

49. $\dfrac{2}{3\sqrt{2} - \sqrt{6}} = \dfrac{2}{3\sqrt{2} - \sqrt{6}} \cdot \dfrac{3\sqrt{2} + \sqrt{6}}{3\sqrt{2} + \sqrt{6}} =$
$\dfrac{2\left(3\sqrt{2} + \sqrt{6}\right)}{18 - 6} = \dfrac{2\left(3\sqrt{2} + \sqrt{6}\right)}{12} =$
$\dfrac{3\sqrt{2} + \sqrt{6}}{6}$

50. $\dfrac{\sqrt{6}}{3\sqrt{7} + 2\sqrt{10}} =$
$\dfrac{\sqrt{6}}{3\sqrt{7} + 2\sqrt{10}} \cdot \dfrac{3\sqrt{7} - 2\sqrt{10}}{3\sqrt{7} - 2\sqrt{10}} =$
$\dfrac{\sqrt{6}\left(3\sqrt{7} - 2\sqrt{10}\right)}{63 - 40} =$
$\dfrac{3\sqrt{42} - 2\sqrt{60}}{23} = \dfrac{3\sqrt{42} - 4\sqrt{15}}{23}$

51. $2\sqrt{50} + 3\sqrt{72} - 5\sqrt{8} =$
$2\sqrt{25}\sqrt{2} + 3\sqrt{36}\sqrt{2} - 5\sqrt{4}\sqrt{2} =$
$10\sqrt{2} + 18\sqrt{2} - 10\sqrt{2} = 18\sqrt{2}$

52. $\sqrt{8x} - 3\sqrt{18x} = \sqrt{4}\sqrt{2x} - 3\sqrt{9}\sqrt{2x} =$
$2\sqrt{2x} - 9\sqrt{2x} = -7\sqrt{2x}$

53. $9\sqrt[3]{2} - 5\sqrt[3]{16} = 9\sqrt[3]{2} - 5\sqrt[3]{8}\sqrt[3]{2} =$
$9\sqrt[3]{2} - 5(2)\sqrt[3]{2} = 9\sqrt[3]{2} - 10\sqrt[3]{2} =$
$-\sqrt[3]{2}$

54. $3\sqrt{10}+\sqrt{\frac{2}{5}}=3\sqrt{10}+\frac{\sqrt{2}}{\sqrt{5}}\cdot\frac{\sqrt{5}}{\sqrt{5}}=$

$3\sqrt{10}+\frac{\sqrt{10}}{5}=\left(\frac{15}{5}+\frac{1}{5}\right)\sqrt{10}=$

$\frac{16}{5}\sqrt{10}$

55. $4\sqrt{20}-\frac{3}{\sqrt{5}}+\sqrt{45}=$

$4\sqrt{4}\sqrt{5}-\frac{3}{\sqrt{5}}\cdot\frac{\sqrt{5}}{\sqrt{5}}+\sqrt{9}\sqrt{5}=$

$8\sqrt{5}-\frac{3\sqrt{5}}{5}+3\sqrt{5}=$

$\left(\frac{40}{5}-\frac{3}{5}+\frac{15}{5}\right)\sqrt{5}=\frac{52}{5}\sqrt{5}$

56. $\sqrt{\frac{2}{3}}-2\sqrt{54}=\frac{\sqrt{2}}{\sqrt{3}}\cdot\frac{\sqrt{3}}{\sqrt{3}}-2\sqrt{9}\sqrt{6}=$

$\frac{\sqrt{6}}{3}-6\sqrt{6}=\left(\frac{1}{3}-\frac{18}{3}\right)\sqrt{6}=$

$-\frac{17}{3}\sqrt{6}$

57. $\sqrt{5x+6}=6$

$\left(\sqrt{5x+6}\right)^2=(6)^2$

$5x+6=36$

$5x=30$

$x=6$

CHECK

$\sqrt{5(6)+6}\stackrel{?}{=}6$

$\sqrt{30+6}\stackrel{?}{=}6$

$\sqrt{36}\stackrel{?}{=}6$

$6=6$

The solution set is $\{6\}$.

58. $\sqrt{6x+1}=\sqrt{3x+13}$

$\left(\sqrt{6x+1}\right)^2=\left(\sqrt{3x+13}\right)^2$

$6x+1=3x+13$

$3x=12$

$x=4$

CHECK

$\sqrt{6(4)+1}\stackrel{?}{=}\sqrt{3(4)+13}$

$\sqrt{24+1}\stackrel{?}{=}\sqrt{12+13}$

$\sqrt{25}\stackrel{?}{=}\sqrt{25}$

$5=5$

The solution set is $\{4\}$.

59. $3\sqrt{n}=n$

$\left(3\sqrt{n}\right)^2=n^2$

$9n=n^2$

$0=n^2-9n$

$0=n(n-9)$

$n=0$ or $n-9=0$

$n=0$ or $n=9$

CHECK

$3\sqrt{0}\stackrel{?}{=}0$

$0=0$

$3\sqrt{9}\stackrel{?}{=}9$

$3(3)=9$

The solution set is $\{0, 9\}$.

60. $\sqrt{y+5}=y+5$

$\left(\sqrt{y+5}\right)^2=(y+5)^2$

$y+5=y^2+10y+25$

$0=y^2+9y+20$

$0=(y+4)(y+5)$

$y+4=0$ or $y+5=0$

$y=-4$ or $y=-5$

CHECK

$$\sqrt{-4+5} \stackrel{?}{=} -4+5$$
$$\sqrt{1}=1$$

$$\sqrt{-5+5} \stackrel{?}{=} -5+5$$
$$\sqrt{0}=0$$

The solution set is $\{-5, -4\}$.

61. $$\sqrt{-3a+10}=a-2$$
$$\left(\sqrt{-3a+10}\right)^2=(a-2)^2$$
$$-3a+10=a^2-4a+4$$
$$0=a^2-a-6$$
$$0=(a+2)(a-3)$$
$$a+2=0 \quad \text{or} \quad a-3=0$$
$$a=-2 \quad \text{or} \quad a=3$$

CHECK

$$\sqrt{-3(-2)+10} \stackrel{?}{=} -2-2$$
$$\sqrt{16} \stackrel{?}{=} -4$$
$$4 \neq -4$$

$$\sqrt{-3(3)+10} \stackrel{?}{=} 3-2$$
$$\sqrt{1} \stackrel{?}{=} 1$$
$$1=1$$

The solution set is $\{3\}$.

62. $$3-\sqrt{2x-1}=2$$
$$-\sqrt{2x-1}=-1$$
$$\left(-\sqrt{2x-1}\right)^2=(-1)^2$$
$$2x-1=1$$
$$2x=2$$
$$x=1$$

CHECK

$$3-\sqrt{2(1)-1} \stackrel{?}{=} 2$$
$$3-\sqrt{1} \stackrel{?}{=} 2$$
$$3-1=2$$
$$2=2$$

The solution set is $\{1\}$.

63. 46

64. $\sqrt{4356}=66$

65. 72

66. $\sqrt{690}=26$ to the nearest whole number.

67. 47

68. $\sqrt{5500}=74$ to the nearest whole number.

CHAPTER 9 Test

1. $-\sqrt{\frac{64}{49}}=-\frac{8}{7}$ because $-\left(\frac{8}{7}\right)^2=-\frac{64}{49}$.

2. $\sqrt{0.0025}=0.05$ because $(0.05)^2=0.0025$.

3. $\sqrt{8}=\sqrt{4}\sqrt{2}=2\sqrt{2}\approx 2(1.41)\approx 2.82\approx 2.8$

4. $-\sqrt{32}=-\sqrt{16}\sqrt{2}=-4\sqrt{2}\approx -4(1.41)\approx -5.6$

5. $\frac{3}{\sqrt{2}}=\frac{3}{\sqrt{2}}\cdot\frac{\sqrt{2}}{\sqrt{2}}=\frac{3\sqrt{2}}{2}\approx \frac{3(1.41)}{2}\approx\frac{4.23}{2}\approx 2.115\approx 2.1$

6. $\sqrt{45}=\sqrt{9}\sqrt{5}=3\sqrt{5}$

7. $-4\sqrt[3]{54} = -4\sqrt[3]{27}\sqrt[3]{2} =$
$-4(3)\sqrt[3]{2} = -12\sqrt[3]{2}$

8. $\frac{2\sqrt{3}}{3\sqrt{6}} = \frac{2\sqrt{1}}{3\sqrt{2}} \cdot \frac{\sqrt{2}}{\sqrt{2}} = \frac{2\sqrt{2}}{6} = \frac{\sqrt{2}}{3}$

9. $\sqrt{\frac{25}{2}} = \frac{5}{\sqrt{2}} \cdot \frac{\sqrt{2}}{\sqrt{2}} = \frac{5\sqrt{2}}{2}$

10. $\frac{\sqrt{24}}{\sqrt{36}} = \frac{\sqrt{4}\sqrt{6}}{6} = \frac{2\sqrt{6}}{6} = \frac{\sqrt{6}}{3}$

11. $\sqrt{\frac{5}{8}} = \frac{\sqrt{5}}{2\sqrt{2}} \cdot \frac{\sqrt{2}}{\sqrt{2}} = \frac{\sqrt{10}}{4}$

12. $\sqrt[3]{-250x^4y^3} = \sqrt[3]{-125x^3y^3}\sqrt[3]{2x} =$
$-5xy\sqrt[3]{2x}$

13. $\frac{\sqrt{3x}}{\sqrt{5y}} = \frac{\sqrt{3x}}{\sqrt{5y}} \cdot \frac{\sqrt{5y}}{\sqrt{5y}} = \frac{\sqrt{15xy}}{5y}$

14. $\frac{3}{4}\sqrt{48x^3y^2} = \frac{3}{4}\sqrt{16x^2y^2}\sqrt{3x} =$
$\frac{3}{4}(4xy)\sqrt{3x} = 3xy\sqrt{3x}$

15. $\left(\sqrt{8}\right)\left(\sqrt{12}\right) = \sqrt{96} = \sqrt{16}\sqrt{6} = 4\sqrt{6}$

16. $\left(6\sqrt[3]{5}\right)\left(4\sqrt[3]{2}\right) = 24\sqrt[3]{10}$

17. $\sqrt{6}\left(2\sqrt{12} - 3\sqrt{8}\right) = 2\sqrt{72} - 3\sqrt{48} =$
$2\sqrt{36}\sqrt{2} - 3\sqrt{16}\sqrt{3} = 12\sqrt{2} - 12\sqrt{3}$

18. $\left(2\sqrt{5} + \sqrt{3}\right)\left(\sqrt{5} - 3\sqrt{3}\right) =$
$2\sqrt{5}\left(\sqrt{5} - 3\sqrt{3}\right) + \sqrt{3}\left(\sqrt{5} - 3\sqrt{3}\right) =$
$2 \cdot 5 - 6\sqrt{15} + \sqrt{15} - 3 \cdot 3 =$
$10 - 5\sqrt{15} - 9 = 1 - 5\sqrt{15}$

19. $\frac{\sqrt{6}}{\sqrt{12} + \sqrt{2}} =$

$\frac{\sqrt{6}}{\sqrt{12} + \sqrt{2}} \cdot \frac{\sqrt{12} - \sqrt{2}}{\sqrt{12} - \sqrt{2}} =$
$\frac{\sqrt{72} - \sqrt{12}}{12 - 2} = \frac{6\sqrt{2} - 2\sqrt{3}}{10} =$
$\frac{2\left(3\sqrt{2} - \sqrt{3}\right)}{10} = \frac{3\sqrt{2} - \sqrt{3}}{5}$

20. $2\sqrt{24} - 4\sqrt{54} + 3\sqrt{96} =$
$2\sqrt{4}\sqrt{6} - 4\sqrt{9}\sqrt{6} + 3\sqrt{16}\sqrt{6} =$
$4\sqrt{6} - 12\sqrt{6} + 12\sqrt{6} = 4\sqrt{6}$

21. $\sqrt{500} = 22$ to the nearest whole number.

22.
$$\begin{aligned} \sqrt{3x+1} &= 4 \\ \left(\sqrt{3x+1}\right)^2 &= (4)^2 \\ 3x + 1 &= 16 \\ 3x &= 15 \\ x &= 5 \end{aligned}$$

CHECK

$$\begin{aligned} \sqrt{3(5)+1} &\stackrel{?}{=} 4 \\ \sqrt{16} &\stackrel{?}{=} 4 \\ 4 &= 4 \end{aligned}$$

The solution set is $\{5\}$.

23.
$$\begin{aligned} \sqrt{2x-5} &= -4 \\ \left(\sqrt{2x-5}\right)^2 &= (-4)^2 \\ 2x - 5 &= 16 \\ 2x &= 21 \\ x &= \frac{21}{2} \end{aligned}$$

CHECK

$$\begin{aligned} \sqrt{2\left(\frac{21}{2}\right) - 5} &\stackrel{?}{=} -4 \\ \sqrt{21 - 5} &\stackrel{?}{=} -4 \\ \sqrt{16} &\stackrel{?}{=} -4 \\ 4 &\neq -4 \end{aligned}$$

The solution set is $\emptyset$.

24. $$\sqrt{n-3} = 3-n$$
$$\left(\sqrt{n-3}\right)^2 = (3-n)^2$$
$$n-3 = 9-6n+n^2$$
$$0 = n^2-7n+12$$
$$0 = (n-3)(n-4)$$
$$n-3=0 \quad \text{or} \quad n-4=0$$
$$n=3 \quad \text{or} \quad n=4$$

CHECK

$$\sqrt{3-3} \stackrel{?}{=} 3-3$$
$$\sqrt{0} = 0$$

$$\sqrt{4-3} \stackrel{?}{=} 3-4$$
$$\sqrt{1} \stackrel{?}{=} -1$$
$$1 \neq -1$$

The solution set is $\{3\}$.

25. $$\sqrt{3x+6} = x+2$$
$$\left(\sqrt{3x+6}\right)^2 = (x+2)^2$$
$$3x+6 = x^2+4x+4$$
$$0 = x^2+x-2$$
$$0 = (x+2)(x-1)$$
$$x+2=0 \quad \text{or} \quad x-1=0$$
$$x=-2 \quad \text{or} \quad x=1$$

CHECK

$$\sqrt{3(-2)+6} \stackrel{?}{=} -2+2$$
$$\sqrt{0} \stackrel{?}{=} 0$$
$$0=0$$

$$\sqrt{3(1)+6} \stackrel{?}{=} 1+2$$
$$\sqrt{9} \stackrel{?}{=} 3$$
$$3=3$$

The solution set is $\{-2, 1\}$.

CHAPTERS 1-9 Cumulative Review

1. $-2^6 = -64$

2. $\left(\frac{1}{4}\right)^{-3} = \frac{1^{-3}}{4^{-3}} = \frac{4^3}{1^3} = 64$

3. $\left(\frac{1}{3}-\frac{1}{4}\right)^{-2} = \left(\frac{4}{12}-\frac{3}{12}\right)^{-2} =$
$\left(\frac{1}{12}\right)^{-2} = 12^2 = 144$

4. $-\sqrt{64} = -8$

5. $\sqrt{\frac{4}{9}} = \frac{2}{3}$

6. $3^0+3^{-1}+3^{-2} =$
$1+\frac{1}{3}+\frac{1}{3^2} = \frac{9}{9}+\frac{3}{9}+\frac{1}{9} = \frac{13}{9}$

7. $3(2x-1)-4(2x+3)-(x+6) =$
$6x-3-8x-12-x-6 =$
$-3x-21 =$
$-3(-4)-21$, when $x=-4$
$= 12-21 = -9$

8. $(3x^2-4x-6)-(3x^2+3x+1) =$
$3x^2-4x-6-3x^2-3x-1 =$
$-7x-7 =$
$-7(6)-7$, when $x=6$
$= -42-7 = -49$

9. $2(a-b)-3(2a+b)+2(a-3b) =$
$2a-2b-6a-3b+2a-6b =$
$-2a-11b =$
$-2(-2)-11(3)$, when $a=-2$, $b=3$
$= 4-33 = -29$

10. $x^2-2xy+y^2 = (x-y)^2 =$
$[5-(-2)]^2$, when $x=5$, $y=-2$
$= (5+2)^2 = 7^2 = 49$

11. $\frac{3}{4x} + \frac{5}{2x} - \frac{7}{x} =$

$\frac{3}{4x} + \frac{5}{2x}\left(\frac{2}{2}\right) - \frac{7}{x}\left(\frac{4}{4}\right) =$

$\frac{3 + 10 - 28}{4x} = -\frac{15}{4x}$

12. $\frac{3}{x-2} - \frac{4}{x+3} =$

$\frac{3}{x-2}\left(\frac{x+3}{x+3}\right) - \frac{4}{x+3}\left(\frac{x-2}{x-2}\right) =$

$\frac{3(x+3) - 4(x-2)}{(x-2)(x+3)} =$

$\frac{3x + 9 - 4x + 8}{(x-2)(x+3)} = \frac{-x + 17}{(x-2)(x+3)}$

13. $\frac{3x}{7y} \div \frac{6x}{35y^2} = \frac{3x}{7y} \cdot \frac{35y^2}{6x} = \frac{5y}{2}$

14. $\frac{x-2}{x^2+x-6} \cdot \frac{x^2+6x+9}{x^2-x-12} =$

$\frac{x-2}{(x+3)(x-2)} \cdot \frac{(x+3)(x+3)}{(x-4)(x+3)} =$

$\frac{1}{x-4}$

15. $\frac{7}{x^2+3x-18} - \frac{8}{x-3} =$

$\frac{7}{(x+6)(x-3)} - \frac{8}{x-3}\left(\frac{x+6}{x+6}\right) =$

$\frac{7 - 8(x+6)}{(x+6)(x-3)} = \frac{7 - 8x - 48}{(x+6)(x-3)} =$

$\frac{-8x - 41}{(x+6)(x-3)}$

16. $(-3xy)(-4y^2)(5x^3y) =$

$(-3)(-4)(5)(x^{1+3})(y^{1+2+1}) =$

$60x^4y^4$

17. $(-4x^{-5})(2x^3) = -8x^{-2} = \frac{-8}{x^2}$

18. $\frac{-12a^{-2}b^3}{4a^{-5}b^4} = -3a^3b^{-1} = \frac{-3a^3}{b}$

19. $(3n^4)^{-1} = 3^{-1}n^{-4} = \frac{1}{3n^4}$

20. $(9x-2)(3x+4) =$

$9x(3x) + (36-6)x - 2(4) =$

$27x^2 + 30x - 8$

21. $(-x-1)(5x+7) =$

$-x(5x) + (-7-5)x - 1(7) =$

$-5x^2 - 12x - 7$

22. $(3x+1)(2x^2-x-4) =$

$3x(2x^2-x-4) + 1(2x^2-x-4) =$

$6x^3 - 3x^2 - 12x + 2x^2 - x - 4 =$

$6x^3 - x^2 - 13x - 4$

23. $\frac{15x^6y^8 - 20x^3y^5}{5x^3y^2} =$

$\frac{15x^6y^8}{5x^3y^2} - \frac{20x^3y^5}{5x^3y^2} =$

$3x^3y^6 - 4y^3$

24.

$$\begin{array}{r} 2x^2 - \ \ 2x - 3 \\ 5x+1 \overline{)10x^3 - 8x^2 - 17x - 3} \\ \underline{10x^3 + 2x^2 \qquad\qquad\quad} \\ -10x^2 - 17x - 3 \\ \underline{-2x^2 - \ \ 2x \qquad} \\ -15x - 3 \\ \underline{-15x - 3} \end{array}$$

25. $\frac{\frac{1}{x} - \frac{1}{y}}{\frac{1}{xy}} = \frac{xy\left(\frac{1}{x} - \frac{1}{y}\right)}{xy\left(\frac{1}{xy}\right)} =$

$$\frac{xy\left(\frac{1}{x}\right) - xy\left(\frac{1}{y}\right)}{1} = y - x$$

26. $\frac{2}{1500} = \frac{x}{3500}$
$2(3500) = 1500x$
$7000 = 1500x$
$4.67 = x$
4.67 gallons of paint will be needed.

27. $18 = x(72)$
$\frac{18}{72} = x$
$0.25 = x$
$0.25 \times 100\% = 25\%$

28. $V = \frac{1}{3}Bh,\ V = 432$ and $h = 12$
$432 = \frac{1}{3}(B)(12)$
$432 = 4B$
$108 = B$

29. $P = 2L + 2W,\ L = 25$ and $W = 40$
$P = 2(25) + 2(40)$
$P = 50 + 80$
$P = 130$
130 feet of fence is needed.

30. Surface Area $= 4\pi r^2$
Surface Area $= 4(3.14)(5)^2$
Surface Area $= 4(3.14)(25)$
Surface Area $= 314$
The surface area is 314 square inches.

31 a. $85000 = (8.5)(10^4)$
b. $0.0009 = (9)(10^{-4})$
c. $0.00000104 = (1.04)(10^{-6})$
d. $53000000 = (5.3)(10^7)$

32. Domain $= \{-2, -1, 0, 1, 2\}$
Range $= \{4, 1, 0, 1, 4\}$
It is a function.

33. $f(x) = \frac{4}{2x-3}$
The denominator can not equal zero.
$2x - 3 = 0$
$2x = 3$
$x = \frac{3}{2}$
So if $x = \frac{3}{2}$, then the denominator will be zero. Therefore $\frac{3}{2}$ is excluded from the domain. Domain is all real numbers except $\frac{3}{2}$.

34. $g(x) = x^2 + 5x - 4$
Domain is all real numbers.

35. $2x - y = -6$

Let $x = 0$	Let $y = 0$
$2(0) - y = -6$	$2x - 0 = -6$
$-y = -6$	$2x = -6$
$y = 6$	$x = -3$
$(0, 6)$	$(-3, 0)$

The x-intercept is -3 and the y-intercept is 6.

36. $f(x) = x^2 - 4$
$f(5) = 5^2 - 4 = 25 - 4 = 21$
$f(1) = 1^2 - 4 = 1 - 4 = -3$
$f(2) = 2^2 - 4 = 4 - 4 = 0$
$f(a) = a^2 - 4$

37. The fourth quadrant, QIV, since all x-coordinates are positive and all y-coordinates are negative.

38. October - 3 movies
December - 1 movie
$3 - 1 = 2$
Men attended 2 more movies in October than in December.

39. None. The difference is the same for all months.

40. October - 5 movies
November - 4 movies
December - 3 movies

$$\text{Average} = \frac{5+4+3}{3} = \frac{12}{3} = 4$$

The average number of movies attended for October, November, and December is 4.

41. $12x^3 + 14x^2 - 40x =$
$2x(6x^2 + 7x - 20) =$
$2x(6x^2 + 15x - 8x - 20) =$
$2x[3x(2x+5) - 4(2x+5)] =$
$2x(2x+5)(3x-4)$

42. $12x^2 - 27 =$
$3(4x^2 - 9) =$
$3(2x+3)(2x-3)$

43. $xy + 3x - 2y - 6 =$
$x(y+3) - 2(y+3) =$
$(y+3)(x-2)$

44. $30 + 19x - 5x^2 =$
$30 + 25x - 6x - 5x^2 =$
$5(6+5x) - x(6+5x) =$
$(6+5x)(5-x)$

45. $4x^4 - 4 =$
$4(x^4 - 1) =$
$4(x^2+1)(x^2-1) =$
$4(x^2+1)(x+1)(x-1)$

46. $21x^2 + 22x - 8 =$
$21x^2 - 6x + 28x - 8 =$
$3x(7x-2) + 4(7x-2) =$
$(7x-2)(3x+4)$

47. $4\sqrt{28} = 4\left(\sqrt{4}\sqrt{7}\right) = 4\left(2\sqrt{7}\right) = 8\sqrt{7}$

48. $-\sqrt{45} = -\sqrt{9}\sqrt{5} = -3\sqrt{5}$

49. $\sqrt{\frac{36}{5}} = \frac{\sqrt{36}}{\sqrt{5}} = \frac{6}{\sqrt{5}} \cdot \frac{\sqrt{5}}{\sqrt{5}} = \frac{6\sqrt{5}}{5}$

50. $\frac{5\sqrt{8}}{6\sqrt{12}} = \frac{5\left(2\sqrt{2}\right)}{6\left(2\sqrt{3}\right)} =$
$\frac{5\sqrt{2}}{6\sqrt{3}} \cdot \frac{\sqrt{3}}{\sqrt{3}} = \frac{5\sqrt{6}}{18}$

51. $\sqrt{72xy^5} = \sqrt{36y^4}\sqrt{2xy} = 6y^2\sqrt{2xy}$

52. $\frac{-2\sqrt{ab^2}}{5\sqrt{b}} = \frac{-2\sqrt{b}\sqrt{ab}}{5\sqrt{b}} = \frac{-2\sqrt{ab}}{5}$

53. $\left(3\sqrt{8}\right)\left(4\sqrt{2}\right) = 12\sqrt{16} = 12 \cdot 4 = 48$

54. $6\sqrt{2}\left(9\sqrt{8} - 3\sqrt{12}\right) =$
$6\sqrt{2}\left(9\sqrt{8}\right) - 6\sqrt{2}\left(3\sqrt{12}\right) =$
$54\sqrt{16} - 18\sqrt{24} =$
$54 \cdot 4 - 18 \cdot 2\sqrt{6} =$
$216 - 36\sqrt{6}$

55. $\left(3\sqrt{2} - \sqrt{7}\right)\left(3\sqrt{2} + \sqrt{7}\right) =$
$\left(3\sqrt{2}\right)^2 - \left(\sqrt{7}\right)^2 =$
$9 \cdot 2 - 7 = 18 - 7 = 11$

56. $\frac{4}{\sqrt{3}+\sqrt{2}} =$
$\frac{4}{\sqrt{3}+\sqrt{2}} \cdot \frac{\sqrt{3}-\sqrt{2}}{\sqrt{3}-\sqrt{2}} =$
$\frac{4\left(\sqrt{3}-\sqrt{2}\right)}{1} = 4\sqrt{3} - 4\sqrt{2}$

57. $\frac{-6}{3\sqrt{5}-\sqrt{6}} =$
$\frac{-6}{3\sqrt{5}-\sqrt{6}} \cdot \frac{3\sqrt{5}+\sqrt{6}}{3\sqrt{5}+\sqrt{6}} =$
$\frac{-6\left(3\sqrt{5}+\sqrt{6}\right)}{39} =$

$$\frac{-2\left(3\sqrt{5}+\sqrt{6}\right)}{13} =$$

$$\frac{-6\sqrt{5}-2\sqrt{6}}{13} = -\frac{6\sqrt{5}+2\sqrt{6}}{13}$$

58. $3\sqrt{50}-7\sqrt{72}+4\sqrt{98} =$
$3\cdot 5\sqrt{2}-7\cdot 6\sqrt{2}+4\cdot 7\sqrt{2} =$
$15\sqrt{2}-42\sqrt{2}+28\sqrt{2} = \sqrt{2}$

59. $\frac{2}{3}\sqrt{20}-\frac{3}{4}\sqrt{45}+\sqrt{80} =$
$\frac{2\cdot 2\sqrt{5}}{3}-\frac{3\cdot 3\sqrt{5}}{4}+4\sqrt{5} =$
$\frac{4\sqrt{5}}{3}\left(\frac{4}{4}\right)-\frac{9\sqrt{5}}{4}\left(\frac{3}{3}\right)+\frac{4\sqrt{5}}{1}\left(\frac{12}{12}\right) =$
$\frac{16\sqrt{5}-27\sqrt{5}+48\sqrt{5}}{12} = \frac{37\sqrt{5}}{12}$

60. $3x-6y = -6$

$x=0:\quad 0-6y = -6$
$y = 1$

$y=0:\quad 3x-0 = -6$
$x = -2$

$x=2:\quad 6-6y = -6$
$-6y = -12$
$y = 2$

Point
(0, 1)
(− 2, 0)
(2, 2)

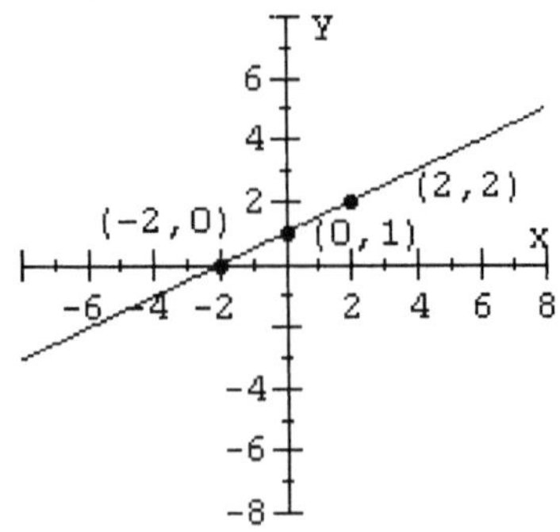

61. $y = \frac{1}{3}x+4$

Let $x = 0$	$x = 3$
$y = \frac{1}{3}(0)+4$	$y = \frac{1}{3}(3)+4$
$y = 4$	$y = 1+4$
(0, 4)	$y = 5$
	(3, 5)

See back of textbook for graph.

62. $y = -\frac{2}{5}x+3$

Let $x = 0$	$x = 5$
$y = -\frac{2}{5}(0)+3$	$y = -\frac{2}{5}(5)+3$
$y = 3$	$y = -2+3$
(0, 3)	$y = 1$
	(5, 1)

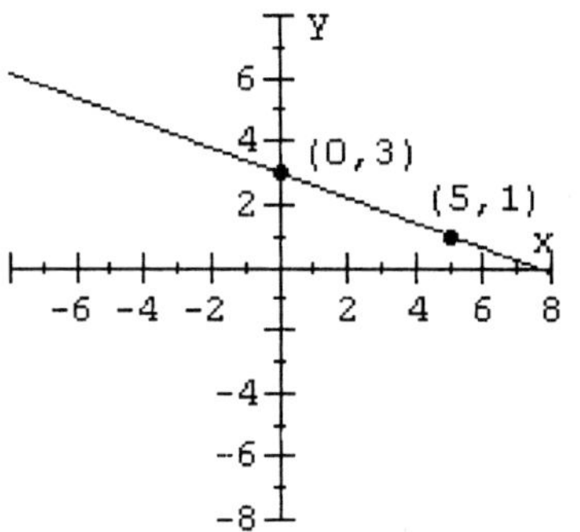

63. $y-2x = 0$
$y = 2x$

Let $x = 0$	$x = 1$
$y = 2(0)$	$y = 2(1)$
$y = 0$	$y = 2$
(0, 0)	(1, 2)

See back of textbook for graph.

64.

$$y = -2x^2 + 1$$

$x = -2$: $y = -2(-2)^2 + 1$
$y = -2(4) + 1$
$y = -7$

$x = -1$: $y = -2(-1)^2 + 1$
$y = -2(1) + 1$
$y = -1$

$x = 0$: $y = -2(0)^2 + 1$
$y = -2(0) + 1$
$y = 1$

$x = 1$: $y = -2(1)^2 + 1$
$y = -2(1) + 1$
$y = -1$

$x = 2$: $y = -2(2)^2 + 1$
$y = -2(4) + 1$
$y = -7$

Point
$(-2, -7)$
$(-1, -1)$
$(0, 1)$
$(1, -1)$
$(2, -7)$

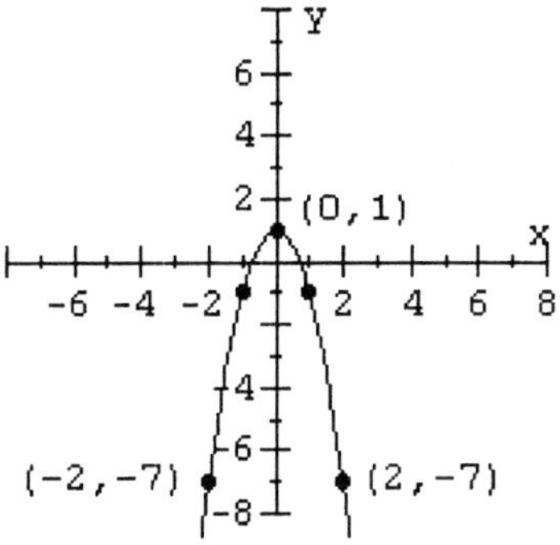

Y
6
4
2 (0,1)
X
-6 -4 -2 2 4 6 8
-4
-6
(-2,-7) (2,-7)
-8

65.

$$y = -2x^3$$

$x = -2$: $y = -2(-2)^3$
$y = -2(-8)$
$y = 16$

$x = -1$: $y = -2(-1)^3$
$y = -2(-1)$
$y = 2$

$x = 0$: $y = -2(0)^3$
$y = -2(0)$
$y = 0$

$x = 1$: $y = -2(1)^3$
$y = -2(1)$
$y = -2$

$x = 2$: $y = -2(2)^3$
$y = -2(8)$
$y = -16$

Point
$(-2, 16)$
$(-1, 2)$
$(0, 0)$
$(1, -2)$
$(2, -16)$

See back of textbook for graph.

66.

$$y = -x$$

$x = -1$: $y = -(-1) = 1$
$x = 0$: $y = 0$
$x = 2$: $y = -(2) = -2$

Point
$(-1, 1)$
$(0, 0)$
$(2, -2)$

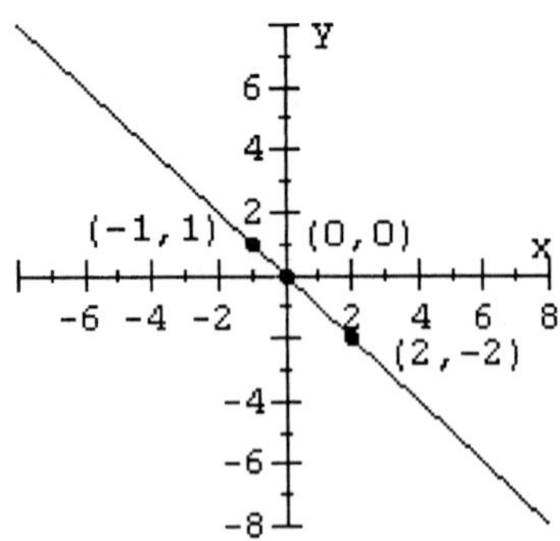

67. Let $P_1 = (-3, 6)$ and $P_2 = (2, -4)$.

$m = \frac{y_2 - y_1}{x_2 - x_1} = \frac{-4-6}{2-(-3)} = \frac{-10}{5} = -2$

68. $4x - 7y = 12$

$-7y = -4x + 12$

$y = \frac{4}{7}x - \frac{12}{7}$

The slope is $\frac{4}{7}$.

69. Use the point-slope form with $m = \frac{2}{3}$ and point $= (7, 2)$.

$y - y_1 = m(x - x_1)$

$y - 2 = \frac{2}{3}(x - 7)$

$3y - 6 = 2(x - 7)$

$3y - 6 = 2x - 14$

$-2x + 3y = -8$ or $2x - 3y = 8$

70. Find the slope.

$m = \frac{y_2 - y_1}{x_2 - x_1} = \frac{-3-1}{-1-(-4)} = \frac{-4}{3} = -\frac{4}{3}$

Use the slope and either point in the point-slope form.

$y - y_1 = m(x - x_1)$

$y - 1 = -\frac{4}{3}[x - (-4)]$

$3y - 3 = -4(x + 4)$

$3y - 3 = -4x - 16$

$4x + 3y = -13$

71. Use the slope-intercept form with $m = -\frac{1}{4}$ and $b = -3$.

$y = mx + b$

$y = -\frac{1}{4}x - 3$

$4y = -x - 12$

$x + 4y = -12$

72. A line perpendicular to the x-axis would be a vertical line. Since it contains the point $(4, -5)$, the line would be $x = 4$.

73. $y = 3x - 5$

$3x + 4y = -5$

Substitute $3x - 5$ for y in $3x + 4y = -5$.

$3x + 4(3x - 5) = -5$

$3x + 12x - 20 = -5$

$15x - 20 = -5$

$15x = 15$

$x = 1$

Substitute 1 for x in $y = 3x - 5$.

$y = 3(1) - 5 = 3 - 5 = -2$

The solution set is $\{(1, -2)\}$.

74. $4x - 3y = -20$ Multiply by 5.

$3x + 5y = 14$ Multiply by 3.

$$\begin{aligned} 20x - 15y &= -100 \\ 9x + 15y &= 42 \\ \hline 29x \quad &= -58 \end{aligned}$$

$x = -2$

Substitute -2 for x in $3x + 5y = 14$.

$3(-2) + 5y = 14$

$-6 + 5y = 14$

$5y = 20$

$y = 4$

The solution set is $\{(-2, 4)\}$.

75. $\frac{1}{2}x - \frac{2}{3}y = -11$ Multiply by 6.

$\frac{1}{3}x + \frac{5}{6}y = 8$ Multiply by 6.

$3x - 4y = -66$ Multiply by 5.
$2x + 5y = 48$ Multiply by 4.

$$\begin{array}{l} 15x - 20y = -330 \\ \underline{8x + 20y = 192} \\ 23x = -138 \\ x = -6 \end{array}$$

Substitute -6 for x in $3x - 4y = -66$.

$$\begin{aligned} 3(-6) - 4y &= -66 \\ -18 - 4y &= -66 \\ -4y &= -48 \\ y &= 12 \end{aligned}$$

The solution set is $\{(6, 12)\}$.

76. $2x + 7y = 22$ Multiply by -2.
$4x - 5y = -13$ Leave alone.

$$\begin{array}{l} -4x - 14y = -44 \\ \underline{4x - 5y = -13} \\ -19y = -57 \\ y = 3 \end{array}$$

Substitute 3 for y in $2x + 7y = 22$.

$$\begin{aligned} 2x + 7(3) &= 22 \\ 2x + 21 &= 22 \\ 2x &= 1 \\ x &= \frac{1}{2} \end{aligned}$$

The solution set is $\left\{\left(\frac{1}{2}, 3\right)\right\}$.

77.
$$\begin{aligned} -2(n-1) + 4(2n-3) &= 4(n+6) \\ -2n + 2 + 8n - 12 &= 4n + 24 \\ 6n - 10 &= 4n + 24 \\ 2n - 10 &= 24 \\ 2n &= 34 \\ n &= 17 \end{aligned}$$

The solution set is $\{17\}$.

78. $\frac{4}{x-1} = \frac{-1}{x+6}$, $x \neq -6$ or $x \neq 1$

Cross products are equal.

$$\begin{aligned} 4(x+6) &= -1(x-1) \\ 4x + 24 &= -x + 1 \\ 5x + 24 &= 1 \\ 5x &= -23 \\ x &= -\frac{23}{5} \end{aligned}$$

The solution set is $\left\{-\frac{23}{5}\right\}$.

79. $\frac{t-1}{3} - \frac{t+2}{4} = -\frac{5}{12}$

Multiply both sides by 12.

$$\begin{aligned} 4(t-1) - 3(t+2) &= -5 \\ 4t - 4 - 3t - 6 &= -5 \\ t - 10 &= -5 \\ t &= 5 \end{aligned}$$

The solution set is $\{5\}$.

80.
$$\begin{aligned} -7 - 2n - 6n &= 7n - 5n + 12 \\ -7 - 8n &= 2n + 12 \\ -7 - 10n &= 12 \\ -10n &= 19 \\ n &= -\frac{19}{10} \end{aligned}$$

The solution set is $\left\{-\frac{19}{10}\right\}$.

81. $\frac{n-5}{2} = 3 - \frac{n+4}{5}$

Multiply both sides by 10.

$$\begin{aligned} 5(n-5) &= 10(3) - 2(n+4) \\ 5n - 25 &= 30 - 2n - 8 \\ 5n - 25 &= 22 - 2n \\ 7n - 25 &= 22 \\ 7n &= 47 \\ n &= \frac{47}{7} \end{aligned}$$

The solution set is $\left\{\frac{47}{7}\right\}$.

82. $0.11x + 0.14(x + 400) = 181$

Multiply by 100.

$$\begin{aligned} 11x + 14(x + 400) &= 18,100 \\ 11x + 14x + 5600 &= 18,100 \\ 25x + 5600 &= 18,100 \\ 25x &= 12,500 \\ x &= 500 \end{aligned}$$

The solution set is $\{500\}$.

83. $$\frac{x}{60 - x} = 7 + \frac{4}{60 - x},\ x \neq 60$$

Multiply by $(60 - x)$.

$$\begin{aligned} (60 - x)\left(\frac{x}{60 - x}\right) &= (60 - x)\left(7 + \frac{4}{60 - x}\right) \\ x &= 7(60 - x) + 4 \\ x &= 420 - 7x + 4 \\ 8x &= 424 \\ x &= 53 \end{aligned}$$

The solution set is $\{53\}$.

84. $$1 + \frac{x + 1}{2x} = \frac{3}{4},\ x \neq 0$$

Multiply both sides by $4x$.

$$\begin{aligned} 4x + 2(x + 1) &= 3x \\ 4x + 2x + 2 &= 3x \\ 6x + 2 &= 3x \\ 2 &= -3x \\ -\frac{2}{3} &= x \end{aligned}$$

The solution set is $\left\{-\frac{2}{3}\right\}$.

85. $x^2 + 4x - 12 = 0$

$(x + 6)(x - 2) = 0$

$x + 6 = 0$ or $x - 2 = 0$

$x = -6$ or $x = 2$

The solution set is $\{-6, 2\}$.

86. $$\begin{aligned} 2x^2 - 8 &= 0 \\ 2(x^2 - 4) &= 0 \\ 2(x + 2)(x - 2) &= 0 \end{aligned}$$

$x + 2 = 0$ or $x - 2 = 0$

$x = -2$ or $x = 2$

The solution set is $\{-2, 2\}$.

87. $$\begin{aligned} \sqrt{3x - 6} &= 9 \\ \left(\sqrt{3x - 6}\right)^2 &= 9^2 \\ 3x - 6 &= 81 \\ 3x &= 87 \\ x &= 29 \end{aligned}$$

CHECK

$$\begin{aligned} \sqrt{3(29) - 6} &\stackrel{?}{=} 9 \\ \sqrt{87 - 6} &\stackrel{?}{=} 9 \\ \sqrt{81} &\stackrel{?}{=} 9 \\ 9 &= 9 \end{aligned}$$

The solution set is $\{29\}$.

88. $$\begin{aligned} \sqrt{3n} - 2 &= 7 \\ \sqrt{3n} &= 9 \\ \left(\sqrt{3n}\right)^2 &= 9^2 \\ 3n &= 81 \\ n &= 27 \end{aligned}$$

CHECK

$$\begin{aligned} \sqrt{3(27)} - 2 &\stackrel{?}{=} 7 \\ \sqrt{81} - 2 &\stackrel{?}{=} 7 \\ 9 - 2 &\stackrel{?}{=} 7 \\ 7 &= 7 \end{aligned}$$

The solution set is $\{27\}$.

89. $$\begin{aligned} -3n - 4 &\leq 11 \\ -3n &\leq 15 \end{aligned}$$

Dividing by -3 reverses the inequality.

$$n \geq -5$$

The solution set is $\{n | n \geq -5\}$.

90. $$\begin{aligned} -5 &> 3n - 4 - 7n \\ -5 &> -4n - 4 \\ -1 &> -4n \end{aligned}$$

Dividing by -4 reverses the inequality.

$$\frac{1}{4} < n$$

The solution set is $\left\{n | n > \frac{1}{4}\right\}$.

91. $2(x-2)+3(x+4)>6$

$2x-4+3x+12>6$

$5x+8>6$

$5x>-2$

$x>-\frac{2}{5}$

The solution set is $\left\{x|x>-\frac{2}{5}\right\}$.

92. $\frac{1}{2}n-\frac{2}{3}n<-1$

Multiply both sides by 6.

$3n-4n<-6$

$-n<-6$

Multiplying by -1 reverses the inequality.

$n>6$

The solution set is $\{n|n>6\}$.

93. $\frac{x+1}{2}+\frac{x-2}{6}<\frac{3}{8}$

$24\left(\frac{x+1}{2}+\frac{x-2}{6}\right)<24\left(\frac{3}{8}\right)$

$12(x+1)+4(x-2)<3(3)$

$12x+12+4x-8<9$

$16x+4<9$

$16x<5$

$x<\frac{5}{16}$

The solution set is $\left\{x|x<\frac{5}{16}\right\}$.

94. $\frac{x-3}{7}-\frac{x-2}{4}\leq\frac{9}{14}$

Multiply both sides by 28.

$4(x-3)-7(x-2)\leq 2(9)$

$4x-12-7x+14\leq 18$

$-3x+2\leq 18$

$-3x\leq 16$

$x\geq-\frac{16}{3}$

The solution set is $\left\{x|x\geq-\frac{16}{3}\right\}$.

95. Let a represent the measure of the smaller angle, then $180-a$ represents the larger angle.

$180-a=2a-15$

$180-3a=-15$

$-3a=-195$

$a=65$

If $a=65$, then $180-a=115$.

The measures of the supplementary angles are 65° and 115°.

96. Let n represent the smaller number, then $3n-2$ represents the larger number.

Their sum is 50.

$n+(3n-2)=50$

$4n-2=50$

$4n=52$

$n=13$

If $n=13$, then $3n-2=3(13)-2=37$.

The numbers are 13 and 37.

97. Let x represent the first odd number.

Then $x+2$ represents the second odd number.

$x^2+(x+2)^2=130$

$x^2+x^2+4x+4=130$

$2x^2+4x-126=0$

$2(x^2+2x-63)=0$

$2(x+9)(x-7)=0$

$x+9=0$ or $x-7=0$

$x=-9$ or $x=7$

Since -9 is not a whole number, the consecutive whole numbers are 7 and 9.

98. Let n represent the number of nickels, then $2n+1$ is the number of dimes, and $3n+4$ is the number of quarters.

$n+(2n+1)+(3n+4)=47$

$6n+5=47$

$6n=42$

$n=7$ nickels

If $n=7$, then $2n+1=2(7)+1=$ $14+1=15$ dimes.

If $n=7$, then $3n+4=3(7)+4=$ $21+4=25$ quarters.

There were 7 nickels, 15 dimes, and 25 quarters.

99. Let t represent the taxes on the $\$90,000$ home. The problem can be solved as a proportion.

$$\frac{\text{taxes}}{\text{value}} \quad \frac{1050}{70,000} = \frac{t}{90,000}$$

Cross products are equal.

$$70,000t = 90,000(1050)$$

Divide both sides by $10,000$.

$$7t = 9(1050)$$
$$7t = 9450$$
$$t = 1350$$

The taxes would be \$1350.

100. Let s represent the selling price. Profit is 60% of cost.

Selling price = Cost + Profit

$$s = 30 + (60\%)(30)$$
$$s = 30 + 0.6(30)$$
$$s = 30 + 18$$
$$s = 48$$

The selling price would be \$48.

101. Let t represent Polly's time, then $t+1$ represents Rosa's time. A diagram of the problem would be as follows:

Start

Rosa —— $45(t + 1)$ ——→

Polly —— $55t$ ——→

Distances are equal.

The two distances are the same.

$$55t = 45(t+1)$$
$$55t = 45t + 45$$
$$10t = 45$$
$$t = \frac{45}{10} = 4\frac{1}{2}$$

It would take Polly $4\frac{1}{2}$ hours to overtake Rosa.

102. Let x represent the amount of pure acid to be added. Then $100 + x$ represents the amount of final solution.

$$\begin{pmatrix}\text{pure}\\\text{acid}\\\text{in 10\%}\\\text{solution}\end{pmatrix} + \begin{pmatrix}\text{pure}\\\text{acid}\\\text{to be}\\\text{added}\end{pmatrix} = \begin{pmatrix}\text{pure}\\\text{acid}\\\text{in final}\\\text{solution}\end{pmatrix}$$

$$(10\%)(100) + x = (20\%)(100 + x)$$
$$0.10(100) + x = 0.20(100 + x)$$
$$10[0.10(100) + x] = 10[0.20(100 + x)]$$
$$1(100) + 10x = 2(100 + x)$$
$$100 + 10x = 200 + 2x$$
$$100 + 8x = 200$$
$$8x = 100$$
$$x = 12.5$$

We must add 12.5 milliliters of pure acid.

103. Let x represent the score on the fourth algebra test.

$$\frac{85 + 90 + 86 + x}{4} \geq 88$$
$$85 + 90 + 86 + x \geq 352$$
$$261 + x \geq 352$$
$$x \geq 91$$

The score on the last test needs to be 91 or better.

104. Let w represent the number of games they must win of the remaining 20 games. They will play a total of $70 + 72 + 20 = 162$ games. To win more than 50% of their games they would need to win more than 81 games.

$$w + 70 > 81$$
$$w > 11$$

They must win more than 11 of the remaining games.

105. Let t represent the time working together. The sum of the individual rates equals the rate working together.

Seth's rate + Butch's rate = Rate together

$$\begin{aligned}
\frac{1}{20}+\frac{1}{30}&=\frac{1}{t}\\
60t\left(\frac{1}{20}+\frac{1}{30}\right)&=60t\left(\frac{1}{t}\right)\\
3t+2t&=60\\
5t&=60\\
t&=12
\end{aligned}$$

It would take 12 minutes to complete the job working together.

Chapter 10 Quadratic Equations

PROBLEM SET 10.1 Quadratic Equations

1. $x^2 + 15x = 0$
$x(x+15) = 0$
$x = 0$ or $x + 15 = 0$
$x = 0$ or $x = -15$
The solution set is $\{-15, 0\}$.

3. $n^2 = 12n$
$n^2 - 12n = 0$
$n(n-12) = 0$
$n = 0$ or $n - 12 = 0$
$n = 0$ or $n = 12$
The solution set is $\{0, 12\}$.

5. $3y^2 = 15y$
$y^2 = 5y$
$y^2 - 5y = 0$
$y(y-5) = 0$
$y = 0$ or $y - 5 = 0$
$y = 0$ or $y = 5$
The solution set is $\{0, 5\}$.

7. $x^2 - 9x + 8 = 0$
$(x-8)(x-1) = 0$
$x - 8 = 0$ or $x - 1 = 0$
$x = 8$ or $x = 1$
The solution set is $\{1, 8\}$.

9. $x^2 - 5x - 14 = 0$
$(x-7)(x+2) = 0$
$x - 7 = 0$ or $x + 2 = 0$
$x = 7$ or $x = -2$
The solution set is $\{-2, 7\}$.

11. $n^2 + 5n - 6 = 0$
$(n+6)(n-1) = 0$
$n + 6 = 0$ or $n - 1 = 0$
$n = -6$ or $n = 1$
The solution set is $\{-6, 1\}$.

13. $6y^2 + 7y - 5 = 0$
$(3y+5)(2y-1) = 0$
$3y + 5 = 0$ or $2y - 1 = 0$
$3y = -5$ or $2y = 1$
$y = -\frac{5}{3}$ or $y = \frac{1}{2}$
The solution set is $\left\{-\frac{5}{3}, \frac{1}{2}\right\}$.

15. $30x^2 - 37x + 10 = 0$
$(5x-2)(6x-5) = 0$
$5x - 2 = 0$ or $6x - 5 = 0$
$5x = 2$ or $6x = 5$
$x = \frac{2}{5}$ or $x = \frac{5}{6}$
The solution set is $\left\{\frac{2}{5}, \frac{5}{6}\right\}$.

17. $4x^2 - 4x + 1 = 0$
$(2x-1)(2x-1) = 0$
$2x - 1 = 0$ or $2x - 1 = 0$
$2x = 1$ or $2x = 1$
$x = \frac{1}{2}$ or $x = \frac{1}{2}$
The solution set is $\left\{\frac{1}{2}\right\}$.

19. $x^2 = 64$
$x = \pm\sqrt{64}$
$x = \pm 8$
The solution set is $\{-8, 8\}$.

21. $x^2 = \frac{25}{9}$
$x = \pm\sqrt{\frac{25}{9}}$
$x = \pm\frac{5}{3}$
The solution set is $\left\{-\frac{5}{3}, \frac{5}{3}\right\}$.

23. $4x^2 = 64$
$x^2 = 16$
$x = \pm\sqrt{16}$
$x = \pm 4$
The solution set is $\{-4, 4\}$.

25. $n^2 = 14$
$n = \pm\sqrt{14}$
The solution set is $\left\{-\sqrt{14}, \sqrt{14}\right\}$.

27. $n^2 + 16 = 0$
$n^2 = -16$
There are no real number solutions because n^2 will always be nonnegative.

29. $y^2 = 32$
$y = \pm\sqrt{32}$
$y = \pm\sqrt{16}\sqrt{2}$
$y = \pm 4\sqrt{2}$
The solution set is $\left\{-4\sqrt{2}, 4\sqrt{2}\right\}$.

31. $3x^2 - 54 = 0$
$x^2 - 18 = 0$
$x^2 = 18$
$x = \pm\sqrt{18}$
$x = \pm\sqrt{9}\sqrt{2}$
$x = \pm 3\sqrt{2}$
The solution set is $\left\{-3\sqrt{2}, 3\sqrt{2}\right\}$.

33. $2x^2 = 9$
$x^2 = \frac{9}{2}$
$x = \pm\sqrt{\frac{9}{2}}$
$x = \pm\frac{\sqrt{9}}{\sqrt{2}}$
$x = \pm\frac{3}{\sqrt{2}} \cdot \frac{\sqrt{2}}{\sqrt{2}}$
$x = \pm\frac{3\sqrt{2}}{2}$
The solution set is $\left\{-\frac{3\sqrt{2}}{2}, \frac{3\sqrt{2}}{2}\right\}$.

35. $8n^2 = 25$
$n^2 = \frac{25}{8}$
$n = \pm\sqrt{\frac{25}{8}}$
$n = \pm\frac{\sqrt{25}}{\sqrt{8}}$
$n = \pm\frac{5}{2\sqrt{2}} \cdot \frac{\sqrt{2}}{\sqrt{2}}$
$n = \pm\frac{5\sqrt{2}}{4}$
The solution set is $\left\{-\frac{5\sqrt{2}}{4}, \frac{5\sqrt{2}}{4}\right\}$.

37. $(x-1)^2 = 4$
$x - 1 = \pm\sqrt{4}$
$x - 1 = \pm 2$
$x - 1 = 2$ or $x - 1 = -2$
$x = 3$ or $x = -1$
The solution set is $\{-1, 3\}$.

39. $(x+3)^2 = 25$
$x + 3 = \pm\sqrt{25}$
$x + 3 = \pm 5$
$x + 3 = 5$ or $x + 3 = -5$
$x = 2$ or $x = -8$
The solution set is $\{-8, 2\}$.

41. $(3x-2)^2 = 49$
$3x - 2 = \pm\sqrt{49}$
$3x - 2 = \pm 7$
$3x - 2 = 7$ or $3x - 2 = -7$
$3x = 9$ or $3x = -5$
$x = 3$ or $x = -\frac{5}{3}$
The solution set is $\left\{-\frac{5}{3}, 3\right\}$.

43. $(x+6)^2 = 5$

$$x+6 = \pm\sqrt{5}$$

$$x+6 = \sqrt{5} \quad \text{or} \quad x+6 = -\sqrt{5}$$

$$x = -6+\sqrt{5} \quad \text{or} \quad x = -6-\sqrt{5}$$

The solution set is

$\left\{-6+\sqrt{5},\ -6-\sqrt{5}\right\}$.

45. $(n-1)^2 = 8$

$$n-1 = \pm\sqrt{8} = \pm 2\sqrt{2}$$

$$n-1 = 2\sqrt{2} \quad \text{or} \quad n-1 = -2\sqrt{2}$$

$$n = 1+2\sqrt{2} \quad \text{or} \quad n = 1-2\sqrt{2}$$

The solution set is

$\left\{1+2\sqrt{2},\ 1-2\sqrt{2}\right\}$.

47. $(2n+3)^2 = 20$

$$2n+3 = \pm\sqrt{20} = \pm 2\sqrt{5}$$

$$2n+3 = 2\sqrt{5} \quad \text{or} \quad 2n+3 = -2\sqrt{5}$$

$$2n = -3+2\sqrt{5} \quad \text{or} \quad 2n = -3-2\sqrt{5}$$

$$n = \frac{-3+2\sqrt{5}}{2} \quad \text{or} \quad n = \frac{-3-2\sqrt{5}}{2}$$

The solution set is

$\left\{\dfrac{-3-2\sqrt{5}}{2}, \dfrac{-3+2\sqrt{5}}{2}\right\}$.

49. $(4x-1)^2 = -2$

The equation has no real number solutions because $(4x-1)^2$ will always be nonnegative.

51. $(3x-5)^2 - 40 = 0$

$$(3x-5)^2 = 40$$

$$3x-5 = \pm\sqrt{40} = \pm 2\sqrt{10}$$

$$3x-5 = 2\sqrt{10} \quad \text{or} \quad 3x-5 = -2\sqrt{10}$$

$$3x = 5+2\sqrt{10} \quad \text{or} \quad 3x = 5-2\sqrt{10}$$

$$x = \frac{5+2\sqrt{10}}{3} \quad \text{or} \quad x = \frac{5-2\sqrt{10}}{3}$$

The solution set is

$\left\{\dfrac{5-2\sqrt{10}}{3}, \dfrac{5+2\sqrt{10}}{3}\right\}$.

53. $2(7x-1)^2 + 5 = 37$

$$2(7x-1)^2 = 32$$

$$(7x-1)^2 = 16$$

$$7x-1 = \pm\sqrt{16} = \pm 4$$

$$7x-1 = 4 \quad \text{or} \quad 7x-1 = -4$$

$$7x = 5 \quad \text{or} \quad 7x = -3$$

$$x = \frac{5}{7} \quad \text{or} \quad x = \frac{-3}{7}$$

The solution set is $\left\{-\dfrac{3}{7}, \dfrac{5}{7}\right\}$.

55. $2(x+8)^2 - 9 = 91$

$$2(x+8)^2 = 100$$

$$(x+8)^2 = 50$$

$$x+8 = \pm\sqrt{50} = \pm 5\sqrt{2}$$

$$x+8 = 5\sqrt{2} \quad \text{or} \quad x+8 = -5\sqrt{2}$$

$$x = -8+5\sqrt{2} \quad \text{or} \quad x = -8-5\sqrt{2}$$

The solution set is

$\left\{-8-5\sqrt{2},\ -8+5\sqrt{2}\right\}$.

57. $c^2 = a^2 + b^2$

$$c^2 = 1^2 + 7^2 = 1 + 49$$

$$c^2 = 50$$

$$c = \sqrt{50} = 5\sqrt{2} \text{ inches}$$

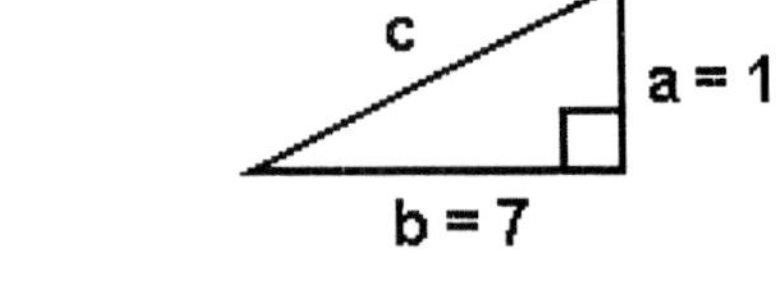

59. $c^2 = a^2 + b^2$

$$8^2 = a^2 + 6^2$$

$$64 = a^2 + 36$$

$$28 = a^2$$

$$a = \sqrt{28} = 2\sqrt{7} \text{ meters}$$

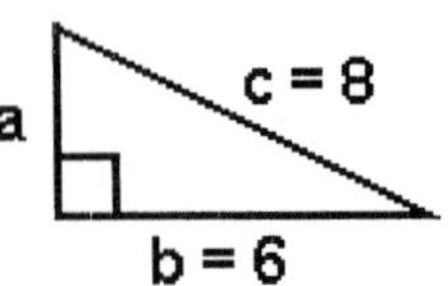

61.
$$a^2 + b^2 = c^2$$
$$10^2 + b^2 = 12^2$$
$$100 + b^2 = 144$$
$$b^2 = 44$$
$$b = \sqrt{44} = 2\sqrt{11} \text{ feet}$$

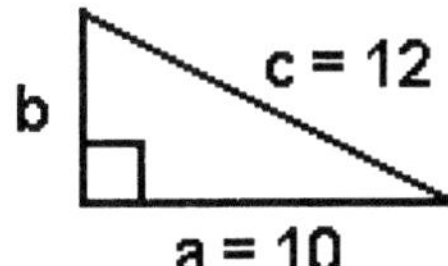

63. If $c = 8$ inches, then
$a = \frac{1}{2}c = \frac{1}{2}(8) = 4$ inches.

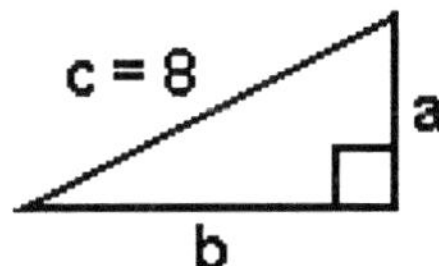

$$c^2 = a^2 + b^2$$
$$8^2 = 4^2 + b^2$$
$$64 = 16 + b^2$$
$$48 = b^2$$
$$b = \sqrt{48} = 4\sqrt{3} \text{ inches}$$

65. If $a = 6$ feet, then
$c = 2a = 2(6) = 12$ feet.

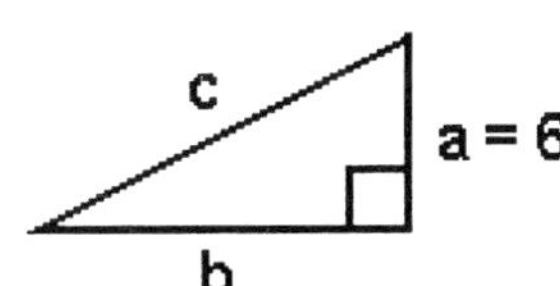

$$c^2 = a^2 + b^2$$
$$12^2 = 6^2 + b^2$$
$$144 = 36 + b^2$$
$$108 = b^2$$
$$b = \sqrt{108} = 6\sqrt{3} \text{ feet}$$

67. Substitute $2a$ for c in the Pythagorean Theorem.

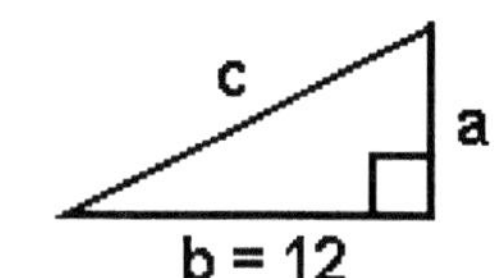

$$c^2 = a^2 + b^2$$
$$(2a)^2 = a^2 + 12^2$$
$$4a^2 = a^2 + 144$$
$$3a^2 = 144$$
$$a^2 = 48$$
$$a = \sqrt{48} = 4\sqrt{3} \text{ meters}$$
$$c = 2a = 2\left(4\sqrt{3}\right) = 8\sqrt{3} \text{ meters}$$

69. If $b = 10$ inches, then $a = 10$ inches.

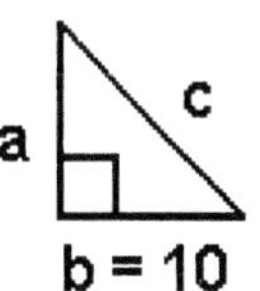

$$c^2 = a^2 + b^2$$
$$c^2 = 10^2 + 10^2$$
$$c^2 = 100 + 100$$
$$c^2 = 200$$
$$c = \sqrt{200} = 10\sqrt{2} \text{ inches}$$

71. Let $b = a$ in the Pythagorean Theorem.

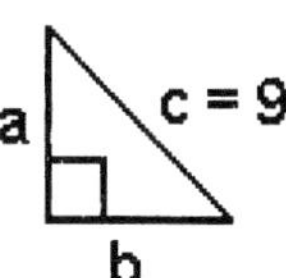

$$a^2 + b^2 = c^2$$
$$a^2 + a^2 = 9^2$$
$$2a^2 = 81$$
$$a^2 = \frac{81}{2}$$
$$a = b = \sqrt{\frac{81}{2}} = \frac{\sqrt{81}}{\sqrt{2}}$$
$$a = b = \frac{9}{\sqrt{2}} \cdot \frac{\sqrt{2}}{\sqrt{2}} = \frac{9\sqrt{2}}{2} \text{ meters}$$

73. Using the Pythagorean Theorem, let c represent the length of the ladder which is 18 feet, then let a represent the height of the windowsill above the ground which is 16 feet.

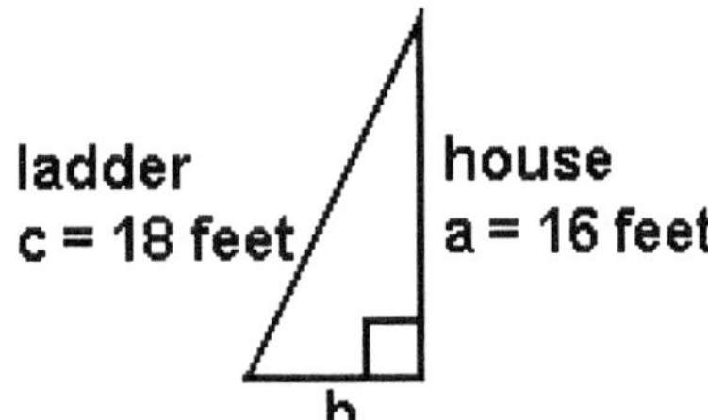

$$a^2 + b^2 = c^2$$
$$16^2 + b^2 = 18^2$$
$$256 + b^2 = 324$$
$$b^2 = 68$$
$$b = \sqrt{68} = 8.2 \text{ to nearest tenth}$$

The ladder is approximately 8.2 feet from the foundation of the house.

75. Let c represent the length of the diagonal of the rectangle.

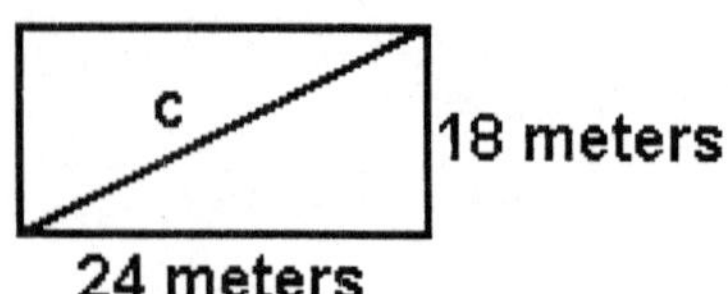

$$c^2 = a^2 + b^2$$
$$c^2 = 18^2 + 24^2$$
$$c^2 = 324 + 576$$
$$c^2 = 900$$
$$c = \sqrt{900} = 30$$

The length of the diagonal is 30 meters.

77. Let x represent a side of the square Parking Lot.

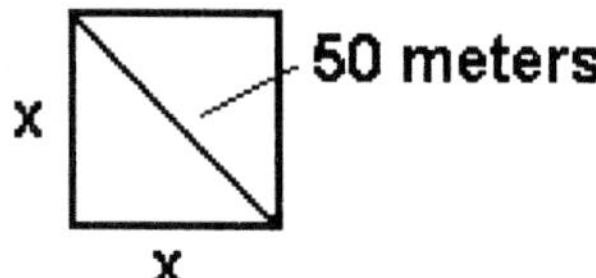

$$a^2 + b^2 = c^2$$
$$x^2 + x^2 = 50^2$$
$$2x^2 = 2500$$
$$x^2 = 1250$$
$$x = \sqrt{1250} = 35 \text{ to the nearest whole number.}$$

The length of a side of the lot is approximately 35 meters.

PROBLEM SET 10.2 Completing the Square

1.
$$x^2 + 8x - 1 = 0$$
$$x^2 + 8x = 1$$
$$x^2 + 8x + 16 = 1 + 16$$
$$(x+4)^2 = 17$$
$$x + 4 = \pm\sqrt{17}$$
$$x + 4 = \sqrt{17} \quad \text{or} \quad x + 4 = -\sqrt{17}$$
$$x = -4 + \sqrt{17} \quad \text{or} \quad x = -4 - \sqrt{17}$$

The solution set is
$$\left\{-4-\sqrt{17},\ -4+\sqrt{17}\right\}.$$

3.
$$x^2 + 10x + 2 = 0$$
$$x^2 + 10x = -2$$
$$x^2 + 10x + 25 = -2 + 25$$
$$(x+5)^2 = 23$$
$$x + 5 = \pm\sqrt{23}$$

$$x + 5 = \sqrt{23} \quad \text{or} \quad x + 5 = -\sqrt{23}$$
$$x = -5 + \sqrt{23} \quad \text{or} \quad x = -5 - \sqrt{23}$$

The solution set is
$\left\{-5 - \sqrt{23},\ -5 + \sqrt{23}\right\}$.

5.
$$x^2 - 4x - 4 = 0$$
$$x^2 - 4x = 4$$
$$x^2 - 4x + 4 = 4 + 4$$
$$(x - 2)^2 = 8$$
$$x - 2 = \pm\sqrt{8} = \pm 2\sqrt{2}$$
$$x - 2 = 2\sqrt{2} \quad \text{or} \quad x - 2 = -2\sqrt{2}$$
$$x = 2 + 2\sqrt{2} \quad \text{or} \quad x = 2 - 2\sqrt{2}$$

The solution set is
$\left\{2 - 2\sqrt{2},\ 2 + 2\sqrt{2}\right\}$.

7.
$$x^2 + 6x + 12 = 0$$
$$x^2 + 6x = -12$$
$$x^2 + 6x + 9 = -12 + 9$$
$$(x + 3)^2 = -3$$

The right side is negative.
There are no real number solutions because $(x + 3)^2$ will always be nonnegative.

9.
$$n^2 + 2n = 17$$
$$n^2 + 2n + 1 = 17 + 1$$
$$(n + 1)^2 = 18$$
$$n + 1 = \pm\sqrt{18} = \pm 3\sqrt{2}$$
$$n + 1 = 3\sqrt{2} \quad \text{or} \quad n + 1 = -3\sqrt{2}$$
$$n = -1 + 3\sqrt{2} \quad \text{or} \quad n = -1 - 3\sqrt{2}$$

The solution set is
$\left\{-1 - 3\sqrt{2},\ -1 + 3\sqrt{2}\right\}$.

11.
$$x^2 + x - 3 = 0$$
$$x^2 + x = 3$$
$$x^2 + x + \frac{1}{4} = 3 + \frac{1}{4}$$
$$\left(x + \frac{1}{2}\right)^2 = \frac{13}{4}$$
$$x + \frac{1}{2} = \pm\sqrt{\frac{13}{4}} = \pm\frac{\sqrt{13}}{2}$$
$$x + \frac{1}{2} = \frac{\sqrt{13}}{2} \quad \text{or} \quad x + \frac{1}{2} = -\frac{\sqrt{13}}{2}$$
$$x = -\frac{1}{2} + \frac{\sqrt{13}}{2} \quad \text{or} \quad x = -\frac{1}{2} - \frac{\sqrt{13}}{2}$$
$$x = \frac{-1 + \sqrt{13}}{2} \quad \text{or} \quad x = \frac{-1 - \sqrt{13}}{2}$$

The solution set is
$\left\{\frac{-1 - \sqrt{13}}{2},\ \frac{-1 + \sqrt{13}}{2}\right\}$.

13.
$$a^2 - 5a = 2$$
$$a^2 - 5a + \frac{25}{4} = 2 + \frac{25}{4}$$
$$\left(a - \frac{5}{2}\right)^2 = \frac{33}{4}$$
$$a - \frac{5}{2} = \pm\sqrt{\frac{33}{4}} = \pm\frac{\sqrt{33}}{2}$$
$$a - \frac{5}{2} = \frac{\sqrt{33}}{2} \quad \text{or} \quad a - \frac{5}{2} = -\frac{\sqrt{33}}{2}$$
$$a = \frac{5}{2} + \frac{\sqrt{33}}{2} \quad \text{or} \quad a = \frac{5}{2} - \frac{\sqrt{33}}{2}$$
$$a = \frac{5 + \sqrt{33}}{2} \quad \text{or} \quad a = \frac{5 - \sqrt{33}}{2}$$

The solution set is
$\left\{\frac{5 - \sqrt{33}}{2},\ \frac{5 + \sqrt{33}}{2}\right\}$.

15. $2x^2 + 8x - 3 = 0$

$$2x^2 + 8x = 3$$
$$x^2 + 4x = \frac{3}{2}$$
$$x^2 + 4x + 4 = \frac{3}{2} + 4$$
$$(x+2)^2 = \frac{11}{2}$$
$$x + 2 = \pm\sqrt{\frac{11}{2}} = \pm\frac{\sqrt{22}}{2}$$
$$x + 2 = \frac{\sqrt{22}}{2} \quad \text{or} \quad x + 2 = -\frac{\sqrt{22}}{2}$$
$$x = -2 + \frac{\sqrt{22}}{2} \quad \text{or} \quad x = -2 - \frac{\sqrt{22}}{2}$$
$$x = \frac{-4 + \sqrt{22}}{2} \quad \text{or} \quad x = \frac{-4 - \sqrt{22}}{2}$$

The solution set is

$$\left\{\frac{-4 - \sqrt{22}}{2}, \frac{-4 + \sqrt{22}}{2}\right\}.$$

17. $3x^2 + 12x - 2 = 0$

$$3x^2 + 12x = 2$$
$$x^2 + 4x = \frac{2}{3}$$
$$x^2 + 4x + 4 = \frac{2}{3} + 4$$
$$(x+2)^2 = \frac{14}{3}$$
$$x + 2 = \pm\sqrt{\frac{14}{3}} = \pm\frac{\sqrt{42}}{3}$$
$$x + 2 = \frac{\sqrt{42}}{3} \quad \text{or} \quad x + 2 = -\frac{\sqrt{42}}{3}$$
$$x = -2 + \frac{\sqrt{42}}{3} \quad \text{or} \quad x = -2 - \frac{\sqrt{42}}{3}$$
$$x = \frac{-6 + \sqrt{42}}{3} \quad \text{or} \quad x = \frac{-6 - \sqrt{42}}{3}$$

The solution set is

$$\left\{\frac{-6 - \sqrt{42}}{3}, \frac{-6 + \sqrt{42}}{3}\right\}.$$

19. $2t^2 - 4t + 1 = 0$

$$2t^2 - 4t = -1$$
$$t^2 - 2t = -\frac{1}{2}$$
$$t^2 - 2t + 1 = -\frac{1}{2} + 1$$
$$(t-1)^2 = \frac{1}{2}$$
$$t - 1 = \pm\sqrt{\frac{1}{2}} = \pm\frac{\sqrt{2}}{2}$$
$$t - 1 = \frac{\sqrt{2}}{2} \quad \text{or} \quad t - 1 = -\frac{\sqrt{2}}{2}$$
$$t = 1 + \frac{\sqrt{2}}{2} \quad \text{or} \quad t = 1 - \frac{\sqrt{2}}{2}$$
$$t = \frac{2 + \sqrt{2}}{2} \quad \text{or} \quad t = \frac{2 - \sqrt{2}}{2}$$

The solution set is

$$\left\{\frac{2 - \sqrt{2}}{2}, \frac{2 + \sqrt{2}}{2}\right\}.$$

21. $5n^2 + 10n + 6 = 0$

$$5n^2 + 10n = -6$$
$$n^2 + 2n = -\frac{6}{5}$$
$$n^2 + 2n + 1 = -\frac{6}{5} + 1$$
$$(n+1)^2 = -\frac{1}{5}$$

The right side is negative.
There are no real number solutions because $(n+1)^2$ will always be nonnegative.

23.

$$-n^2 + 9n = 4$$
$$n^2 - 9n = -4$$
$$n^2 - 9n + \frac{81}{4} = -4 + \frac{81}{4}$$
$$\left(n - \frac{9}{2}\right)^2 = \frac{65}{4}$$
$$n - \frac{9}{2} = \pm\sqrt{\frac{65}{4}} = \pm\frac{\sqrt{65}}{2}$$

$n - \frac{9}{2} = \frac{\sqrt{65}}{2}$ or $n - \frac{9}{2} = -\frac{\sqrt{65}}{2}$

$n = \frac{9}{2} + \frac{\sqrt{65}}{2}$ or $n = \frac{9}{2} - \frac{\sqrt{65}}{2}$

$n = \frac{9+\sqrt{65}}{2}$ or $n = \frac{9-\sqrt{65}}{2}$

The solution set is

$\left\{\frac{9-\sqrt{65}}{2}, \frac{9+\sqrt{65}}{2}\right\}.$

25. $2x^2 + 3x - 1 = 0$

$2x^2 + 3x = 1$

$x^2 + \frac{3}{2}x = \frac{1}{2}$

$x^2 + \frac{3}{2}x + \frac{9}{16} = \frac{1}{2} + \frac{9}{16}$

$\left(x + \frac{3}{4}\right)^2 = \frac{17}{16}$

$x + \frac{3}{4} = \pm\sqrt{\frac{17}{16}} = \pm\frac{\sqrt{17}}{4}$

$x + \frac{3}{4} = \frac{\sqrt{17}}{4}$ or $x + \frac{3}{4} = -\frac{\sqrt{17}}{4}$

$x = -\frac{3}{4} + \frac{\sqrt{17}}{4}$ or $x = -\frac{3}{4} - \frac{\sqrt{17}}{4}$

$x = \frac{-3+\sqrt{17}}{4}$ or $x = \frac{-3-\sqrt{17}}{4}$

The solution set is

$\left\{\frac{-3-\sqrt{17}}{4}, \frac{-3+\sqrt{17}}{4}\right\}.$

27. $3x^2 + 2x - 2 = 0$

$3x^2 + 2x = 2$

$x^2 + \frac{2}{3}x = \frac{2}{3}$

$x^2 + \frac{2}{3}x + \frac{1}{9} = \frac{2}{3} + \frac{1}{9}$

$\left(x + \frac{1}{3}\right)^2 = \frac{7}{9}$

$x + \frac{1}{3} = \pm\sqrt{\frac{7}{9}} = \pm\frac{\sqrt{7}}{3}$

$x + \frac{1}{3} = \frac{\sqrt{7}}{3}$ or $x + \frac{1}{3} = -\frac{\sqrt{7}}{3}$

$x = -\frac{1}{3} + \frac{\sqrt{7}}{3}$ or $x = -\frac{1}{3} - \frac{\sqrt{7}}{3}$

$x = \frac{-1+\sqrt{7}}{3}$ or $x = \frac{-1-\sqrt{7}}{3}$

The solution set is

$\left\{\frac{-1-\sqrt{7}}{3}, \frac{-1+\sqrt{7}}{3}\right\}.$

29. $n(n+2) = 168$

$n^2 + 2n = 168$

$n^2 + 2n + 1 = 168 + 1$

$(n+1)^2 = 169$

$n + 1 = \pm\sqrt{169} = \pm 13$

$n + 1 = 13$ or $n + 1 = -13$

$n = -1 + 13$ or $n = -1 - 13$

$n = 12$ or $n = -14$

The solution set is $\{-14, 12\}$.

31. $n(n-4) = 165$

$n^2 - 4n = 165$

$n^2 - 4n + 4 = 165 + 4$

$(n-2)^2 = 169$

$n - 2 = \pm\sqrt{169} = \pm 13$

$n - 2 = 13$ or $n - 2 = -13$

$n = 2 + 13$ or $n = 2 - 13$

$n = 15$ or $n = -11$

The solution set is $\{-11, 15\}$.

33 a. $x^2 + 4x - 12 = 0$

$(x+6)(x-2) = 0$

$x + 6 = 0$ or $x - 2 = 0$

$x = -6$ or $x = 2$

The solution set is $\{-6, 2\}$.

b. $x^2+4x-12=0$

$$x^2+4x=12$$
$$x^2+4x+4=12+4$$
$$(x+2)^2=16$$
$$x+2=\pm\sqrt{16}=\pm 4$$
$$x+2=4 \quad\text{or}\quad x+2=-4$$
$$x=-2+4 \quad\text{or}\quad x=-2-4$$
$$x=2 \quad\text{or}\quad x=-6$$

The solution set is $\{-6,\ 2\}$.

35 a. $x^2+12x+27=0$

$$(x+9)(x+3)=0$$
$$x+9=0 \quad\text{or}\quad x+3=0$$
$$x=-9 \quad\text{or}\quad x=-3$$

The solution set is $\{-9,\ -3\}$.

b. $x^2+12x+27=0$

$$x^2+12x=-27$$
$$x^2+12x+36=-27+36$$
$$(x+6)^2=9$$
$$x+6=\pm\sqrt{9}=\pm 3$$
$$x+6=3 \quad\text{or}\quad x+6=-3$$
$$x=-6+3 \quad\text{or}\quad x=-6-3$$
$$x=-3 \quad\text{or}\quad x=-9$$

The solution set is $\{-9,\ -3\}$.

37 a. $n^2-3n-40=0$

$$(n+5)(n-8)=0$$
$$n+5=0 \quad\text{or}\quad n-8=0$$
$$n=-5 \quad\text{or}\quad n=8$$

The solution set is $\{-5,\ 8\}$.

b. $n^2-3n-40=0$

$$n^2-3n=40$$
$$n^2-3n+\frac{9}{4}=40+\frac{9}{4}$$
$$\left(n-\frac{3}{2}\right)^2=\frac{169}{4}$$
$$n-\frac{3}{2}=\pm\sqrt{\frac{169}{4}}=\pm\frac{13}{2}$$
$$n-\frac{3}{2}=\frac{13}{2} \quad\text{or}\quad n-\frac{3}{2}=-\frac{13}{2}$$
$$n=\frac{3}{2}+\frac{13}{2} \quad\text{or}\quad n=\frac{3}{2}-\frac{13}{2}$$
$$n=\frac{16}{2}=8 \quad\text{or}\quad n=\frac{-10}{2}=-5$$

The solution set is $\{-5,\ 8\}$.

39a. $2n^2-9n+4=0$

$$(2n-1)(n-4)=0$$
$$2n-1=0 \quad\text{or}\quad n-4=0$$
$$2n=1 \quad\text{or}\quad n=4$$
$$n=\frac{1}{2} \quad\text{or}\quad n=4$$

The solution set is $\left\{\frac{1}{2},\ 4\right\}$.

b. $2n^2-9n+4=0$

$$2n^2-9n=-4$$
$$n^2-\frac{9}{2}n=-2$$
$$n^2-\frac{9}{2}n+\frac{81}{16}=-2+\frac{81}{16}$$
$$\left(n-\frac{9}{4}\right)^2=\frac{49}{16}$$
$$n-\frac{9}{4}=\pm\sqrt{\frac{49}{16}}=\pm\frac{7}{4}$$
$$n-\frac{9}{4}=\frac{7}{4} \quad\text{or}\quad n-\frac{9}{4}=-\frac{7}{4}$$
$$n=\frac{9}{4}+\frac{7}{4} \quad\text{or}\quad n=\frac{9}{4}-\frac{7}{4}$$
$$n=\frac{16}{4}=4 \quad\text{or}\quad n=\frac{2}{4}=\frac{1}{2}$$

The solution set is $\left\{\frac{1}{2},\ 4\right\}$.

41 a. $4n^2+4n-15=0$

$$(2n-3)(2n+5)=0$$
$$2n-3=0 \quad\text{or}\quad 2n+5=0$$
$$2n=3 \quad\text{or}\quad 2n=-5$$
$$n=\frac{3}{2} \quad\text{or}\quad n=-\frac{5}{2}$$

The solution set is $\left\{-\frac{5}{2},\ \frac{3}{2}\right\}$.

b. $4n^2 + 4n - 15 = 0$

$$4n^2 + 4n = 15$$
$$n^2 + n = \frac{15}{4}$$
$$n^2 + n + \frac{1}{4} = \frac{15}{4} + \frac{1}{4}$$
$$\left(n + \frac{1}{2}\right)^2 = \frac{16}{4} = 4$$
$$n + \frac{1}{2} = \pm\sqrt{4} = \pm 2$$

$n + \frac{1}{2} = 2$ or $n + \frac{1}{2} = -2$

$n = -\frac{1}{2} + 2$ or $n = -\frac{1}{2} - 2$

$n = \frac{3}{2}$ or $n = -\frac{5}{2}$

The solution set is $\left\{-\frac{5}{2}, \frac{3}{2}\right\}$.

PROBLEM SET 10.3 The Quadratic Formula

1. $x^2 - 5x - 6 = 0$

$$x = \frac{-(-5) \pm \sqrt{(-5)^2 - 4(1)(-6)}}{2(1)}$$
$$x = \frac{5 \pm \sqrt{49}}{2}$$
$$x = \frac{5 \pm 7}{2}$$

$x = \frac{5-7}{2}$ or $x = \frac{5+7}{2}$

$x = \frac{-2}{2} = -1$ or $x = \frac{12}{2} = 6$

The solution set is $\{-1, 6\}$.

3. $x^2 + 5x = 36$

$$x^2 + 5x - 36 = 0$$
$$x = \frac{-(5) \pm \sqrt{(5)^2 - 4(1)(-36)}}{2(1)}$$
$$x = \frac{-5 \pm \sqrt{169}}{2}$$
$$x = \frac{-5 \pm 13}{2}$$

$x = \frac{-5-13}{2}$ or $x = \frac{-5+13}{2}$

$x = \frac{-18}{2} = -9$ or $x = \frac{8}{2} = 4$

The solution set is $\{-9, 4\}$.

5. $n^2 - 2n - 5 = 0$

$$n = \frac{-(-2) \pm \sqrt{(-2)^2 - 4(1)(-5)}}{2(1)}$$
$$n = \frac{2 \pm \sqrt{24}}{2}$$
$$n = \frac{2 \pm 2\sqrt{6}}{2}$$
$$n = 1 \pm \sqrt{6}$$

The solution set is $\left\{1 - \sqrt{6}, 1 + \sqrt{6}\right\}$.

7. $a^2 - 5a - 2 = 0$

$$a = \frac{-(-5) \pm \sqrt{(-5)^2 - 4(1)(-2)}}{2(1)}$$
$$a = \frac{5 \pm \sqrt{33}}{2}$$

The solution set is

$$\left\{\frac{5 - \sqrt{33}}{2}, \frac{5 + \sqrt{33}}{2}\right\}.$$

9. $x^2 - 2x + 6 = 0$

$$x = \frac{-(-2) \pm \sqrt{(-2)^2 - 4(1)(6)}}{2(1)}$$
$$x = \frac{2 \pm \sqrt{-20}}{2}$$

Since $\sqrt{-20}$ is not a real number, this equation has no real number solutions.

Problem Set 10.3

11. $y^2+4y+2=0$

$$y=\frac{-(4)\pm\sqrt{(4)^2-4(1)(2)}}{2(1)}$$

$$y=\frac{-4\pm\sqrt{8}}{2}$$

$$y=\frac{-4\pm 2\sqrt{2}}{2}$$

$$y=-2\pm\sqrt{2}$$

The solution set is

$\left\{-2-\sqrt{2},\ -2+\sqrt{2}\right\}$.

13. $x^2-6x=0$ (Note: $c=0$)

$$x=\frac{-(-6)\pm\sqrt{(-6)^2-4(1)(0)}}{2(1)}$$

$$x=\frac{6\pm\sqrt{36}}{2}$$

$$x=\frac{6\pm 6}{2}$$

$$x=\frac{6-6}{2} \quad \text{or} \quad x=\frac{6+6}{2}$$

$$x=\frac{0}{2}=0 \quad \text{or} \quad x=\frac{12}{2}=6$$

The solution set is $\{0,\ 6\}$.

15. $2x^2=7x$

$2x^2-7x=0$ (Note: $c=0$)

$$x=\frac{-(-7)\pm\sqrt{(-7)^2-4(2)(0)}}{2(2)}$$

$$x=\frac{7\pm\sqrt{49}}{4}$$

$$x=\frac{7\pm 7}{4}$$

$$x=\frac{7-7}{4} \quad \text{or} \quad x=\frac{7+7}{4}$$

$$x=\frac{0}{4}=0 \quad \text{or} \quad x=\frac{14}{4}=\frac{7}{2}$$

The solution set is $\left\{0,\ \frac{7}{2}\right\}$.

17. $n^2-34n+288=0$

$$n=\frac{-(-34)\pm\sqrt{(-34)^2-4(1)(288)}}{2(1)}$$

$$n=\frac{34\pm\sqrt{4}}{2}$$

$$n=\frac{34\pm 2}{2}$$

$$n=\frac{34-2}{2} \quad \text{or} \quad n=\frac{34+2}{2}$$

$$n=\frac{32}{2}=16 \quad \text{or} \quad n=\frac{36}{2}=18$$

The solution set is $\{16,\ 18\}$.

19. $x^2+2x-80=0$

$$x=\frac{-(2)\pm\sqrt{(2)^2-4(1)(-80)}}{2(1)}$$

$$x=\frac{-2\pm\sqrt{324}}{2}$$

$$x=\frac{-2\pm 18}{2}$$

$$x=\frac{-2-18}{2} \quad \text{or} \quad x=\frac{-2+18}{2}$$

$$x=\frac{-20}{2}=-10 \quad \text{or} \quad x=\frac{16}{2}=8$$

The solution set is $\{-10,\ 8\}$.

21. $t^2+4t+4=0$

$$t=\frac{-(4)\pm\sqrt{(4)^2-4(1)(4)}}{2(1)}$$

$$t=\frac{-4\pm\sqrt{0}}{2}$$

$$t=\frac{-4\pm 0}{2}$$

$$t=\frac{-4}{2}=-2$$

The solution set is $\{-2\}$.

23. $6x^2 + x - 2 = 0$

$$x = \frac{-(1) \pm \sqrt{(1)^2 - 4(6)(-2)}}{2(6)}$$

$$x = \frac{-1 \pm \sqrt{49}}{12}$$

$$x = \frac{-1 \pm 7}{12}$$

$$x = \frac{-1-7}{12} \quad \text{or} \quad x = \frac{-1+7}{12}$$

$$x = \frac{-8}{12} = -\frac{2}{3} \quad \text{or} \quad x = \frac{6}{12} = \frac{1}{2}$$

The solution set is $\left\{-\frac{2}{3}, \frac{1}{2}\right\}$.

25. $5x^2 + 3x - 2 = 0$

$$x = \frac{-(3) \pm \sqrt{(3)^2 - 4(5)(-2)}}{2(5)}$$

$$x = \frac{-3 \pm \sqrt{49}}{10}$$

$$x = \frac{-3 \pm 7}{10}$$

$$x = \frac{-3-7}{10} \quad \text{or} \quad x = \frac{-3+7}{10}$$

$$x = \frac{-10}{10} = -1 \quad \text{or} \quad x = \frac{4}{10} = \frac{2}{5}$$

The solution set is $\left\{-1, \frac{2}{5}\right\}$.

27. $12x^2 + 19x = -5$

$12x^2 + 19x + 5 = 0$

$$x = \frac{-(19) \pm \sqrt{(19)^2 - 4(12)(5)}}{2(12)}$$

$$x = \frac{-19 \pm \sqrt{121}}{24}$$

$$x = \frac{-19 \pm 11}{24}$$

$$x = \frac{-19-11}{24} \quad \text{or} \quad x = \frac{-19+11}{24}$$

$$x = \frac{-30}{24} = -\frac{5}{4} \quad \text{or} \quad x = \frac{-8}{24} = -\frac{1}{3}$$

The solution set is $\left\{-\frac{5}{4}, -\frac{1}{3}\right\}$.

29. $2x^2 + 5x - 6 = 0$

$$x = \frac{-(5) \pm \sqrt{(5)^2 - 4(2)(-6)}}{2(2)}$$

$$x = \frac{-5 \pm \sqrt{73}}{4}$$

The solution set is

$$\left\{\frac{-5-\sqrt{73}}{4}, \frac{-5+\sqrt{73}}{4}\right\}.$$

31. $3x^2 + 4x - 1 = 0$

$$x = \frac{-(4) \pm \sqrt{(4)^2 - 4(3)(-1)}}{2(3)}$$

$$x = \frac{-4 \pm \sqrt{28}}{6}$$

$$x = \frac{-4 \pm 2\sqrt{7}}{6}$$

$$x = \frac{-2 \pm \sqrt{7}}{3}$$

The solution set is

$$\left\{\frac{-2-\sqrt{7}}{3}, \frac{-2+\sqrt{7}}{3}\right\}.$$

33. $16x^2 + 24x + 9 = 0$

$$x = \frac{-(24) \pm \sqrt{(24)^2 - 4(16)(9)}}{2(16)}$$

$$x = \frac{-24 \pm \sqrt{0}}{32}$$

$$x = \frac{-24 \pm 0}{32} = -\frac{3}{4}$$

The solution set is $\left\{-\frac{3}{4}\right\}$.

35. $4n^2 + 8n - 1 = 0$

$$n = \frac{-(8) \pm \sqrt{(8)^2 - 4(4)(-1)}}{2(4)}$$

$$n = \frac{-8 \pm \sqrt{80}}{8}$$

$$n = \frac{-8 \pm 4\sqrt{5}}{8}$$

$$n = \frac{-2 \pm \sqrt{5}}{2}$$

The solution set is

$$\left\{\frac{-2-\sqrt{5}}{2}, \frac{-2+\sqrt{5}}{2}\right\}.$$

37. $6n^2 + 9n + 1 = 0$

$$n = \frac{-(9) \pm \sqrt{(9)^2 - 4(6)(1)}}{2(6)}$$

$$n = \frac{-9 \pm \sqrt{57}}{12}$$

The solution set is

$$\left\{\frac{-9-\sqrt{57}}{12}, \frac{-9+\sqrt{57}}{12}\right\}.$$

39. $2y^2 - y - 4 = 0$

$$y = \frac{-(-1) \pm \sqrt{(-1)^2 - 4(2)(-4)}}{2(2)}$$

$$y = \frac{1 \pm \sqrt{33}}{4}$$

The solution set is

$$\left\{\frac{1-\sqrt{33}}{4}, \frac{1+\sqrt{33}}{4}\right\}.$$

41. $4t^2 + 5t + 3 = 0$

$$t = \frac{-(5) \pm \sqrt{(5)^2 - 4(4)(3)}}{2(4)}$$

$$t = \frac{-5 \pm \sqrt{-23}}{8}$$

Since $\sqrt{-23}$ is not a real number, this equation has no real number solutions.

43. $7x^2 + 5x - 4 = 0$

$$x = \frac{-(5) \pm \sqrt{(5)^2 - 4(7)(-4)}}{2(7)}$$

$$x = \frac{-5 \pm \sqrt{137}}{14}$$

The solution set is

$$\left\{\frac{-5-\sqrt{137}}{14}, \frac{-5+\sqrt{137}}{14}\right\}.$$

45. $7 = 3x^2 - x$

$0 = 3x^2 - x - 7$

$$x = \frac{-(-1) \pm \sqrt{(-1)^2 - 4(3)(-7)}}{2(3)}$$

$$x = \frac{1 \pm \sqrt{85}}{6}$$

The solution set is

$$\left\{\frac{1-\sqrt{85}}{6}, \frac{1+\sqrt{85}}{6}\right\}.$$

47. $n^2 + 23n = -126$

$n^2 + 23n + 126 = 0$

$$n = \frac{-(23) \pm \sqrt{(23)^2 - 4(1)(126)}}{2(1)}$$

$$n = \frac{-23 \pm \sqrt{25}}{2}$$

$$n = \frac{-23 \pm 5}{2}$$

$$n = \frac{-23-5}{2} \quad \text{or} \quad n = \frac{-23+5}{2}$$

$$n = \frac{-28}{2} = -14 \quad \text{or} \quad n = \frac{-18}{2} = -9$$

The solution set is $\{-14, -9\}$.

PROBLEM SET 10.4 Solving Quadratic Equations - Which Method?

For the problems in this set, the method chosen to solve the problem is not the only possible method. If a different method is used, the solution sets obtained should agree.

1. $x^2 + 4x = 45$

$x^2 + 4x - 45 = 0$

$(x+9)(x-5) = 0$

$x + 9 = 0$ or $x - 5 = 0$

$x = -9$ or $x = 5$

The solution set is $\{-9, 5\}$.

3. $(5n+6)^2 = 49$

$5n + 6 = \pm\sqrt{49} = \pm 7$

$5n + 6 = -7$ or $5n + 6 = 7$

$5n = -13$ or $5n = 1$

$n = -\frac{13}{5}$ or $n = \frac{1}{5}$

The solution set is $\left\{-\frac{13}{5}, \frac{1}{5}\right\}$.

5. $t^2 - t - 2 = 0$

$(t-2)(t+1) = 0$

$t - 2 = 0$ or $t + 1 = 0$

$t = 2$ or $t = -1$

The solution set is $\{-1, 2\}$.

7. $8x = 3x^2$

$0 = 3x^2 - 8x$

$0 = x(3x - 8)$

$x = 0$ or $3x - 8 = 0$

$x = 0$ or $3x = 8$

$x = 0$ or $x = \frac{8}{3}$

The solution set is $\left\{0, \frac{8}{3}\right\}$.

9. $9x^2 - 6x + 1 = 0$

$(3x-1)^2 = 0$

$3x - 1 = 0$

$3x = 1$

$x = \frac{1}{3}$

The solution set is $\left\{\frac{1}{3}\right\}$.

11. $5n^2 = \sqrt{8}n$

$5n^2 - \sqrt{8}n = 0$

$n\left(5n - \sqrt{8}\right) = 0$

$n = 0$ or $5n - \sqrt{8} = 0$

$n = 0$ or $5n = \sqrt{8} = 2\sqrt{2}$

$n = 0$ or $n = \frac{2\sqrt{2}}{5}$

The solution set is $\left\{0, \frac{2\sqrt{2}}{5}\right\}$.

13. $n^2 - 14n = 19$

$n^2 - 14n + 49 = 19 + 49$

$(n-7)^2 = 68$

$n - 7 = \pm\sqrt{68} = \pm 2\sqrt{17}$

$n - 7 = -2\sqrt{17}$ or $n - 7 = 2\sqrt{17}$

$n = 7 - 2\sqrt{17}$ or $n = 7 + 2\sqrt{17}$

The solution set is

$\left\{7 - 2\sqrt{17}, 7 + 2\sqrt{17}\right\}$.

15. $5x^2 - 2x - 7 = 0$

$(5x-7)(x+1) = 0$

$5x - 7 = 0$ or $x + 1 = 0$

$5x = 7$ or $x = -1$

$x = \frac{7}{5}$ or $x = -1$

The solution set is $\left\{-1, \frac{7}{5}\right\}$.

Problem Set 10.4

17.
$$15x^2 + 28x + 5 = 0$$
$$(5x+1)(3x+5) = 0$$
$$5x + 1 = 0 \quad \text{or} \quad 3x + 5 = 0$$
$$5x = -1 \quad \text{or} \quad 3x = -5$$
$$x = -\frac{1}{5} \quad \text{or} \quad x = -\frac{5}{3}$$
The solution set is $\left\{ -\frac{5}{3}, -\frac{1}{5} \right\}$.

19.
$$x^2 - \sqrt{8}x - 7 = 0$$
$$x^2 - 2\sqrt{2x} = 7$$
$$x^2 - 2\sqrt{2x} + 2 = 7 + 2$$
$$\left(x - \sqrt{2}\right)^2 = 9$$
$$x - \sqrt{2} = \pm\sqrt{9} = \pm 3$$
$$x = \sqrt{2} \pm 3$$
The solution set is $\left\{ \sqrt{2} - 3, \sqrt{2} + 3 \right\}$.

21.
$$y^2 + 5y = 84$$
$$y^2 + 5y - 84 = 0$$
$$(y+12)(y-7) = 0$$
$$y + 12 = 0 \quad \text{or} \quad y - 7 = 0$$
$$y = -12 \quad \text{or} \quad y = 7$$
The solution set is $\{-12, 7\}$.

23.
$$2n = 3 + \frac{3}{n}$$
Multiply both sides by n.
$$2n^2 = 3n + 3$$
$$2n^2 - 3n - 3 = 0$$
$$n = \frac{-(-3) \pm \sqrt{(-3)^2 - 4(2)(-3)}}{2(2)}$$
$$n = \frac{3 \pm \sqrt{33}}{4}$$
The solution set is
$\left\{ \frac{3 - \sqrt{33}}{4}, \frac{3 + \sqrt{33}}{4} \right\}$.

25.
$$3x^2 - 9x - 12 = 0$$
$$x^2 - 3x - 4 = 0$$
$$(x-4)(x+1) = 0$$
$$x - 4 = 0 \quad \text{or} \quad x + 1 = 0$$
$$x = 4 \quad \text{or} \quad x = -1$$
The solution set is $\{-1, 4\}$.

27.
$$2x^2 - 3x + 7 = 0$$
$$x = \frac{-(-3) \pm \sqrt{(-3)^2 - 4(2)(7)}}{2(2)}$$
$$x = \frac{3 \pm \sqrt{-47}}{4}$$
Since $\sqrt{-47}$ is not a real number, this equation has no real number solutions.

29.
$$n(n - 46) = -480$$
$$n^2 - 46n = -480$$
$$n^2 - 46n + 480 = 0$$
$$n = \frac{-(-46) \pm \sqrt{(-46)^2 - 4(1)(480)}}{2(1)}$$
$$n = \frac{46 \pm \sqrt{196}}{2}$$
$$n = \frac{46 \pm 14}{2}$$
$$n = \frac{46 - 14}{2} \quad \text{or} \quad n = \frac{46 + 14}{2}$$
$$n = \frac{32}{2} = 16 \quad \text{or} \quad n = \frac{60}{2} = 30$$
The solution set is $\{16, 30\}$.

31.
$$n - \frac{3}{n} = -1$$
Multiply both sides by n.
$$n^2 - 3 = -n$$
$$n^2 + n - 3 = 0$$
$$n = \frac{-(1) \pm \sqrt{(1)^2 - 4(1)(-3)}}{2(1)}$$
$$n = \frac{-1 \pm \sqrt{13}}{2}$$
The solution set is
$\left\{ \frac{-1 - \sqrt{13}}{2}, \frac{-1 + \sqrt{13}}{2} \right\}$.

33. $x + \frac{1}{x} = \frac{25}{12}$

Multiply both sides by $12x$.

$$12x^2 + 12 = 25x$$
$$12x^2 - 25x + 12 = 0$$
$$(3x - 4)(4x - 3) = 0$$
$$3x - 4 = 0 \quad \text{or} \quad 4x - 3 = 0$$
$$3x = 4 \quad \text{or} \quad 4x = 3$$
$$x = \frac{4}{3} \quad \text{or} \quad x = \frac{3}{4}$$

The solution set is $\left\{\frac{3}{4}, \frac{4}{3}\right\}$.

35. $t^2 + 12t + 36 = 49$

$$(t + 6)^2 = 49$$
$$t + 6 = \pm\sqrt{49} = \pm 7$$
$$t + 6 = -7 \quad \text{or} \quad t + 6 = 7$$
$$t = -13 \quad \text{or} \quad t = 1$$

The solution set is $\{-13, 1\}$.

37. $x^2 - 28x + 187 = 0$

$$x = \frac{-(-28) \pm \sqrt{(-28)^2 - 4(1)(187)}}{2(1)}$$
$$x = \frac{28 \pm \sqrt{36}}{2}$$
$$x = \frac{28 \pm 6}{2}$$
$$x = \frac{28 - 6}{2} \quad \text{or} \quad x = \frac{28 + 6}{2}$$
$$x = \frac{22}{2} = 11 \quad \text{or} \quad x = \frac{34}{2} = 17$$

The solution set is $\{11, 17\}$.

39. $\frac{x^2}{3} - x = -\frac{1}{2}$

Multiply both sides by 6.

$$2x^2 - 6x = -3$$
$$2x^2 - 6x + 3 = 0$$
$$x = \frac{-(-6) \pm \sqrt{(-6)^2 - 4(2)(3)}}{2(2)}$$
$$x = \frac{6 \pm \sqrt{12}}{4} = \frac{6 \pm 2\sqrt{3}}{4}$$
$$x = \frac{3 \pm \sqrt{3}}{2}$$

The solution set is $\left\{\frac{3 - \sqrt{3}}{2}, \frac{3 + \sqrt{3}}{2}\right\}$.

41. $\frac{2}{x + 2} - \frac{1}{x} = 3$

Multiply both sides by $x(x + 2)$.

$$2x - (x + 2) = 3x(x + 2)$$
$$x - 2 = 3x^2 + 6x$$
$$0 = 3x^2 + 5x + 2$$
$$0 = (3x + 2)(x + 1)$$
$$3x + 2 = 0 \quad \text{or} \quad x + 1 = 0$$
$$3x = -2 \quad \text{or} \quad x = -1$$
$$x = -\frac{2}{3} \quad \text{or} \quad x = -1$$

The solution set is $\left\{-1, -\frac{2}{3}\right\}$.

43. $\frac{2}{3n - 1} = \frac{n + 2}{6}$

Cross products are equal.

$$(3n - 1)(n + 2) = 2(6)$$
$$3n^2 + 5n - 2 = 12$$
$$3n^2 + 5n - 14 = 0$$
$$n = \frac{-(5) \pm \sqrt{(5)^2 - 4(3)(-14)}}{2(3)}$$
$$n = \frac{-5 \pm \sqrt{193}}{6}$$

The solution set is

$\left\{\frac{-5 - \sqrt{193}}{6}, \frac{-5 + \sqrt{193}}{6}\right\}$.

45. $(n - 2)(n + 4) = 7$

$$n^2 + 2n - 8 = 7$$
$$n^2 + 2n - 15 = 0$$
$$(n - 3)(n + 5) = 0$$
$$n - 3 = 0 \quad \text{or} \quad n + 5 = 0$$
$$n = 3 \quad \text{or} \quad n = -5$$

The solution set is $\{-5, 3\}$.

PROBLEM SET 10.5 Solving Problems Using Quadratic Equations

1. Let n and $n+1$ represent the consecutive whole numbers.

$$n(n+1) = 306$$
$$n^2 + n = 306$$
$$n^2 + n - 306 = 0$$
$$(n+18)(n-17) = 0$$
$$n + 18 = 0 \quad \text{or} \quad n - 17 = 0$$
$$n = -18 \quad \text{or} \quad n = 17$$

Discard the negative number since whole numbers are wanted. The numbers are 17 and $17 + 1 = 18$.

3. Let x and y represent the two positive integers.

$x + y = 44$ Their sum is 44.

$xy = 475$ Their product is 475.

Solve $x + y = 44$ for y.

$$x + y = 44$$
$$y = 44 - x$$

Substitute $44 - x$ for y in $xy = 475$.

$$x(44 - x) = 475$$
$$44x - x^2 = 475$$
$$0 = x^2 - 44x + 475$$
$$0 = (x - 25)(x - 19)$$
$$x - 25 = 0 \quad \text{or} \quad x - 19 = 0$$
$$x = 25 \quad \text{or} \quad x = 19$$

If $x = 25$, then $y = 44 - 25 = 19$.
If $x = 19$, then $y = 44 - 19 = 25$.
The numbers would be 25 and 19.

5. Let x and y represent the numbers.

$x + y = 6$ Their sum is 6.

$xy = 4$ Their product is 4.

Solve $x + y = 6$ for y.

$$x + y = 6$$
$$y = 6 - x$$

Substitute $6 - x$ for y in $xy = 4$.

$$x(6 - x) = 4$$
$$6x - x^2 = 4$$
$$0 = x^2 - 6x + 4$$
$$x = \frac{-(-6) \pm \sqrt{(-6)^2 - 4(1)(4)}}{2(1)}$$
$$x = \frac{6 \pm \sqrt{20}}{2}$$
$$x = \frac{6 \pm 2\sqrt{5}}{2}$$
$$x = 3 \pm \sqrt{5}$$

If $x = 3 + \sqrt{5}$, then $y = 6 - \left(3 + \sqrt{5}\right) = 3 - \sqrt{5}$.

If $x = 3 - \sqrt{5}$, then $y = 6 - \left(3 - \sqrt{5}\right) = 3 + \sqrt{5}$.

The numbers would be $3 - \sqrt{5}$ and $3 + \sqrt{5}$.

7. Let x represent the number, then $\frac{1}{x}$ represents its reciprocal.

$$x + \frac{1}{x} = \frac{3\sqrt{2}}{2}$$

Their sum is $\frac{3\sqrt{2}}{2}$.

$$x^2 + 1 = \frac{3\sqrt{2}}{2}x$$

Multiplied both sides by x.

$$x^2 - \frac{3\sqrt{2}}{2}x = -1$$

$$x^2 - \frac{3\sqrt{2}}{2}x + \frac{18}{16} = -1 + \frac{18}{16}$$

$$\left(x - \frac{3\sqrt{2}}{4}\right)^2 = \frac{2}{16}$$

$$x - \frac{3\sqrt{2}}{4} = \pm\sqrt{\frac{2}{16}} = \pm\frac{\sqrt{2}}{4}$$

$$x = \frac{3\sqrt{2}}{4} \pm \frac{\sqrt{2}}{4}$$

$$x = \frac{3\sqrt{2} \pm \sqrt{2}}{4}$$

$$x = \frac{3\sqrt{2} + \sqrt{2}}{4} \quad \text{or} \quad x = \frac{3\sqrt{2} - \sqrt{2}}{4}$$

$$x = \frac{4\sqrt{2}}{4} = \sqrt{2} \quad \text{or} \quad x = \frac{2\sqrt{2}}{4} = \frac{\sqrt{2}}{2}$$

The number could be $\sqrt{2}$ or $\frac{\sqrt{2}}{2}$.

9. Let n, $n+2$ and $n+4$ represent the three consecutive even whole numbers.

$$n^2 + (n+2)^2 + (n+4)^2 = 596$$

The sum of their squares is 596.

$$n^2 + n^2 + 4n + 4 + n^2 + 8n + 16 = 596$$

$$3n^2 + 12n + 20 = 596$$

$$3n^2 + 12n - 576 = 0$$

$$n^2 + 4n - 192 = 0$$

$$(n+16)(n-12) = 0$$

$$n + 16 = 0 \quad \text{or} \quad n - 12 = 0$$

$$n = -16 \quad \text{or} \quad n = 12$$

Since whole numbers are needed, the negative number is discarded. Thus, the numbers are 12, $12 + 2 = 14$, and $12 + 4 = 16$.

11. Let n represent the number, then $\frac{1}{2}n$ represents one-half of the number.

$$x^2 + \left(\frac{1}{2}x\right)^2 = 80$$

The sum of their squares is 80.

$$x^2 + \frac{1}{4}x^2 = 80$$

Multiply both sides by 4.

$$4x^2 + x^2 = 320$$

$$5x^2 = 320$$

$$x^2 = 64$$

$$x = \pm\sqrt{64} = \pm 8$$

Discard the negative solution.

The number could be either -8 or 8.

13. Let w represent the width of the rectangle, then $2w - 4$ represents the length of the rectangle.

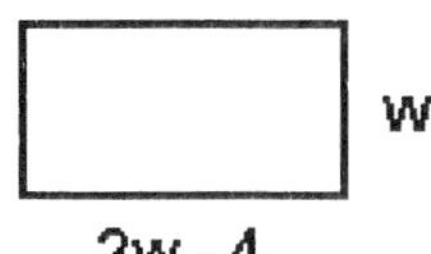

$$w(2w-4) = 96$$

$$2w^2 - 4w = 96$$

$$2w^2 - 4w - 96 = 0$$

$$w^2 - 2w - 48 = 0$$

$$(w-8)(w+6) = 0$$

$$w - 8 = 0 \quad \text{or} \quad w + 6 = 0$$

$$w = 8 \quad \text{or} \quad w = -6$$

The negative solution must be discarded. Thus, the rectangle is 8 meters by $2(8) - 4 = 12$ meters.

Problem Set 10.5

15. Let l represent the length and w represent the width of the rectangle.

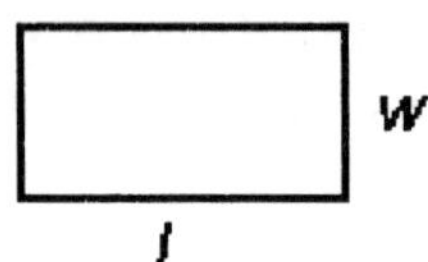

$2l + 2w = 80$ The perimeter is 80 centimeters.
$lw = 375$ The area is 375 square centimeters.

Solve $2l + 2w = 80$ for l.
$2l + 2w = 80$
$l + w = 40$
$l = 40 - w$
Substitute $40 - w$ for l in $lw = 375$.
$w(40 - w) = 375$
$40w - w^2 = 375$
$0 = w^2 - 40w + 375$
$0 = (w - 25)(w - 15)$
$w - 25 = 0$ or $w - 15 = 0$
$w = 25$ or $w = 15$
If $w = 25$, then $l = 40 - 25 = 15$.
If $w = 15$, then $l = 40 - 15 = 25$.
Thus, the rectangle is 15 centimeters by 25 centimeters.

17. Let w represent the width and $\frac{26}{9}w$ the length of the tennis court.

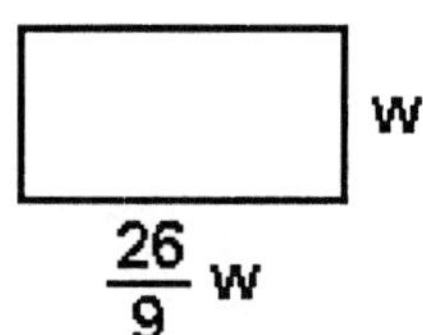

$w\left(\frac{26}{9}w\right) = 2106$
$\frac{26}{9}w^2 = 2106$
$26w^2 = 18,954$
$w^2 = 729$
$w = \pm\sqrt{729} = \pm 27$
Discard the negative solution.
If $w = 27$, then $l = \frac{26}{9}(27) = 78$.
The tennis court would be 27 feet by 78 feet.

19. Let s represent the number of seats per row, then $s - 5$ represents the number of rows.

$s(s - 5) = 300$ The product of the seats per row and the number of rows is 300.

$s^2 - 5s = 300$
$s^2 - 5s - 300 = 0$
$(s - 20)(s + 15) = 0$
$s - 20 = 0$ or $s + 15 = 0$
$s = 20$ or $s = -15$
Discard the negative solution.
There are 20 seats per row and $20 - 5 = 15$ rows in the auditorium.

21. Let l represent the length and w represent the width of the original rectangle. Then $l + 3$ and $w + 3$ represent the lenght and width, respectively, of the larger rectangle.

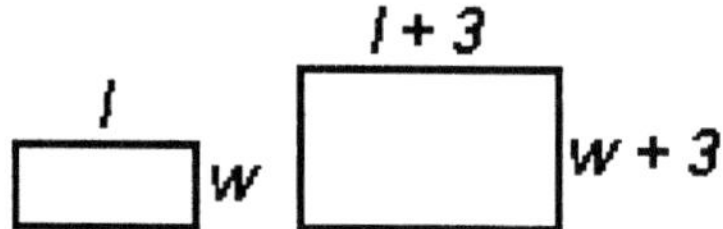

$lw = 63$
The original area is 63 square feet.
$(l + 3)(w + 3) = 63 + 57 = 120$
If the length and width are each increased by 3, the area is increased by 57.

Solve $lw = 63$ for l.
$lw = 63$
$l = \frac{63}{w}$

Substitute $\frac{63}{w}$ for l in the second equation.

$$\left(\frac{63}{w}+3\right)(w+3)=120$$
$$63+\frac{189}{w}+3w+9=120$$
$$63w+189+3w^2+9w=120w$$
$$3w^2+72w+189=120w$$
$$3w^2-48w+189=0$$
$$w^2-16w+63=0$$
$$(w-9)(w-7)=0$$
$$w-9=0 \quad \text{or} \quad w-7=0$$
$$w=9 \quad \text{or} \quad w=7$$

If $w=9$, then $l=\frac{63}{9}=7$.

If $w=7$, then $l=\frac{63}{7}=9$.

The original rectangle is 7feet by 9 feet.

23. Let a and b represent the length of the legs of the right triangle.

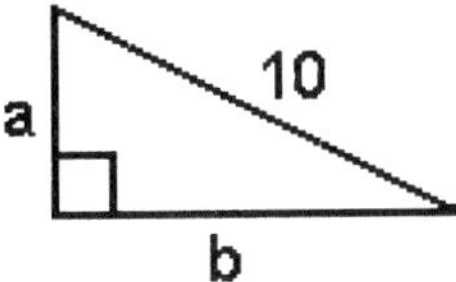

$a+b=14$ The sum of the lengths is 14 inches.

$a^2+b^2=10^2$ The Pythagorean Theorem using $c=10$ inches.

Solve $a+b=14$ for a.

$$a+b=14$$
$$a=14-b$$

Substitute $14-b$ for a in $a^2+b^2=100$.

$$(14-b)^2+b^2=100$$
$$196-28b+b^2+b^2=100$$
$$196-28b+2b^2=100$$
$$2b^2-28b+96=0$$
$$b^2-14b+48=0$$
$$(b-8)(b-6)=0$$
$$b-8=0 \quad \text{or} \quad b-6=0$$
$$b=8 \quad \text{or} \quad b=6$$

If $b=8$, then $a=14-8=6$.
If $b=6$, then $a=14-6=8$.
The lengths of the legs would be 6 inches and 8 inches.

25. Let x represent the uniform width of the frame. Then $(5+2x)$ and $(7+2x)$ represent the dimensions of the picture and frame together. The area is 80 square inches.

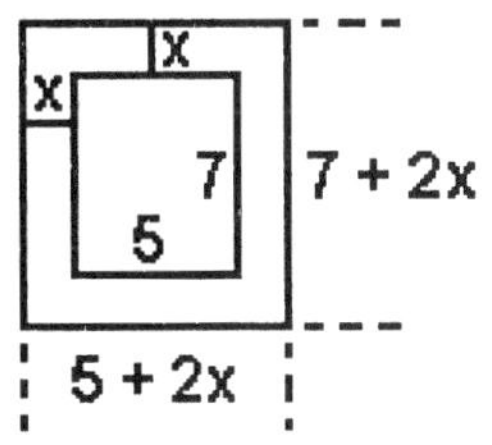

$$(5+2x)(7+2x)=80$$
$$35+24x+4x^2=80$$
$$4x^2+24x-45=0$$
$$(2x+15)(2x-3)=0$$
$$2x+15=0 \quad \text{or} \quad 2x-3=0$$
$$2x=-15 \quad \text{or} \quad 2x=3$$
$$x=-\frac{15}{2} \quad \text{or} \quad x=\frac{3}{2}=1\frac{1}{2}$$

Discard the negative solution. The frame width would be $1\frac{1}{2}$ inches.

27. Let s represent the number of students. Let c represent the cost per student of the trip.

$cs=3000$ The cost of the trip was \$3000.

$(c-25)(s+10)=3000$ With ten more students, the cost was \$25 less per student.

Solve $cs=3000$ for c.

$$cs=3000$$
$$c=\frac{3000}{s}$$

Substitute $\frac{3000}{s}$ for c in the second equation.

$$\left(\frac{3000}{s} - 25\right)(s+10) = 3000$$
$$3000 + \frac{30,000}{s} - 25s - 250 = 3000$$
$$3000s + 30,000 - 25s^2 - 250s = 3000s$$
$$-25s^2 + 2750s + 30,000 = 3000s$$
$$-25s^2 - 250s + 30,000 = 0$$

Divided by -25.

$$s^2 + 10s - 1200 = 0$$
$$(s-30)(s+40) = 0$$
$$s - 30 = 0 \quad \text{or} \quad s + 40 = 0$$
$$s = 30 \quad \text{or} \quad s = -40$$

Discard the negative solution.
There were 30 students on the trip.

29. Let x represent the length of a side of the first square. Let y represent the length of a side of the second square. Then $4x$ represents the length of the wire used for the first square and $4y$ represents the length of the wire used for the second square.

$4x + 4y = 56$ The sum of the lengths of the two wires is 56 inches.

$x^2 + y^2 = 100$ The sum of the areas of the two squares is 100 square inches.

4x 4y
56 inches
x x
y y

Solve $4x + 4y = 56$ for y.

$$4x + 4y = 56$$
$$x + y = 14$$
$$y = 14 - x$$

Substitute $14 - x$ for y in $x^2 + y^2 = 100$.

$$x^2 + (14-x)^2 = 100$$
$$x^2 + 196 - 28x + x^2 = 100$$
$$2x^2 - 28x + 196 = 100$$
$$2x^2 - 28x + 96 = 0$$
$$x^2 - 14x + 48 = 0$$
$$(x-6)(x-8) = 0$$
$$x - 6 = 0 \quad \text{or} \quad x - 8 = 0$$
$$x = 6 \quad \text{or} \quad x = 8$$

If $x = 6$, then $y = 14 - 6 = 8$.
If $x = 8$, then $y = 14 - 8 = 6$.
Thus, a side of one square is 6 inches for a wire length of $4(6) = 24$ inches and a side of the other square is 8 inches for a wire length of $4(8) = 32$ inches.

31. Let r represent her rate for the first 36 miles, then $r - 4$ represents the rate for the last 14 miles.
Let t represent her time for the first 36 miles, then $3 - t$ represents the time for the last 14 miles.

$rt = 36$ The first part of the trip.
$(r-4)(3-t) = 14$ The second part of the trip.

Solve $rt = 36$ for t.

$$rt = 36$$
$$t = \frac{36}{r}$$

Substitute $\frac{36}{r}$ for t in the second equation.

$$(r-4)\left(3 - \frac{36}{r}\right) = 14$$
$$3r - 36 - 12 + \frac{144}{r} = 14$$
$$3r^2 - 36r - 12r + 144 = 14r$$
$$3r^2 - 48r + 144 = 14r$$
$$3r^2 - 62r + 144 = 0$$
$$(3r-8)(r-18) = 0$$
$$3r - 8 = 0 \quad \text{or} \quad r - 18 = 0$$
$$3r = 8 \quad \text{or} \quad r = 18$$
$$r = \frac{8}{3} \quad \text{or} \quad r = 18$$

If $r = \frac{8}{3}$, then $r - 4 = \frac{8}{3} - 4 = -\frac{4}{3}$ which is not reasonable.
If $r = 18$, then $r - 4 = 18 - 4 = 14$.
The rate for the first 36 miles was 18 miles per hour.

CHAPTER 10 **Review Problem Set**

1. $(2x+7)^2 = 25$

$2x+7 = \pm\sqrt{25} = \pm 5$

$2x+7 = 5$ or $2x+7 = -5$

$2x = -2$ or $2x = -12$

$x = -1$ or $x = -6$

The solution set is $\{-6, -1\}$.

2. $x^2 + 8x = -3$

$x^2 + 8x + 16 = -3 + 16$

$(x+4)^2 = 13$

$x + 4 = \pm\sqrt{13}$

$x + 4 = -\sqrt{13}$ or $x + 4 = \sqrt{13}$

$x = -4 - \sqrt{13}$ or $x = -4 + \sqrt{13}$

The solution set is

$\{-4-\sqrt{13}, -4+\sqrt{13}\}$.

3. $21x^2 - 13x + 2 = 0$

$(3x-1)(7x-2) = 0$

$3x - 1 = 0$ or $7x - 2 = 0$

$3x = 1$ or $7x = 2$

$x = \frac{1}{3}$ or $x = \frac{2}{7}$

The solution set is $\left\{\frac{2}{7}, \frac{1}{3}\right\}$.

4. $x^2 = 17x$

$x^2 - 17x = 0$

$x(x-17) = 0$

$x = 0$ or $x - 17 = 0$

$x = 0$ or $x = 17$

The solution set is $\{0, 17\}$.

5. $n - \frac{4}{n} = -3$

$n^2 - 4 = -3n$

$n^2 + 3n - 4 = 0$

$(n+4)(n-1) = 0$

$n + 4 = 0$ or $n - 1 = 0$

$n = -4$ or $n = 1$

The solution set is $\{-4, 1\}$.

6. $n^2 - 26n + 165 = 0$

$$n = \frac{-(-26) \pm \sqrt{(-26)^2 - 4(1)(165)}}{2(1)}$$

$$n = \frac{26 \pm \sqrt{16}}{2}$$

$$n = \frac{26 \pm 4}{2} = 13 \pm 2$$

$n = 13 - 2 = 11$ or $n = 13 + 2 = 15$

The solution set is $\{11, 15\}$.

7. $3a^2 + 7a - 1 = 0$

$$a = \frac{-(7) \pm \sqrt{(7)^2 - 4(3)(-1)}}{2(3)}$$

$$a = \frac{-7 \pm \sqrt{61}}{6}$$

The solution set is

$\left\{\frac{-7-\sqrt{61}}{6}, \frac{-7+\sqrt{61}}{6}\right\}$.

8. $4x^2 - 4x + 1 = 0$

$(2x-1)^2 = 0$

$2x - 1 = 0$

$2x = 1$

$x = \frac{1}{2}$

The solution set is $\left\{\frac{1}{2}\right\}$.

9. $5x^2 + 6x + 7 = 0$

$$x = \frac{-(6) \pm \sqrt{(6)^2 - 4(5)(7)}}{2(5)}$$

$$x = \frac{-6 \pm \sqrt{-104}}{10}$$

Since $\sqrt{-104}$ is not a real number, this equation has no real number solutions.

10. $3x^2+18x+15=0$

$x^2+6x+5=0$

$(x+5)(x+1)=0$

$x+5=0$ or $x+1=0$

$x=-5$ or $x=-1$

The solution set is $\{-5, -1\}$.

11. $3(x-2)^2-2=4$

$3(x-2)^2=6$

$(x-2)^2=2$

$x-2=\pm\sqrt{2}$

$x=2\pm\sqrt{2}$

The solution set is

$\{2-\sqrt{2}, 2+\sqrt{2}\}$.

12. $x^2+4x-14=0$

$$x=\frac{-(4)\pm\sqrt{(4)^2-4(1)(-14)}}{2(1)}$$

$$x=\frac{-4\pm\sqrt{72}}{2}$$

$$x=\frac{-4\pm6\sqrt{2}}{2}=-2\pm3\sqrt{2}$$

The solution set is

$\{-2-3\sqrt{2}, -2+3\sqrt{2}\}$.

13. $y^2=45$

$y=\pm\sqrt{45}$

$y=\pm3\sqrt{5}$

The solution set is $\{-3\sqrt{5}, 3\sqrt{5}\}$.

14. $x(x-6)=27$

$x^2-6x=27$

$x^2-6x-27=0$

$(x-9)(x+3)=0$

$x-9=0$ or $x+3=0$

$x=9$ or $x=-3$

The solution set is $\{-3, 9\}$.

15. $x^2=x$

$x^2-x=0$

$x(x-1)=0$

$x=0$ or $x-1=0$

$x=0$ or $x=1$

The solution set is $\{0, 1\}$.

16. $n^2-4n-3=6$

$n^2-4n-9=0$

$$n=\frac{-(-4)\pm\sqrt{(-4)^2-4(1)(-9)}}{2(1)}$$

$$n=\frac{4\pm\sqrt{52}}{2}$$

$$n=\frac{4\pm2\sqrt{13}}{2}=2\pm\sqrt{13}$$

The solution set is

$\{2-\sqrt{13}, 2+\sqrt{13}\}$.

17. $n^2-44n+480=0$

$(n-24)(n-20)=0$

$n-24=0$ or $n-20=0$

$n=24$ or $n=20$

The solution set is $\{20, 24\}$.

18. $\frac{x^2}{4}=x+1$

$x^2=4x+4$

$x^2-4x-4=0$

$$x=\frac{-(-4)\pm\sqrt{(-4)^2-4(1)(-4)}}{2(1)}$$

$$x=\frac{4\pm\sqrt{32}}{2}$$

$$x=\frac{4\pm4\sqrt{2}}{2}=2\pm2\sqrt{2}$$

The solution set is

$\{2-2\sqrt{2}, 2+2\sqrt{2}\}$.

19. $\frac{5x-2}{3}=\frac{2}{x+1}$

Cross products are equal.

$(5x-2)(x+1)=2(3)$

$5x^2+3x-2=6$

$5x^2+3x-8=0$

$(5x+8)(x-1)=0$

$$5x + 8 = 0 \quad \text{or} \quad x - 1 = 0$$
$$5x = -8 \quad \text{or} \quad x = 1$$
$$x = -\frac{8}{5} \quad \text{or} \quad x = 1$$

The solution set is $\left\{-\frac{8}{5}, 1\right\}$.

20. $\frac{-1}{3x - 1} = \frac{2x + 1}{-2}$

Cross products are equal.

$$(3x - 1)(2x + 1) = -1(-2)$$
$$6x^2 + x - 1 = 2$$
$$6x^2 + x - 3 = 0$$
$$x = \frac{-(1) \pm \sqrt{(1)^2 - 4(6)(-3)}}{2(6)}$$
$$x = \frac{-1 \pm \sqrt{73}}{12}$$

The solution set is

$\left\{\frac{-1 - \sqrt{73}}{12}, \frac{-1 + \sqrt{73}}{12}\right\}$.

21. $\frac{5}{x - 3} + \frac{4}{x} = 6$

Multiply both sides by $x(x - 3)$.

$$5x + 4(x - 3) = 6x(x - 3)$$
$$5x + 4x - 12 = 6x^2 - 18x$$
$$9x - 12 = 6x^2 - 18x$$
$$0 = 6x^2 - 27x + 12$$
$$0 = 2x^2 - 9x + 4$$
$$0 = (2x - 1)(x - 4)$$
$$2x - 1 = 0 \quad \text{or} \quad x - 4 = 0$$
$$2x = 1 \quad \text{or} \quad x = 4$$
$$x = \frac{1}{2} \quad \text{or} \quad x = 4$$

The solution set is $\left\{\frac{1}{2}, 4\right\}$.

22. $\frac{1}{x + 2} - \frac{2}{x} = 3$

Multiplied both sides by $x(x + 2)$.

$$x - 2(x + 2) = 3x(x + 2)$$
$$x - 2x - 4 = 3x^2 + 6x$$
$$-x - 4 = 3x^2 + 6x$$
$$0 = 3x^2 + 7x + 4$$
$$0 = (3x + 4)(x + 1)$$
$$3x + 4 = 0 \quad \text{or} \quad x + 1 = 0$$
$$3x = -4 \quad \text{or} \quad x = -1$$
$$x = -\frac{4}{3} \quad \text{or} \quad x = -1$$

The solution set is $\left\{-\frac{4}{3}, -1\right\}$.

23. Let l represent the length and w represent the width of the rectangle.

$2l + 2w = 42$ The perimeter is 42 inches.

$lw = 108$ The area is 108 square inches.

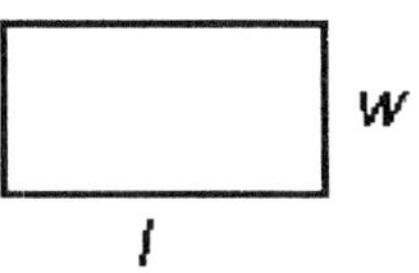

Solve $2l + 2w = 42$ for l.

$$2l + 2w = 42$$
$$l + w = 21$$
$$l = 21 - w$$

Substitute $21 - w$ for l in $lw = 108$.

$$w(21 - w) = 108$$
$$21w - w^2 = 108$$
$$0 = w^2 - 21w + 108$$
$$0 = (w - 9)(w - 12)$$
$$w - 9 = 0 \quad \text{or} \quad w - 12 = 0$$
$$w = 9 \quad \text{or} \quad w = 12$$

If $w = 9$, then $l = 21 - 9 = 12$.
If $w = 12$, then $l = 21 - 12 = 9$.
Thus, the rectangle is 9 inches by 12 inches.

24. Let n and $n+1$ represent the two consecutive whole numbers.

$n(n+1) = 342$ Their product is 342.

$$n^2 + n = 342$$
$$n^2 + n - 342 = 0$$
$$(n+19)(n-18) = 0$$
$$n + 19 = 0 \quad \text{or} \quad n - 18 = 0$$
$$n = -19 \quad \text{or} \quad n = 18$$

Since whole numbers must be positive the negative solution will not satisfy the condition. Thus, the numbers are 18 and $18 + 1 = 19$.

25. Let n, $n+2$ and $n+4$ represent the three consecutive odd numbers.

$$n^2 + (n+2)^2 + (n+4)^2 = 251$$

The sum of their squares is 251.

$$n^2 + n^2 + 4n + 4 + n^2 + 8n + 16 = 251$$
$$3n^2 + 12n + 20 = 251$$
$$3n^2 + 12n - 231 = 0$$
$$n^2 + 4n - 77 = 0$$
$$(n+11)(n-7) = 0$$
$$n + 11 = 0 \quad \text{or} \quad n - 7 = 0$$
$$n = -11 \quad \text{or} \quad n = 7$$

If $n = -11$, $n + 2 = -11 + 2 = -9$, and $n + 4 = 7 + 4 = -7$.

If $n = 7$, $n + 2 = 7 + 2 = 9$, and $n + 4 = 7 + 4 = 11$.

Since a negative number squared is positive, both groups of numbers will satisfy the condition. Thus, the numbers are -11, -9, and -7 or 7, 9, and 11.

26. Let x represent a side of the smaller square, then $3x$ represents a side of the larger square.

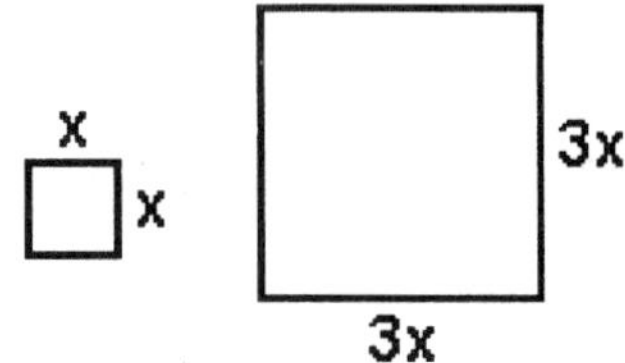

$x^2 + (3x)^2 = 50$ The combined area is 50 square meters.

$$x^2 + 9x^2 = 50$$
$$10x^2 = 50$$
$$x^2 = 5$$
$$x = \pm\sqrt{5}$$

Discard the negative solutions. The lengths of the sides are $\sqrt{5}$ meters and $3\sqrt{5}$ meters.

27. Let a represent the length of one leg of the triangle, then $a - 2$ represents the length of the other leg.

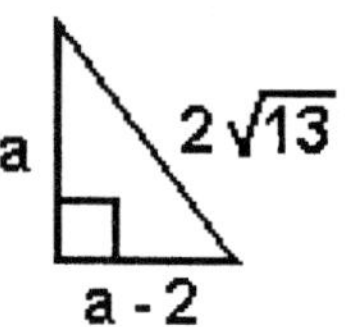

$$a^2 + (a-2)^2 = \left(2\sqrt{13}\right)^2$$

The Pythagorean Theorem.

$$a^2 + a^2 - 4a + 4 = 4(13)$$
$$2a^2 - 4a + 4 = 52$$
$$2a^2 - 4a - 48 = 0$$
$$a^2 - 2a - 24 = 0$$
$$(a-6)(a+4) = 0$$
$$a - 6 = 0 \quad \text{or} \quad a + 4 = 0$$
$$a = 6 \quad \text{or} \quad a = -4$$

Discard the negative solution. The lengths of the legs are 6 yards and $6 - 2 = 4$ yards.

28. Let n represent the number of shares purchased. Let v represent the purchase price per share. The $n - 20$ represents the number of shares sold and $v + 8$ represents the selling price per share.

$nv = 720$ The original purchase was \$720.

$(n-20)(v+8) = 800$

The sale price was \$720 + \$80 = \$800

Solve $nv = 720$ for v.

$$v = \frac{720}{n}$$

Substitute $\frac{720}{n}$ for v in the second equation.

$$(n-20)\left(\frac{720}{n}+8\right)=800$$
$$720+8n-\frac{14,400}{n}-160=800$$
$$720n+8n^2-14,400-160n=800n$$
$$8n^2+560n-14,400=800n$$
$$8n^2-240n-14,400=0$$
$$n^2-30n-1800=0$$
$$(n-60)(n+30)=0$$
$$n-60=0 \quad \text{or} \quad n+30=0$$
$$n=60 \quad \text{or} \quad n=-30$$

Discard the negative solution.

If $n=60$, then $v=\frac{720}{60}=12$.

The number of shares sold was $60-20=40$ shares at $12+8=\$20$ per share.

29. Let x represent the width of the strip to be added. Then $(40+x)$ and $(60+x)$ represent the dimensions of the parking lot after the addition. The original area is 2400 square meters.

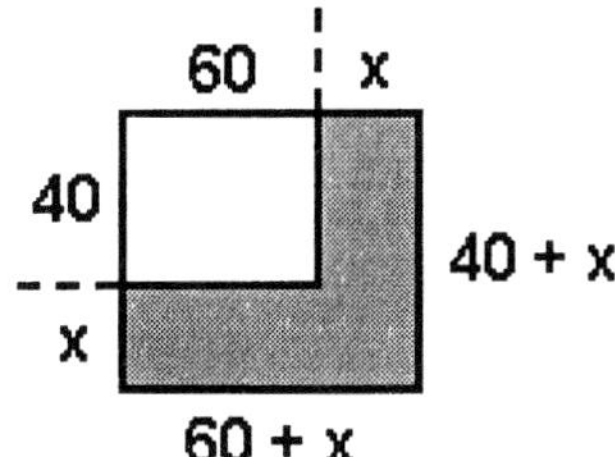

$$(40+x)(60+x)=2400+1100$$
$$2400+100x+x^2=3500$$
$$x^2+100x-1100=0$$
$$(x+110)(x-10)=0$$
$$x+110=0 \quad \text{or} \quad x-10=0$$
$$x=-110 \quad \text{or} \quad x=10$$

Discard the negative solution. The width of the strip would be 10 meters.

30. Let r represent Jean's rate, then $r-3$ represents Jay's rate. Let t represent Jean's time, then $t-2$ represents Jay's time.

$rt=336$ Jean's trip was 336 miles.

$(r-3)(t-2)=225$ Jay's trip was 225 miles.

Solve $rt=336$ for t.

$$t=\frac{336}{r}$$

Substitute $\frac{336}{r}$ for t in the second equation.

$$(r-3)\left(\frac{336}{r}-2\right)=225$$
$$336-2r-\frac{1008}{r}+6=225$$
$$336r-2r^2-1008+6r=225r$$
$$-2r^2+342r-1008=225r$$
$$-2r^2+117r-1008=0$$
$$2r^2-117r+1008=0$$
$$r=\frac{-(-117)\pm\sqrt{(-117)^2-4(2)(1008)}}{2(2)}$$
$$r=\frac{117\pm\sqrt{5625}}{4}$$
$$r=\frac{117\pm 75}{4}$$
$$r=\frac{117-75}{4} \quad \text{or} \quad r=\frac{117+75}{4}$$
$$r=\frac{42}{4}=\frac{21}{2} \quad \text{or} \quad r=\frac{192}{4}=48$$

If $r=\frac{21}{2}=10\frac{1}{2}$, then $r-3=7\frac{1}{2}$.

If $r=48$, then $r-3=45$.

If Jean's rate is $10\frac{1}{2}$ miles per hour, then Jay's rate is $7\frac{1}{2}$ miles per hour. Or, if Jean's rate is 48 miles per hour, then Jay's rate is 45 miles per hour.

31. Let a represent the length of the two equal legs.

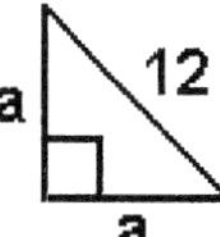

$a^2 + a^2 = 12^2$
The Pythagorean Theorem
$2a^2 = 144$
$a^2 = 72$
$a = \pm\sqrt{72} = \pm 6\sqrt{2}$
Discard the negative solution. The length of each leg is $6\sqrt{2}$ inches.

32. Let c represent the length of the hypotenuse. Then $\frac{1}{2}c$ represents the length of the other leg.

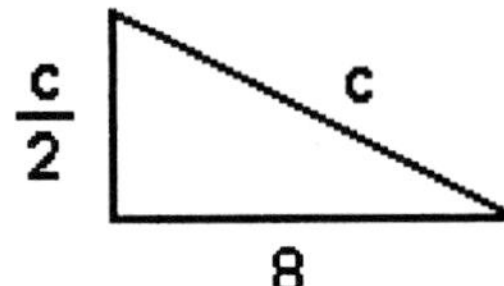

$8^2 + \left(\frac{1}{2}c\right)^2 = c^2$
The Pythagorean Theorem
$64 + \frac{1}{4}c^2 = c^2$
$256 + c^2 = 4c^2$
$-3c^2 = -256$
$c^2 = \frac{256}{3}$
$c = \pm\sqrt{\frac{256}{3}} = \pm\frac{16}{\sqrt{3}} = \pm\frac{16\sqrt{3}}{3}$
Discard the negative solution. The length of the hypotenuse is $\frac{16\sqrt{3}}{3}$ centimeters.

CHAPTER 10 Test

1. Let $a = 4$ and $b = 6$ in the Pythagorean Theorem.

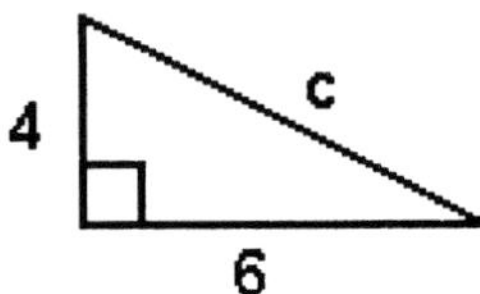

$c^2 = a^2 + b^2$
$c^2 = 4^2 + 6^2$
$c^2 = 16 + 36$
$c^2 = 52$
$c = 2\sqrt{13}$ inches

2. Let $c = 14$ and $a = 5$ in the Pythagorean Theorem.

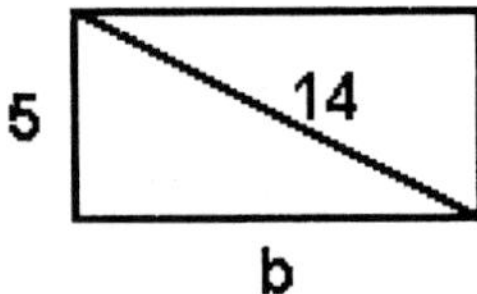

$c^2 = a^2 + b^2$
$14^2 = 5^2 + b^2$
$196 = 25 + b^2$
$171 = b^2$
$b = \sqrt{171} = 13$ to the nearest whole number.
The length of the rectangle is approximately 13 meters.

3. Let x represent a side of the square. Use the Pythagorean Theorem.

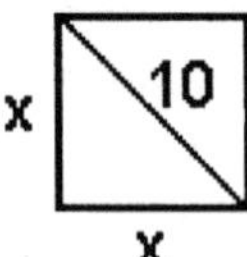

$x^2 + x^2 = 10^2$
$2x^2 = 100$
$x^2 = 50$
$x = \sqrt{50} = 7$ to the nearest whole number.
The length of a side of the square would be approximately 7 inches.

4. Let a represent the side opposite the 60° angle. The hypotenuse is 8 centimeters which is twice the side opposite the 30° angle.

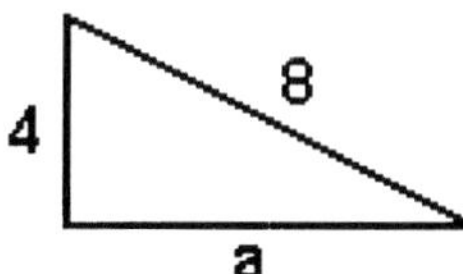

$$a^2 + 4^2 = 8^2$$
$$a^2 + 16 = 64$$
$$a^2 = 48$$
$$a = \sqrt{48} = 4\sqrt{3}$$

The length of a side opposite the 60° angle is $4\sqrt{3}$ centimeters.

5. $(3x+2)^2 = 49$

$$3x + 2 = \pm\sqrt{49} = \pm 7$$
$$3x + 2 = -7 \quad \text{or} \quad 3x + 2 = 7$$
$$3x = -9 \quad \text{or} \quad 3x = 5$$
$$x = -3 \quad \text{or} \quad x = \frac{5}{3}$$

The solution set is $\left\{-3, \frac{5}{3}\right\}$.

6. $4x^2 = 64$

$$x^2 = 16$$
$$x = \pm\sqrt{16} = \pm 4$$

The solution set is $\{-4, 4\}$.

7. $8x^2 - 10x + 3 = 0$

$$(2x-1)(4x-3) = 0$$
$$2x - 1 = 0 \quad \text{or} \quad 4x - 3 = 0$$
$$2x = 1 \quad \text{or} \quad 4x = 3$$
$$x = \frac{1}{2} \quad \text{or} \quad x = \frac{3}{4}$$

The solution set is $\left\{\frac{1}{2}, \frac{3}{4}\right\}$.

8. $x^2 - 3x - 5 = 0$

$$x = \frac{-(-3) \pm \sqrt{(-3)^2 - 4(1)(-5)}}{2(1)}$$
$$x = \frac{3 \pm \sqrt{29}}{2}$$

The solution set is $\left\{\frac{3-\sqrt{29}}{2}, \frac{3+\sqrt{29}}{2}\right\}$.

9. $n^2 + 2n = 9$

$$n^2 + 2n + 1 = 9 + 1$$
$$(n+1)^2 = 10$$
$$n + 1 = \pm\sqrt{10}$$
$$n + 1 = -\sqrt{10} \quad \text{or} \quad n + 1 = \sqrt{10}$$
$$n = -1 - \sqrt{10} \quad \text{or} \quad n = -1 + \sqrt{10}$$

The solution set is $\{-1-\sqrt{10}, -1+\sqrt{10}\}$.

10. $(2x-1)^2 = -16$

$$2x - 1 = \pm\sqrt{-16}$$

Since $\sqrt{-16}$ is not a real number, this equation has no real number solutions.

11. $y^2 + 10y = 24$

$$y^2 + 10y - 24 = 0$$
$$(y+12)(y-2) = 0$$
$$y + 12 = 0 \quad \text{or} \quad y - 2 = 0$$
$$y = -12 \quad \text{or} \quad y = 2$$

The solution set is $\{-12, 2\}$.

12. $2x^2 - 3x - 4 = 0$

$$x = \frac{-(-3) \pm \sqrt{(-3)^2 - 4(2)(-4)}}{2(2)}$$
$$x = \frac{3 \pm \sqrt{41}}{4}$$

The solution set is $\left\{\frac{3-\sqrt{41}}{4}, \frac{3+\sqrt{41}}{4}\right\}$.

13. $\frac{x-2}{3} = \frac{4}{x+1}$

Cross products are equal.

$(x-2)(x+1) = 3(4)$

$x^2 - x - 2 = 12$

$x^2 - x - 14 = 0$

$$x = \frac{-(-1) \pm \sqrt{(-1)^2 - 4(1)(-14)}}{2(1)}$$

$$x = \frac{1 \pm \sqrt{57}}{2}$$

The solution set is

$\left\{\frac{1-\sqrt{57}}{2}, \frac{1+\sqrt{57}}{2}\right\}$.

14. $\frac{2}{x-1} + \frac{1}{x} = \frac{5}{2}$

Multiply both sides by $2x(x-1)$.

$4x + 2(x-1) = 5x(x-1)$

$4x + 2x - 2 = 5x^2 - 5x$

$6x - 2 = 5x^2 - 5x$

$0 = 5x^2 - 11x + 2$

$0 = (5x-1)(x-2)$

$5x - 1 = 0$ or $x - 2 = 0$

$5x = 1$ or $x = 2$

$x = \frac{1}{5}$ or $x = 2$

The solution set is $\left\{\frac{1}{5}, 2\right\}$.

15. $n(n-28) = -195$

$n^2 - 28n = -195$

$n^2 - 28n + 195 = 0$

$(n-15)(n-13) = 0$

$n - 15 = 0$ or $n - 13 = 0$

$n = 15$ or $n = 13$

The solution set is $\{13, 15\}$.

16. $n + \frac{3}{n} = \frac{19}{4}$

Multiply both sides by $4n$.

$4n^2 + 12 = 19n$

$4n^2 - 19n + 12 = 0$

$(4n-3)(n-4) = 0$

$4n - 3 = 0$ or $n - 4 = 0$

$4n = 3$ or $n = 4$

$n = \frac{3}{4}$ or $n = 4$

The solution set is $\left\{\frac{3}{4}, 4\right\}$.

17. $(2x+1)(3x-2) = -2$

$6x^2 - x - 2 = -2$

$6x^2 - x = 0$

$x(6x-1) = 0$

$x = 0$ or $6x - 1 = 0$

$x = 0$ or $6x = 1$

$x = 0$ or $x = \frac{1}{6}$

The solution set is $\left\{0, \frac{1}{6}\right\}$.

18. $(7x+2)^2 - 4 = 21$

$(7x+2)^2 = 25$

$7x + 2 = \pm\sqrt{25} = \pm 5$

$7x + 2 = -5$ or $7x + 2 = 5$

$7x = -7$ or $7x = 3$

$x = -1$ or $x = \frac{3}{7}$

The solution set is $\left\{-1, \frac{3}{7}\right\}$.

19. $(4x-1)^2 = 27$

$4x - 1 = \pm\sqrt{27} = \pm 3\sqrt{3}$

$4x - 1 = -3\sqrt{3}$ or $4x - 1 = 3\sqrt{3}$

$4x = 1 - 3\sqrt{3}$ or $4x = 1 + 3\sqrt{3}$

$x = \frac{1-3\sqrt{3}}{4}$ or $x = \frac{1+3\sqrt{3}}{4}$

The solution set is

$\left\{\frac{1-3\sqrt{3}}{4}, \frac{1+3\sqrt{3}}{4}\right\}$.

20. $n^2 - 5n + 7 = 0$

$$n = \frac{-(-5) \pm \sqrt{(-5)^2 - 4(1)(7)}}{2(1)}$$

$$n = \frac{5 \pm \sqrt{-3}}{2}$$

Since $\sqrt{-3}$ is not a real number, this equation has no real number solutions.

21. Let r represent the number of rows, then $2r-1$ represents the number of seats per row.

$r(2r-1)=120$ The product of the seats per row and the number of rows is 120.

$$2r^2-r=120$$
$$2r^2-r-120=0$$
$$(2r+15)(r-8)=0$$
$$2r+15=0 \quad \text{or} \quad r-8=0$$
$$2r=-15 \quad \text{or} \quad r=8$$
$$r=-\frac{15}{2} \quad \text{or} \quad r=8$$

Discard the negative solution. There are $2r-1=2(8)-1=15$ seats per row.

22. Let r represent Stan's rate, then $r+2$ represents Abu's rate. Let t represent Stan's time, then $t-2$ represents Abu's time.

$rt=72$ Stan's trip was 72 miles.

$(r+2)(t-2)=56$ Abu's trip was 56 miles.

Solve $rt=72$ for t.

$$t=\frac{72}{r}$$

Substitute $\frac{72}{r}$ for t in the second equation.

$$(r+2)\left(\frac{72}{r}-2\right)=56$$
$$72-2r+\frac{144}{r}-4=56$$
$$72r-2r^2+144-4r=56r$$
$$-2r^2+68r+144=56r$$
$$-2r^2+12r+144=0$$
$$r^2-6r-72=0$$
$$(r-12)(r+6)=0$$
$$r-12=0 \quad \text{or} \quad r+6=0$$
$$r=12 \quad \text{or} \quad r=-6$$

If $r=12$, then $r+2=14$.
Abu's rate is 14 miles per hour.

23. Let n and $n+2$ represent the two consecutive odd whole numbers.

$n(n+2)=255$ Their product is 255.

$$n^2+2n=255$$
$$n^2+2n-255=0$$
$$(n+17)(n-15)=0$$
$$n+17=0 \quad \text{or} \quad n-15=0$$
$$n=-17 \quad \text{or} \quad n=15$$

If $n=15$, $n+2=15+2=17$.
Discard the negative solution.
The numbers are 15 and 17.

24. Let x represent a side of the smaller square, then $2x+1$ represents a side of the larger square.

$$x^2+(2x+1)^2=97$$

The combined area is 97 square feet.

$$x^2+4x^2+4x+1=97$$
$$5x^2+4x+1=97$$
$$5x^2+4x-96=0$$
$$(5x+24)(x-4)=0$$
$$5x+24=0 \quad \text{or} \quad x-4=0$$
$$5x=-24 \quad \text{or} \quad x=4$$
$$x=-\frac{24}{5} \quad \text{or} \quad x=4$$

Discard the negative solution. The lengths of the sides of the larger square are $2(4)+1=9$ feet.

25. Let n represent the number of shares of stock she purchased. Then $n-4$ represents the number of shares she sold. Let x represent the original price of a single share, then $x+2$ represents the selling price per share.

$nx=160$ The original purchase.

$(x+2)(n-4)=160$ The sale of the stock.

Solve $nx=160$ for x.

$$nx=160$$
$$x=\frac{160}{n}$$

Substitute $\frac{160}{n}$ for x in the second equation.

$$\left(\frac{160}{n}+2\right)(n-4)=160$$
$$160-\frac{640}{n}+2n-8=160$$
$$160n-640+2n^2-8n=160n$$
$$2n^2+152n-640=160n$$
$$2n^2-8n-640=0$$
$$n^2-4n-320=0$$
$$(n+16)(n-20)=0$$
$$n+16=0 \quad \text{or} \quad n-20=0$$
$$n=-16 \quad \text{or} \quad n=20$$

Discard the negative solution. The original purchase was 20 shares.

Chapter 11 Additional Topics

PROBLEM SET 11.1 Equations and Inequalities Involving Absolute Value

In Problems 1 – 25, *see the text for a graph of the solution.*

1. $|x| = 4$ is equivalent to $x = -4$ or $x = 4$. The solution set is $\{-4, 4\}$.

3. $|x| < 1$ means that x must be less than one unit away from zero. Therefore, $|x| < 1$ is equivalent to $x > -1$ and $x < 1$. The solution set is $\{x|x > -1 \text{ and } x < 1\}$ or $(-1, 1)$.

5. $|x| \geq 2$ means that x must be equal to or more than two unit away from zero. Therefore, $|x| \geq 2$ is equivalent to $x \leq -2$ or $x \geq 2$. The solution set is $\{x|x \leq -2 \text{ or x} \geq 2\}$ or $(-\infty, -2] \cup [2, \infty)$.

7. $|x+2| = 1$ is equivalent to $x + 2 = -1$ or $x + 2 = 1$.

$x + 2 = -1$ or $x + 2 = 1$

$x = -3$ or $x = -1$

The solution set is $\{-3, -1\}$.

9. $|x-1| = 2$ is equivalent to $x - 1 = -2$ or $x - 1 = 2$.

$x - 1 = -2$ or $x - 1 = 2$

$x = -1$ or $x = 3$

The solution set is $\{-1, 3\}$.

11. $|x-2| \leq 2$ is equivalent to $x - 2 \geq -2$ and $x - 2 \leq 2$.

$x - 2 \geq -2$ and $x - 2 \leq 2$

$x \geq 0$ and $x \leq 4$

The solution set is $\{x|x \geq 0 \text{ and } x \leq 4\}$ or $[0, 4]$.

13. $|x+1| > 3$ is equivalent to $x + 1 < -3$ or $x + 1 > 3$.

$x + 1 < -3$ or $x + 1 > 3$

$x < -4$ or $x > 2$

The solution set is $\{x|x < -4 \text{ or } x > 2\}$ or $(-\infty, -4) \cup (2, \infty)$.

15. $|2x+1| = 3$ is equivalent to $2x + 1 = -3$ or $2x + 1 = 3$.

$2x + 1 = -3$ or $2x + 1 = 3$

$2x = -4$ or $2x = 2$

$x = -2$ or $x = 1$

The solution set is $\{-2, 1\}$.

17. $|5x-2| = 4$ is equivalent to $5x - 2 = -4$ or $5x - 2 = 4$.

$5x - 2 = -4$ or $5x - 2 = 4$

$5x = -2$ or $5x = 6$

$x = -\frac{2}{5}$ or $x = \frac{6}{5}$

The solution set is $\left\{-\frac{2}{5}, \frac{6}{5}\right\}$.

19. $|2x-3| \geq 1$ is equivalent to $2x - 3 \leq -1$ or $2x - 3 \geq 1$.

$2x - 3 \leq -1$ or $2x - 3 \geq 1$

$2x \leq 2$ or $2x \geq 4$

$x \leq 1$ or $x \geq 2$

The solution set is $\{x|x \leq 1 \text{ or } x \geq 2\}$ or $(-\infty, 1] \cup [2, \infty)$.

21. $|4x+3| < 2$ is equivalent to $4x + 3 > -2$ and $4x + 3 < 2$.

$4x + 3 > -2$ and $4x + 3 < 2$

$4x > -5$ and $4x < -1$

$x > -\frac{5}{4}$ and $x < -\frac{1}{4}$

The solution set is

$\left\{x|x > -\frac{5}{4} \text{ and } x < -\frac{1}{4}\right\}$ or $\left(-\frac{5}{4}, -\frac{1}{4}\right)$.

23. $|3x+6|=0$ is equivalent to $3x+6=0$.
$$3x+6=0$$
$$3x=-6$$
$$x=-2$$
The solution set is $\{-2\}$.

25. $|3x-2|>0$ is equivalent to $3x-2\neq 0$.
$$3x-2\neq 0$$
$$3x\neq 2$$
$$x\neq \frac{2}{3}$$
The solution set is
$\left\{x|x\neq \frac{2}{3}\right\}$ or $\left(-\infty,\ \frac{2}{3}\right)\cup\left(\frac{2}{3},\ \infty\right)$.

27. $|3x-1|=17$ is equivalent to
$3x-1=-17$ or $3x-1=17$.
$$3x-1=-17 \quad \text{or} \quad 3x-1=17$$
$$3x=-16 \quad \text{or} \quad 3x=18$$
$$x=-\frac{16}{3} \quad \text{or} \quad x=6$$
The solution set is $\left\{-\frac{16}{3},\ 6\right\}$.

29. $|2x+1|>9$ is equivalent to
$2x+1<-9$ or $2x+1>9$.
$$2x+1<-9 \quad \text{or} \quad 2x+1>9$$
$$2x<-10 \quad \text{or} \quad 2x>8$$
$$x<-5 \quad \text{or} \quad x>4$$
The solution set is
$\{x|x<-5 \text{ or } x>4\}$ or
$(-\infty,\ -5)\cup(4,\ \infty)$.

31. $|3x-5|<19$ is equivalent to
$3x-5>-19$ and $3x-5<19$.
$$3x-5>-19 \quad \text{and} \quad 3x-5<19$$
$$3x>-14 \quad \text{and} \quad 3x<24$$
$$x>-\frac{14}{3} \quad \text{and} \quad x<8$$
The solution set is
$\left\{x|x>-\frac{14}{3} \text{ and } x<8\right\}$
or $\left(-\frac{14}{3},\ 8\right)$.

33. $|-3x-1|=17$ is equivalent to
$-3x-1=-17$ or $-3x-1=17$.
$$-3x-1=-17 \quad \text{or} \quad -3x-1=17$$
$$-3x=-16 \quad \text{or} \quad -3x=18$$
$$x=\frac{16}{3} \quad \text{or} \quad x=-6$$
The solution set is $\left\{-6,\ \frac{16}{3}\right\}$.

35. $|4x-7|\leq 31$ is equivalent to
$4x-7\geq -31$ and $4x-7\leq 31$.
$$4x-7\geq -31 \quad \text{and} \quad 4x-7\leq 31$$
$$4x\geq -24 \quad \text{and} \quad 4x\leq 38$$
$$x\geq -6 \quad \text{and} \quad x\leq \frac{38}{4}=\frac{19}{2}$$
The solution set is
$\left\{x|x\geq -6 \text{ and } x\leq \frac{19}{2}\right\}$ or $\left[-6,\ \frac{19}{2}\right]$.

37. $|5x+3|\geq 18$ is equivalent to
$5x+3\leq -18$ or $5x+3\geq 18$.
$$5x+3\leq -18 \quad \text{or} \quad 5x+3\geq 18$$
$$5x\leq -21 \quad \text{or} \quad 5x\geq 15$$
$$x\leq -\frac{21}{5} \quad \text{or} \quad x\geq 3$$
The solution set is $\left\{x|x\leq -\frac{21}{5} \text{ or } x\geq 3\right\}$
or $\left(-\infty,\ -\frac{21}{5}\right]\cup[3,\ \infty)$.

39. $|-x-2|<4$ is equivalent to
$-x-2>-4$ and $-x-2<4$.
$$-x-2>-4 \quad \text{and} \quad -x-2<4$$
$$-x>-2 \quad \text{and} \quad -x<6$$
$$x<2 \quad \text{and} \quad x>-6$$
The solution set is
$\{x|x>-6 \text{ and } x<2\}$ or $(-6,\ 2)$.

41. $|-2x+1|>6$ is equivalent to
$-2x+1<-6$ or $-2x+1>6$.
$$-2x+1<-6 \quad \text{or} \quad -2x+1>6$$
$$-2x<-7 \quad \text{or} \quad -2x>5$$
$$x>\frac{7}{2} \quad \text{or} \quad x<-\frac{5}{2}$$
The solution set is $\left\{x|x<-\frac{5}{2} \text{ or } x>\frac{7}{2}\right\}$
or $\left(-\infty,\ -\frac{5}{2}\right)\cup\left(\frac{7}{2},\ \infty\right)$.

43. $|7x| = 0$ has only one solution since absolute value is defined to be the distance between the number and zero on a number line. In this case the distance is zero. The solution set is $\{0\}$.

45. $|x-6| > -4$ will be true for all real numbers since the absolute value of $x-6$, regardless of what number is substituted for x, will always be greater than -4. The solution set is $\{x|x$ is a real number$\}$.

47. $|x+4| < -7$ has no solutions since we cannot obtain an absolute value less than -7. The solution set is $\emptyset$.

49. $|x+6| \leq 0$ has only one solution since absolute value is defined to be the distance between the number and zero on a number line. In this case the distance is zero and the x would be -6. The solution set is $\{-6\}$.

PROBLEM SET 11.2 Special Functions

In this section, a few points are given to help in graphing the function. See the text for a graph of the equation.

1. $f(x) = 4$
Points $f(0) = 4$, $f(2) = 4$, and $f(-2) = 4$ are on the horizontal line.

3. $f(x) = -2$
Points $f(0) = -2$, $f(2) = -2$, and $f(-2) = -2$ are on the horizontal line.

5. $f(x) = \frac{1}{2}$
Points $f(0) = \frac{1}{2}$, $f(2) = \frac{1}{2}$, and $f(-2) = \frac{1}{2}$ are on the horizontal line.

7. $f(x) = 3x + 2$
The points $(0, 2)$, $(1, 5)$, and $(-1, -1)$ are on the straight line.

9. $f(x) = 2x - 5$
The points $(0, -5)$, $(1, -3)$, and $(-1, -7)$ are on the straight line.

11. $f(x) = -4x + 1$
The points $(0, 1)$, $(1, -3)$, and $(-1, 5)$ are on the straight line.

13. $f(x) = -5x - 1$
The points $(0, -1)$, $(1, -6)$, and $(-1, 4)$ are on the straight line.

15. $f(x) = 3x$
The points $(0, 0)$, $(1, 3)$, and $(-1, -3)$ are on the straight line.

17. $f(x) = -x$
The points $(0, 0)$, $(1, -1)$, and $(-1, 1)$ are on the straight line.

19. $f(x) = 3x^2$
The points $(0, 0)$, $(1, 3)$, and $(-1, 3)$ are on this parabola that opens upward with its vertex at the origin.

21. $f(x) = \frac{1}{4}x^2$
The points $(0, 0)$, $(4, 4)$, and $(-4, 4)$ are on this parabola that opens upward with its vertex at the origin.

23. $f(x) = -3x^2$
The points $(0, 0)$, $(1, -3)$, and $(-1, -3)$ are on this parabola that opens downward with its vertex at the origin.

25. $f(x) = -\frac{1}{4}x^2$
The points $(0, 0)$, $(4, -4)$, and $(-4, -4)$ are on this parabola that opens downward with its vertex at the origin.

PROBLEM SET 11.3 Factional Exponents

1. $\sqrt{81} = 9$ because $9^2 = 81$.

3. $-\sqrt{100} = -10$ because $-(10^2) = -100$.

5. $\sqrt[3]{125} = 5$ because $(5)^3 = 125$

7. $\sqrt[3]{-64} = -4$ because $(-4)^3 = -64$.

9. $\frac{\sqrt[3]{64}}{\sqrt{49}} = \frac{4}{7}$ because $\frac{4^3}{7^2} = \frac{64}{49}$.

11. $\sqrt[4]{81} = 3$ because $3^4 = 81$.

13. $\sqrt[5]{-243} = -3$ because $(-3)^5 = -243$.

15. $64^{\frac{1}{2}} = \sqrt{64} = 8$

17. $64^{\frac{2}{3}} = \left(\sqrt[3]{64}\right)^2 = 4^2 = 16$

19. $(-64)^{\frac{2}{3}} = \left(\sqrt[3]{-64}\right)^2 = (-4)^2 = 16$

21. $4^{\frac{5}{2}} = \left(\sqrt{4}\right)^5 = 2^5 = 32$

23. $32^{-\frac{1}{5}} = \left(\sqrt[5]{32}\right)^{-1} = 2^{-1} = \frac{1}{2}$

25. $-27^{\frac{1}{3}} = -\sqrt[3]{27} = -3$

27. $16^{-\frac{3}{4}} = \left(\sqrt[4]{16}\right)^{-3} = 2^{-3} = \frac{1}{2^3} = \frac{1}{8}$

29. $\left(\frac{2}{3}\right)^{-3} = \frac{1}{\left(\frac{2}{3}\right)^3} = \frac{1}{\frac{8}{27}} = \frac{27}{8}$

31. $\left(\frac{16}{64}\right)^{-\frac{1}{2}} = \left(\frac{1}{4}\right)^{-\frac{1}{2}} = 4^{\frac{1}{2}} = \sqrt{4} = 2$

33. $125^{\frac{4}{3}} = \left(\sqrt[3]{125}\right)^4 = 5^4 = 625$

35. $-16^{\frac{5}{4}} = -\left(\sqrt[4]{16}\right)^5 = -(2)^5 = -32$

37. $\left(\frac{1}{32}\right)^{\frac{3}{5}} = \left(\sqrt[5]{\frac{1}{32}}\right)^3 = \left(\frac{1}{2}\right)^3 = \frac{1}{8}$

39. $2^{\frac{1}{3}} \cdot 2^{\frac{2}{3}} = 2^{\frac{1}{3}+\frac{2}{3}} = 2^{\frac{3}{3}} = 2^1 = 2$

41. $3^{\frac{4}{3}} \cdot 3^{\frac{5}{3}} = 3^{\frac{4}{3}+\frac{5}{3}} = 3^{\frac{9}{3}} = 3^3 = 27$

43. $\frac{2^{\frac{1}{2}}}{2^{\frac{1}{2}}} = 1$

45. $\frac{3^{-\frac{2}{3}}}{3^{\frac{1}{3}}} = 3^{-\frac{1}{3}-\frac{2}{3}} = 3^{-\frac{3}{3}} = 3^{-1} = \frac{1}{3}$

47. $\frac{2^{\frac{9}{4}}}{2^{\frac{1}{4}}} = 2^{\frac{9}{4}-\frac{1}{4}} = 2^{\frac{8}{4}} = 2^2 = 4$

49. $\frac{7^{\frac{4}{3}}}{7^{-\frac{2}{3}}} = 7^{\frac{4}{3}-\left(-\frac{2}{3}\right)} = 7^{\frac{4}{3}+\frac{2}{3}} = 7^{\frac{6}{3}} = 7^2 = 49$

51. $x^{\frac{1}{2}} \cdot x^{\frac{1}{4}} = x^{\frac{1}{2}+\frac{1}{4}} = x^{\frac{2}{4}+\frac{1}{4}} = x^{\frac{3}{4}}$

53. $a^{\frac{2}{3}} \cdot a^{\frac{3}{4}} = a^{\frac{2}{3}+\frac{3}{4}} = a^{\frac{8}{12}+\frac{9}{12}} = a^{\frac{17}{12}}$

55. $\left(3x^{\frac{1}{4}}\right)\left(5x^{\frac{1}{3}}\right)=3\cdot 5\cdot x^{\frac{1}{4}}\cdot x^{\frac{1}{3}}=$
$15x^{\frac{1}{4}+\frac{1}{3}}=15x^{\frac{3}{12}+\frac{4}{12}}=15x^{\frac{7}{12}}$

57. $\left(4x^{\frac{2}{3}}\right)\left(6x^{\frac{1}{4}}\right)=4\cdot 6\cdot x^{\frac{2}{3}}\cdot x^{\frac{1}{4}}=$
$24x^{\frac{2}{3}+\frac{1}{4}}=24x^{\frac{8}{12}+\frac{3}{12}}=24x^{\frac{11}{12}}$

59. $\left(2y^{\frac{2}{3}}\right)\left(y^{-\frac{1}{4}}\right)=2\cdot y^{\frac{2}{3}}\cdot y^{-\frac{1}{4}}=$
$2y^{\frac{2}{3}-\frac{1}{4}}=2y^{\frac{8}{12}-\frac{3}{12}}=2y^{\frac{5}{12}}$

61. $\left(5n^{\frac{3}{4}}\right)\left(2n^{-\frac{1}{2}}\right)=5\cdot 2\cdot n^{\frac{3}{4}}\cdot n^{-\frac{1}{2}}=$
$10n^{\frac{3}{4}-\frac{1}{2}}=10n^{\frac{3}{4}-\frac{2}{4}}=10n^{\frac{1}{4}}$

63. $\left(2x^{\frac{1}{3}}\right)\left(x^{-\frac{1}{2}}\right)=2\cdot x^{\frac{1}{3}}\cdot x^{-\frac{1}{2}}=$
$2x^{\frac{1}{3}-\frac{1}{2}}=2x^{\frac{2}{6}-\frac{3}{6}}=2x^{-\frac{1}{6}}=\dfrac{2}{x^{\frac{1}{6}}}$

65. $\left(5x^{\frac{1}{2}}y\right)^2=(5)^2\left(x^{\frac{1}{2}}\right)^2(y)^2=25xy^2$

67. $\left(4x^{\frac{1}{4}}y^{\frac{1}{2}}\right)^3=(4)^3\left(x^{\frac{1}{4}}\right)^3\left(y^{\frac{1}{2}}\right)^3=64x^{\frac{3}{4}}y^{\frac{3}{2}}$

69. $(8x^6y^3)^{\frac{1}{3}}=(8)^{\frac{1}{3}}(x^6)^{\frac{1}{3}}(y^3)^{\frac{1}{3}}=$
$\sqrt[3]{8}x^{\frac{6}{3}}y^{\frac{3}{3}}=2x^2y$

71. $\dfrac{24x^{\frac{3}{5}}}{6x^{\frac{1}{3}}}=\dfrac{24}{6}x^{\frac{3}{5}-\frac{1}{3}}=4x^{\frac{9}{15}-\frac{5}{15}}=4x^{\frac{4}{15}}$

73. $\dfrac{48b^{\frac{1}{3}}}{12b^{\frac{3}{4}}}=\dfrac{48}{12}b^{\frac{1}{3}-\frac{3}{4}}=4b^{\frac{4}{12}-\frac{9}{12}}=4b^{-\frac{5}{12}}=\dfrac{4}{b^{\frac{5}{12}}}$

75. $\dfrac{27n^{-\frac{1}{3}}}{9n^{-\frac{1}{3}}}=\dfrac{27}{9}(1)=3$

77. $\left[\dfrac{3x^{\frac{1}{3}}}{2x^{\frac{1}{2}}}\right]^2=\dfrac{(3)^2\left(x^{\frac{1}{3}}\right)^2}{(2)^2\left(x^{\frac{1}{2}}\right)^2}=\dfrac{9x^{\frac{2}{3}}}{4x^{\frac{2}{2}}}=$
$\dfrac{9}{4}x^{\frac{2}{3}-1}=\dfrac{9}{4}x^{-\frac{1}{3}}=\dfrac{9}{4x^{\frac{1}{3}}}$

79. $\left[\dfrac{5x^{\frac{1}{2}}}{6y^{\frac{1}{3}}}\right]^3=\dfrac{(5)^3\left(x^{\frac{1}{2}}\right)^3}{(6)^3\left(y^{\frac{1}{3}}\right)^3}=\dfrac{125x^{\frac{3}{2}}}{216y^{\frac{3}{3}}}=\dfrac{125x^{\frac{3}{2}}}{216y}$

PROBLEM SET 11.4 Complex Numbers

1. $\sqrt{-64}=i\sqrt{64}=8i$

3. $\sqrt{-\dfrac{25}{9}}=i\sqrt{\dfrac{25}{9}}=\dfrac{5}{3}i$

5. $\sqrt{-11}=i\sqrt{11}$

7. $\sqrt{-50}=i\sqrt{25}\sqrt{2}=5i\sqrt{2}$

9. $\sqrt{-48}=i\sqrt{16}\sqrt{3}=4i\sqrt{3}$

11. $\sqrt{-54}=i\sqrt{9}\sqrt{6}=3i\sqrt{6}$

13. $(3+8i)+(5+9i)=$
$(3+5)+(8+9)i=$
$8+17i$

15. $(7-6i)+(3-4i)=$
$(7+3)+(-6-4)i=$
$10-10i$

17. $(10+4i)-(6+2i)=$
$(10-6)+(4-2)i=$
$4+2i$

19. $(5+2i)-(7+8i)=$
$(5-7)+(2-8)i=$
$-2-6i$

21. $(-2-i)-(3-4i)=$
$(-2-3)+(-1+4)i=$
$-5+3i$

23. $(-4-7i)+(-8-9i)=$
$(-4-8)+(-7-9)i=$
$-12-16i$

25. $(0-6i)+(-10+2i)=$
$(0-10)+(-6+2)i=$
$-10-4i$

27. $(-9+7i)-(-8-5i)=$
$(-9+8)+(7+5)i=$
$-1+12i$

29. $(-10-4i)-(10+4i)=$
$(-10-10)+(-4-4)i=$
$-20-8i$

31. $\left(\frac{1}{2}+\frac{2}{3}i\right)+\left(\frac{1}{3}-\frac{1}{4}i\right)=$
$\left(\frac{1}{2}+\frac{1}{3}\right)+\left(\frac{2}{3}-\frac{1}{4}\right)i=$
$\left(\frac{3}{6}+\frac{2}{6}\right)+\left(\frac{8}{12}-\frac{3}{12}\right)i=$
$\frac{5}{6}+\frac{5}{12}i$

33. $\left(\frac{3}{5}-\frac{1}{4}i\right)-\left(\frac{2}{3}-\frac{5}{6}i\right)=$
$\left(\frac{3}{5}-\frac{2}{3}\right)+\left(-\frac{1}{4}+\frac{5}{6}\right)i=$
$\left(\frac{9}{15}-\frac{10}{15}\right)+\left(-\frac{3}{12}+\frac{10}{12}\right)i=$
$-\frac{1}{15}+\frac{7}{12}i$

35. $(7i)(8i)=56i^2=$
$56(-1)=-56+0i$

37. $2i(6+3i)=12i+6i^2=$
$12i+6(-1)=12i-6=$
$-6+12i$

39. $-4i(-5-6i)=20i+24i^2=$
$20i+24(-1)=-24+20i$

41. $(2+3i)(5+4i)=$
$10+8i+15i+12i^2=$
$10+23i+12(-1)=$
$-2+23i$

43. $(7-3i)(8+i)=$
$56+7i-24i-3i^2=$
$56-17i-3(-1)=$
$59-17i$

45. $(-2-3i)(6-3i)=$
$-12+6i-18i+9i^2=$
$-12-12i+9(-1)=$
$-21-12i$

47. $(-1-4i)(-2-7i)=$
$2+7i+8i+28i^2=$
$2+15i+28(-1)=$
$-26+15i$

49. $(4+5i)^2=$
$(4+5i)(4+5i)=$
$16+20i+20i+25i^2=$
$16+40i+25(-1)=$
$-9+40i$

51. $(5-6i)(5+6i)=$
$25-30i+30i-36i^2=$
$25-36(-1)=$
$61+0i$

53. $(-2+i)(-2-i)=$
$4+2i-2i-i^2=$
$4-(-1)=$
$5+0i$

PROBLEM SET 11.5 Complex Solutions of Quadratic Equations

1. $x^2 = -64$

$x = \pm\sqrt{-64}$

$x = \pm i\sqrt{64} = \pm 8i$

The solution set is $\{-8i, 8i\}$.

3. $(x-2)^2 = -1$

$x - 2 = \pm\sqrt{-1} = \pm i$

$x - 2 = -i$ or $x - 2 = i$

$x = 2 - i$ or $x = 2 + i$

The solution set is $\{2 - i, 2 + i\}$.

5. $(x+5)^2 = -13$

$x + 5 = \pm\sqrt{-13} = \pm i\sqrt{13}$

$x + 5 = -i\sqrt{13}$ or $x + 5 = i\sqrt{13}$

$x = -5 - i\sqrt{13}$ or $x = -5 + i\sqrt{13}$

The solution set is

$\left\{-5 - i\sqrt{13}, -5 + i\sqrt{13}\right\}$.

7. $(x-3)^2 = -18$

$x - 3 = \pm\sqrt{-18} = \pm 3i\sqrt{2}$

$x - 3 = -3i\sqrt{2}$ or $x - 3 = 3i\sqrt{2}$

$x = 3 - 3i\sqrt{2}$ or $x = 3 + 3i\sqrt{2}$

The solution set is

$\left\{3 - 3i\sqrt{2}, 3 + 3i\sqrt{2}\right\}$.

9. $(5x-1)^2 = 9$

$5x - 1 = \pm\sqrt{9} = \pm 3$

$5x - 1 = -3$ or $5x - 1 = 3$

$5x = -2$ or $5x = 4$

$x = -\frac{2}{5}$ or $x = \frac{4}{5}$

The solution set is $\left\{-\frac{2}{5}, \frac{4}{5}\right\}$.

11. $a^2 - 3a - 4 = 0$

$(a+1)(a-4) = 0$

$a + 1 = 0$ or $a - 4 = 0$

$a = -1$ or $a = 4$

The solution set is $\{-1, 4\}$.

13. $t^2 + 6t = -12$

$t^2 + 6t + 9 = -12 + 9$

$(t+3)^2 = -3$

$t + 3 = \pm\sqrt{-3} = \pm i\sqrt{3}$

$t + 3 = -i\sqrt{3}$ or $t + 3 = i\sqrt{3}$

$t = -3 - i\sqrt{3}$ or $t = -3 + i\sqrt{3}$

The solution set is

$\left\{-3 - i\sqrt{3}, -3 + i\sqrt{3}\right\}$.

15. $n^2 - 6n + 13 = 0$

$$n = \frac{-(-6) \pm \sqrt{(-6)^2 - 4(1)(13)}}{2(1)}$$

$$n = \frac{6 \pm \sqrt{-16}}{2}$$

$$n = \frac{6 \pm 4i}{2} = 3 \pm 2i$$

The solution set is $\{3 - 2i, 3 + 2i\}$.

17. $x^2 - 4x + 20 = 0$

$$x = \frac{-(-4) \pm \sqrt{(-4)^2 - 4(1)(20)}}{2(1)}$$

$$x = \frac{4 \pm \sqrt{-64}}{2}$$

$$x = \frac{4 \pm 8i}{2} = 2 \pm 4i$$

The solution set is $\{2 - 4i, 2 + 4i\}$.

Problem Set 11.5

19. $3x^2 - 2x + 1 = 0$

$$x = \frac{-(-2) \pm \sqrt{(-2)^2 - 4(3)(1)}}{2(3)}$$

$$x = \frac{2 \pm \sqrt{-8}}{6}$$

$$x = \frac{2 \pm 2i\sqrt{2}}{6} = \frac{1 \pm i\sqrt{2}}{3}$$

The solution set is

$$\left\{\frac{1 - i\sqrt{2}}{3}, \frac{1 + i\sqrt{2}}{3}\right\}.$$

21. $2x^2 - 3x - 5 = 0$

$(2x - 5)(x + 1) = 0$

$2x - 5 = 0$ or $x + 1 = 0$

$2x = 5$ or $x = -1$

$x = \frac{5}{2}$ or $x = -1$

The solution set is $\left\{-1, \frac{5}{2}\right\}$.

23. $y^2 - 2y = -19$

$y^2 - 2y + 1 = -19 + 1$

$(y - 1)^2 = -18$

$y - 1 = \pm\sqrt{-18} = \pm 3i\sqrt{2}$

$y - 1 = -3i\sqrt{2}$ or $y - 1 = 3i\sqrt{2}$

$y = 1 - 3i\sqrt{2}$ or $y = 1 + 3i\sqrt{2}$

The solution set is

$\left\{1 - 3i\sqrt{2}, 1 + 3i\sqrt{2}\right\}$.

25. $x^2 - 4x + 7 = 0$

$$x = \frac{-(-4) \pm \sqrt{(-4)^2 - 4(1)(7)}}{2(1)}$$

$$x = \frac{4 \pm \sqrt{-12}}{2}$$

$$x = \frac{4 \pm 2i\sqrt{3}}{2} = 2 \pm i\sqrt{3}$$

The solution set is $\left\{2 - i\sqrt{3}, 2 + i\sqrt{3}\right\}$.

27. $4x^2 - x + 2 = 0$

$$x = \frac{-(-1) \pm \sqrt{(-1)^2 - 4(4)(2)}}{2(4)}$$

$$x = \frac{1 \pm \sqrt{-31}}{8}$$

$$x = \frac{1 \pm i\sqrt{31}}{8}$$

The solution set is

$$\left\{\frac{1 - i\sqrt{31}}{8}, \frac{1 + i\sqrt{31}}{8}\right\}.$$

29. $6x^2 + 2x + 1 = 0$

$$x = \frac{-(2) \pm \sqrt{(2)^2 - 4(6)(1)}}{2(6)}$$

$$x = \frac{-2 \pm \sqrt{-20}}{12}$$

$$x = \frac{-2 \pm 2i\sqrt{5}}{12} = \frac{-1 \pm i\sqrt{5}}{6}$$

The solution set is

$$\left\{\frac{-1 - i\sqrt{5}}{6}, \frac{-1 + i\sqrt{5}}{6}\right\}.$$

CHAPTER 11 Review Problem Set

1. $\sqrt{-64} = i\sqrt{64} = 8i$

2. $\sqrt{-29} = i\sqrt{29}$

3. $\sqrt{-54} = i\sqrt{9}\sqrt{6} = 3i\sqrt{6}$

4. $\sqrt{-\frac{9}{4}} = \frac{3}{2}i$

5. $\sqrt{-108} = i\sqrt{36}\sqrt{3} = 6i\sqrt{3}$

6. $\sqrt{-96} = i\sqrt{16}\sqrt{6} = 4i\sqrt{6}$

7. $(5-7i)+(-4+9i) =$
$(5-4)+(-7+9)i =$
$1+2i$

8. $(-3+2i)+(-4-7i) =$
$(-3-4)+(2-7)i =$
$-7-5i$

9. $(6-9i)-(4-5i) =$
$(6-4)+(-9+5)i =$
$2-4i$

10. $(-5+3i)-(-8+7i) =$
$(-5+8)+(3-7)i =$
$3-4i$

11. $(7-2i)-(6-4i)+(-2+i) =$
$(7-6-2)+(-2+4+1)i =$
$-1+3i$

12. $(-4+i)-(-4-i)-(6-8i) =$
$(-4+4-6)+(1+1+8)i =$
$-6+10i$

13. $(2+5i)(3+8i) =$
$6+16i+15i+40i^2 =$
$6+31i+40(-1) =$
$-34+31i$

14. $(4-3i)(1-2i) =$
$4-8i-3i+6i^2 =$
$4-11i+6(-1) =$
$-2-11i$

15. $(-1+i)(-2+6i) =$
$2-6i-2i+6i^2 =$
$2-8i+6(-1) =$
$-4-8i$

16. $(-3-3i)(7+8i) =$
$-21-24i-21i-24i^2 =$
$-21-45i-24(-1) =$
$3-45i$

17. $(2+9i)(2-9i) =$
$4-18i+18i-81i^2 =$
$4-81(-1) = 85$

18. $(-3+7i)(-3-7i) =$
$9+21i-21i-49i^2 =$
$9-49(-1) = 58$

19. $(-3-8i)(3+8i) =$
$-9-24i-24i-64i^2 =$
$-9-48i-64(-1) =$
$55-48i$

20. $(6+9i)(-1-i) =$
$-6-6i-9i-9i^2 =$
$-6-15i-9(-1) =$
$3-15i$

21. $(x-6)^2 = -25$
$x-6 = \pm\sqrt{-25} = \pm 5i$
$x-6 = -5i$ or $x-6 = 5i$
$x = 6-5i$ or $x = 6+5i$
The solution set is $\{6-5i, 6+5i\}$.

22. $n^2+2n = -7$
$n^2+2n+1 = -7+1$
$(n+1)^2 = -6$
$n+1 = \pm\sqrt{-6} = \pm i\sqrt{6}$
$n+1 = -i\sqrt{6}$ or $n+1 = i\sqrt{6}$
$n = -1-i\sqrt{6}$ or $n = -1+i\sqrt{6}$
The solution set is
$\left\{-1-i\sqrt{6}, -1+i\sqrt{6}\right\}$.

23. $x^2-2x+17 = 0$
$$x = \frac{-(-2)\pm\sqrt{(-2)^2-4(1)(17)}}{2(1)}$$
$$x = \frac{2\pm\sqrt{-64}}{2}$$
$$x = \frac{2\pm 8i}{2} = 1\pm 4i$$
The solution set is $\{1-4i, 1+4i\}$.

24. $x^2 - x + 7 = 0$

$$x = \frac{-(-1) \pm \sqrt{(-1)^2 - 4(1)(7)}}{2(1)}$$
$$x = \frac{1 \pm \sqrt{-27}}{2}$$
$$x = \frac{1 \pm 3i\sqrt{3}}{2}$$

The solution set is
$$\left\{\frac{1 - 3i\sqrt{3}}{2}, \frac{1 + 3i\sqrt{3}}{2}\right\}.$$

25. $2x^2 - x + 3 = 0$

$$x = \frac{-(-1) \pm \sqrt{(-1)^2 - 4(2)(3)}}{2(2)}$$
$$x = \frac{1 \pm \sqrt{-23}}{4}$$
$$x = \frac{1 \pm i\sqrt{23}}{4}$$

The solution set is
$$\left\{\frac{1 - i\sqrt{23}}{4}, \frac{1 + i\sqrt{23}}{4}\right\}.$$

26. $6x^2 - 11x + 3 = 0$

$(3x - 1)(2x - 3) = 0$

$3x - 1 = 0$ or $2x - 3 = 0$

$3x = 1$ or $2x = 3$

$x = \frac{1}{3}$ or $x = \frac{3}{2}$

The solution set is $\left\{\frac{1}{3}, \frac{3}{2}\right\}$.

27. $-x^2 + 5x - 7 = 0$

$x^2 - 5x + 7 = 0$

$$x = \frac{-(-5) \pm \sqrt{(-5)^2 - 4(1)(7)}}{2(1)}$$
$$x = \frac{5 \pm \sqrt{-3}}{2}$$
$$x = \frac{5 \pm i\sqrt{3}}{2}$$

The solution set is
$$\left\{\frac{5 - i\sqrt{3}}{2}, \frac{5 + i\sqrt{3}}{2}\right\}.$$

28. $-2x^2 - 3x - 6 = 0$

$2x^2 + 3x + 6 = 0$

$$x = \frac{-(3) \pm \sqrt{(3)^2 - 4(2)(6)}}{2(2)}$$
$$x = \frac{-3 \pm \sqrt{-39}}{4}$$
$$x = \frac{-3 \pm i\sqrt{39}}{4}$$

The solution set is
$$\left\{\frac{-3 - i\sqrt{39}}{4}, \frac{-3 + i\sqrt{39}}{4}\right\}.$$

29. $3x^2 + x + 5 = 0$

$$x = \frac{-(1) \pm \sqrt{(1)^2 - 4(3)(5)}}{2(3)}$$
$$x = \frac{-1 \pm \sqrt{-59}}{6}$$
$$x = \frac{-1 \pm i\sqrt{59}}{6}$$

The solution set is
$$\left\{\frac{-1 - i\sqrt{59}}{6}, \frac{-1 + i\sqrt{59}}{6}\right\}.$$

30. $x(4x + 1) = -3$

$4x^2 + x = -3$

$4x^2 + x + 3 = 0$

$$x = \frac{-(1) \pm \sqrt{(1)^2 - 4(4)(3)}}{2(4)}$$
$$x = \frac{-1 \pm \sqrt{-47}}{8}$$
$$x = \frac{-1 \pm i\sqrt{47}}{8}$$

The solution set is
$$\left\{\frac{-1 - i\sqrt{47}}{8}, \frac{-1 + i\sqrt{47}}{8}\right\}.$$

31. $\sqrt{\frac{64}{36}} = \frac{8}{6} = \frac{4}{3}$ because $\left(\frac{2 \cdot 4}{2 \cdot 3}\right)^2 = \frac{64}{36}$

32. $-\sqrt{1} = -1$

33. $\sqrt[3]{\frac{27}{64}} = \frac{3}{4}$ because $\left(\frac{3}{4}\right)^3 = \frac{27}{64}$

34. $\sqrt[3]{-125} = -5$

35. $\sqrt[4]{\frac{81}{16}} = \frac{3}{2}$ because $\left(\frac{3}{2}\right)^4 = \frac{81}{16}$

36. $25^{\frac{3}{2}} = \left(\sqrt{25}\right)^3 = 5^3 = 125$

37. $8^{\frac{5}{3}} = \left(\sqrt[3]{8}\right)^5 = 2^5 = 32$

38. $(-8)^{\frac{5}{3}} = \left(\sqrt[3]{-8}\right)^5 = (-2)^5 = -32$

39. $4^{-2} = \frac{1}{4^2} = \frac{1}{16}$

40. $4^{-\frac{1}{2}} = \frac{1}{4^{\frac{1}{2}}} = \frac{1}{\sqrt{4}} = \frac{1}{2}$

41. $32^{-\frac{2}{5}} = \left(\sqrt[5]{32}\right)^{-2} = 2^{-2} = \frac{1}{2^2} = \frac{1}{4}$

42. $\left(\frac{2}{3}\right)^{-1} = \frac{3}{2}$

43. $2^{\frac{7}{4}} \cdot 2^{\frac{5}{4}} = 2^{\frac{7}{4}+\frac{5}{4}} = 2^{\frac{12}{4}} = 2^3 = 8$

44. $3^{\frac{1}{3}} \cdot 3^{\frac{5}{3}} = 3^{\frac{1}{3}+\frac{5}{3}} = 3^{\frac{6}{3}} = 3^2 = 9$

45. $\frac{3^{\frac{1}{3}}}{3^{\frac{4}{3}}} = 3^{\frac{1}{3}-\frac{4}{3}} = 3^{-\frac{3}{3}} = 3^{-1} = \frac{1}{3}$

46. $x^{\frac{5}{6}} \cdot x^{\frac{5}{6}} = x^{\frac{5}{6}+\frac{5}{6}} = x^{\frac{10}{6}} = x^{\frac{5}{3}}$

47. $\left(3x^{\frac{1}{4}}\right)\left(2x^{\frac{3}{5}}\right) = 3 \cdot 2 \cdot x^{\frac{1}{4}} \cdot x^{\frac{3}{5}} =$
$6x^{\frac{1}{4}+\frac{3}{5}} = 6x^{\frac{5}{20}+\frac{12}{20}} = 6x^{\frac{17}{20}}$

48. $\left(9a^{\frac{1}{2}}\right)\left(4a^{-\frac{1}{3}}\right) = 9 \cdot 4 \cdot a^{\frac{1}{2}} \cdot a^{-\frac{1}{3}} =$
$36a^{\frac{3}{6}-\frac{2}{6}} = 36a^{\frac{1}{6}}$

49. $\left(3x^{\frac{1}{3}}y^{\frac{2}{3}}\right)^3 = (3)^3\left(x^{\frac{1}{3}}\right)^3\left(y^{\frac{2}{3}}\right)^3 =$
$27x^{\frac{3}{3}}y^{\frac{6}{3}} = 27xy^2$

50. $(25x^4y^6)^{\frac{1}{2}} = (25)^{\frac{1}{2}}(x^4)^{\frac{1}{2}}(y^6)^{\frac{1}{2}} =$
$\sqrt{25}x^{\frac{4}{2}}y^{\frac{6}{2}} = 5x^2y^3$

51. $\frac{39n^{\frac{3}{5}}}{3n^{\frac{1}{4}}} = \frac{39}{3}n^{\frac{3}{5}-\frac{1}{4}} = 13n^{\frac{12}{20}-\frac{5}{20}} = 13n^{\frac{7}{20}}$

52. $\frac{64n^{\frac{5}{8}}}{16n^{\frac{7}{8}}} = \frac{64}{16}n^{\frac{5}{8}-\frac{7}{8}} = 4n^{-\frac{2}{8}} = 4n^{-\frac{1}{4}} = \frac{4}{n^{\frac{1}{4}}}$

53. $\left[\frac{6x^{\frac{2}{7}}}{3x^{-\frac{5}{7}}}\right]^3 = \left[\frac{6}{3}x^{\frac{2}{7}-\left(-\frac{5}{7}\right)}\right]^3 =$

$\left(2x^{\frac{7}{7}}\right)^3 = (2x)^3 = 8x^3$

54.

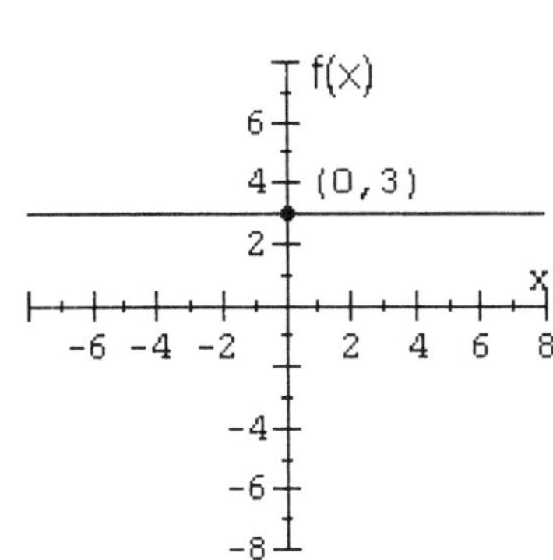

55.

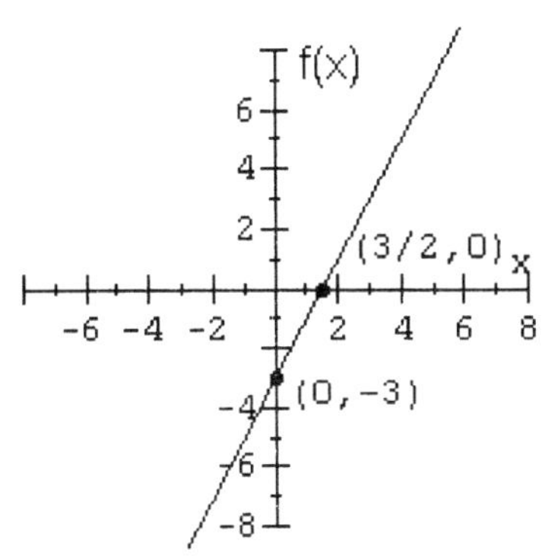

56.

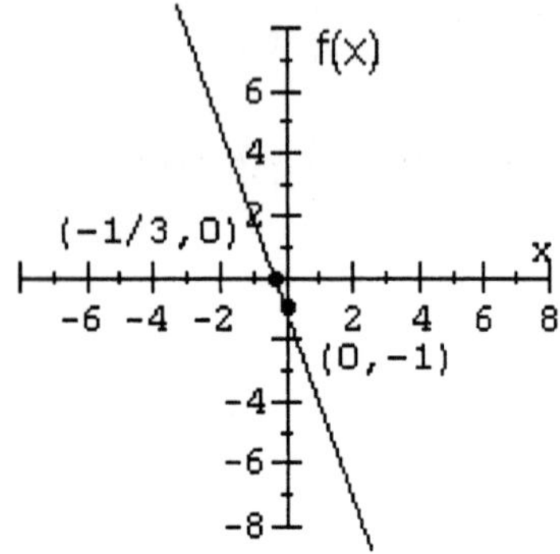

57.

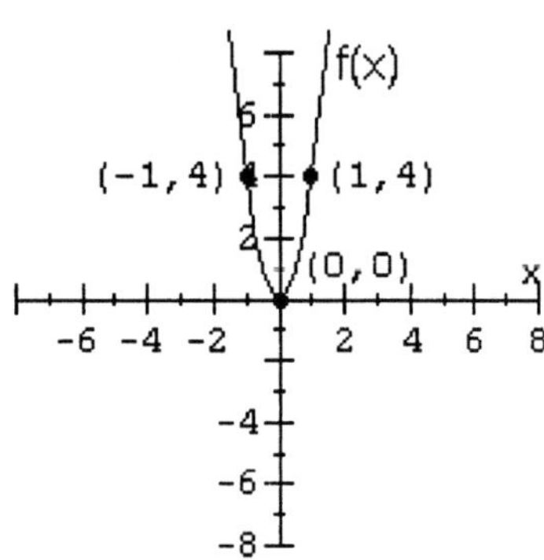

58.

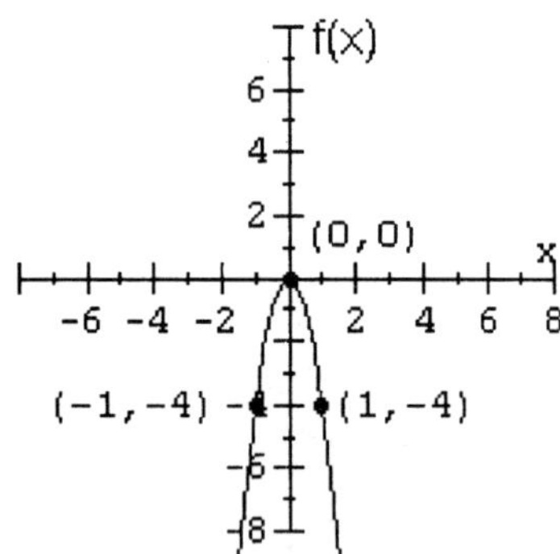

59. $|3x-5|=7$

$3x-5=7 \quad \text{or} \quad 3x-5=-7$

$3x=12 \quad \text{or} \quad 3x=-2$

$x=4 \quad \text{or} \quad x=-\frac{2}{3}$

The solution set is $\left\{-\frac{2}{3}, 4\right\}$.

60. $|x-4|<1$

$-1<x-4<1$

$3<x<5$

The solution set is $\{x|3<x<5\}$ or $(3, 5)$.

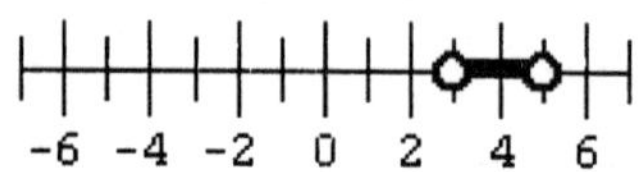

61. $|2x-1|\geq 3$

$2x-1\leq -3 \quad \text{or} \quad 2x-1\geq 3$

$2x\leq -2 \quad \text{or} \quad 2x\geq 4$

$x\leq -1 \quad \text{or} \quad x\geq 2$

The solution set is $\{x|x\leq -1 \text{ or } x\geq 2\}$ or $(-\infty, -1]\cup[2, \infty)$.

62. $|3x-2|\leq 4$

$-4\leq 3x-2\leq 4$

$-2\leq 3x\leq 6$

$-\frac{2}{3}\leq x\leq 2$

The solution set is $\left\{x\middle|-\frac{2}{3}\leq x\leq 2\right\}$ or $\left[-\frac{2}{3}, 2\right]$.

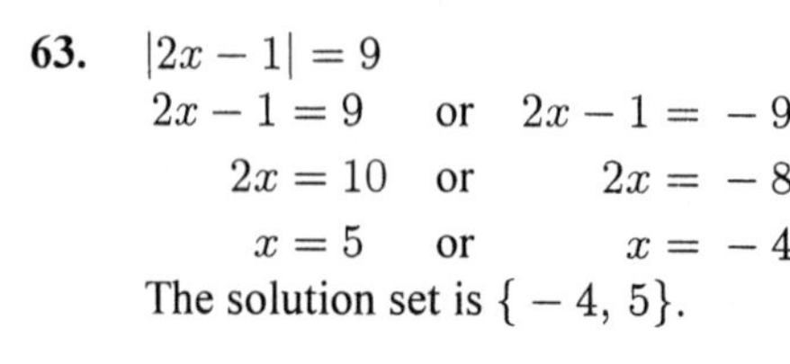

63. $|2x-1|=9$

$2x-1=9 \quad \text{or} \quad 2x-1=-9$

$2x=10 \quad \text{or} \quad 2x=-8$

$x=5 \quad \text{or} \quad x=-4$

The solution set is $\{-4, 5\}$.

64. $|5x-2| \geq 6$

$5x-2 \leq -6$ or $5x-2 \geq 6$

$5x \leq -4$ or $5x \geq 8$

$x \leq -\frac{4}{5}$ or $x \geq \frac{8}{5}$

The solution set is $\left\{x \mid x \leq -\frac{4}{5} \text{ or } x \geq \frac{8}{5}\right\}$ or $\left(-\infty, -\frac{4}{5}\right] \cup \left[\frac{8}{5}, \infty\right)$.

-6 -4 -2 0 2 4 6

CHAPTER 11 Test

1. $\sqrt{-75} = i\sqrt{25}\sqrt{3} = 5i\sqrt{3}$

2. $9^{-\frac{1}{2}} = \frac{1}{9^{\frac{1}{2}}} = \frac{1}{\sqrt{9}} = \frac{1}{3}$

3. $9^{-\frac{1}{2}} = \left(\sqrt{9}\right)^{-1} = 3^{-1} = \frac{1}{3}$

4. $\left(\frac{2}{3}\right)^{-3} = \frac{2^{-3}}{3^{-3}} = \frac{3^3}{2^3} = \frac{27}{8}$

5. $16^{\frac{5}{4}} = \left(\sqrt[4]{16}\right)^5 = 2^5 = 32$

6. $(-8)^{-\frac{1}{3}} = \frac{1}{(-8)^{\frac{1}{3}}} = \frac{1}{\sqrt[3]{-8}} = \frac{1}{-2} = -\frac{1}{2}$

7. $(-9)^{\frac{1}{2}} = \sqrt{-9} = \sqrt{-1}\sqrt{9} = 3i$

8. $\left(2x^{\frac{1}{4}}\right)\left(5x^{\frac{2}{3}}\right) = 10x^{\frac{1}{4}+\frac{2}{3}} = 10x^{\frac{11}{12}}$

9. $\frac{30n^{\frac{1}{2}}}{6n^{\frac{2}{5}}} = \frac{30}{6}n^{\frac{1}{2}-\frac{2}{5}} = 5n^{\frac{5}{10}-\frac{4}{10}} = 5n^{\frac{1}{10}}$

10. $(x-2)^2 = -16$

$x-2 = \pm\sqrt{-16}$

$x-2 = \pm 4i$

$x = 2 \pm 4i$

The solution set is $\{2-4i, 2+4i\}$.

11. $x^2 - 2x + 3 = 0$

$$x = \frac{-(-2) \pm \sqrt{(-2)^2 - 4(1)(3)}}{2(1)}$$

$$x = \frac{2 \pm \sqrt{-8}}{2}$$

$$x = \frac{2 \pm 2i\sqrt{2}}{2} = 1 \pm i\sqrt{2}$$

The solution set is $\left\{1 - i\sqrt{2}, 1 + i\sqrt{2}\right\}$.

12. $x^2 + 6x = -21$

$x^2 + 6x + 9 = -21 + 9$

$(x+3)^2 = -12$

$x + 3 = \pm\sqrt{-12}$

$x + 3 = \pm 2i\sqrt{3}$

$x = -3 \pm 2i\sqrt{3}$

The solution set is $\left\{-3 - 2i\sqrt{3}, -3 + 2i\sqrt{3}\right\}$.

13. $x^2 - 3x + 5 = 0$

$$x = \frac{-(-3) \pm \sqrt{(-3)^2 - 4(1)(5)}}{2(1)}$$

$$x = \frac{3 \pm \sqrt{-11}}{2}$$

$$x = \frac{3 \pm i\sqrt{11}}{2}$$

The solution set is $\left\{\frac{3 - i\sqrt{11}}{2}, \frac{3 + i\sqrt{11}}{2}\right\}$.

14. $|x-2|=6$

$x-2=-6$ or $x-2=6$

$x=-4$ or $x=8$

The solution set is $\{-4, 8\}$.

15. $|4x+5|=2$ is equivalent to $4x+5=-2$ or $4x+5=2$.

$4x+5=-2$ or $4x+5=2$

$4x=-7$ or $4x=-3$

$x=-\frac{7}{4}$ or $x=-\frac{3}{4}$

The solution set is $\left\{-\frac{7}{4}, -\frac{3}{4}\right\}$.

16. $|3x-1|=-4$

No Solution. Absolute value can not equal a negative number. The solution set is $\emptyset$.

17. $2x^2-x+1=0$

$$x=\frac{-(-1)\pm\sqrt{(-1)^2-4(2)(1)}}{2(2)}$$

$$x=\frac{1\pm\sqrt{-7}}{4}$$

$$x=\frac{1\pm i\sqrt{7}}{4}$$

The solution set is

$$\left\{\frac{1-i\sqrt{7}}{4}, \frac{1+i\sqrt{7}}{4}\right\}.$$

18. $3x^2+5x-28=0$

$(3x-7)(x+4)=0$

$3x-7=0$ or $x+4=0$

$3x=7$ or $x=-4$

$x=\frac{7}{3}$ or $x=-4$

The solution set is $\left\{-4, \frac{7}{3}\right\}$.

19. $|x+3|\geq 2$ is equivalent to $x+3\leq -2$ or $x+3\geq 2$.

$x+3\leq -2$ or $x+3\geq 2$

$x\leq -5$ or $x\geq -1$

The solution set is $\{x|x\leq -5 \text{ or } x\geq -1\}$ or $(-\infty, -5]\cup[-1, \infty)$.

20. $|2x-1|<7$

$-7<2x-1<7$

$-6<2x<8$

$-3<x<4$

The solution set is $\{x|-3<x<4\}$ or $(-3, 4)$.

21. $|5x-2|>-3$ will be true for all real numbers since the absolute value of $5x-2$, regardless of what number is substituted for x, will always be greater than -3. The solution set is $\{x|x \text{ is a real number}\}$.

22.

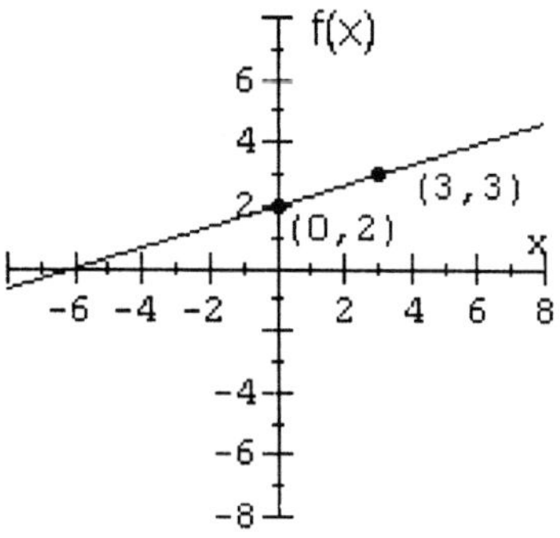

f(x)

6, 4, 2, -4, -6, -8

(3,3)

(0,2)

x

-6 -4 -2 2 4 6 8

23.

f(x)

6, 4, 2, -4, -6, -8

(0,4)

(3,4)

x

-6 -4 -2 2 4 6 8

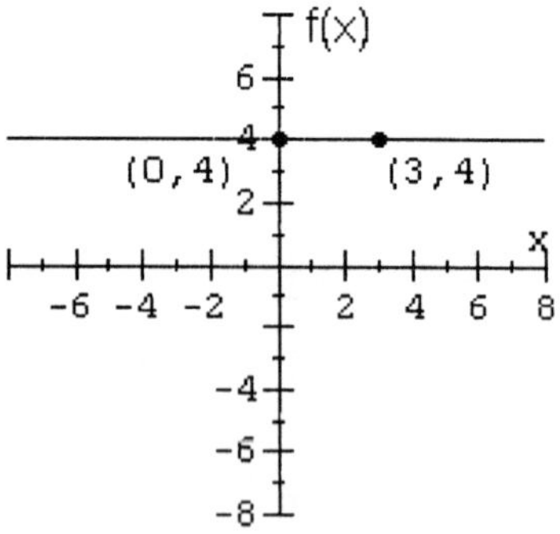

24.

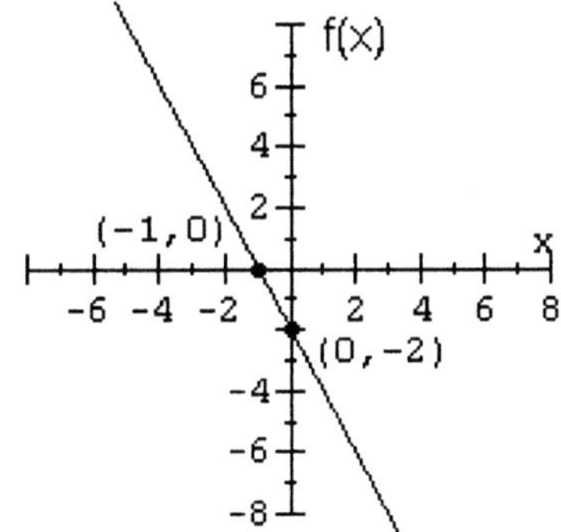

25.

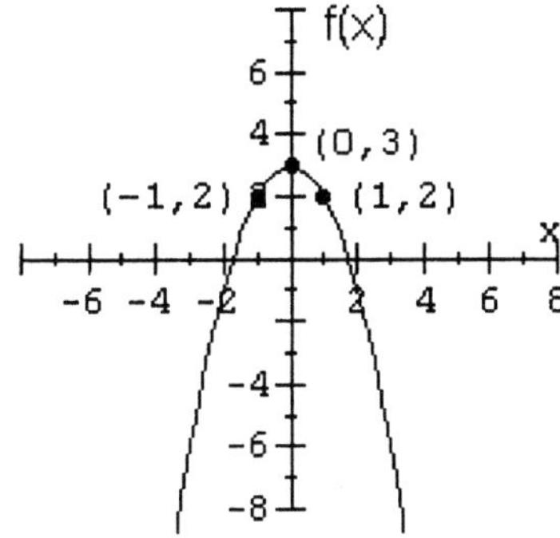